Probabil...

RA(*cond Edition*
RIO
ROF   d Mathematical
St
SCH
WIL   al Statistics
WIL
ZA(

Applied F...
BAI   plications to the
N
BAF   s, *Second Edition*
BEN   emistry and the
Cl
BHA
BO>   ical Method for
Pr
BR(ce and Engineer-
in
BUf
CHI
CH(ms
CLf   · Baisc Statistics
w
CO(
CO(*dition*
CO>

COX and MILLER · The Theory of Stochastic Processes, *Second Edition*
DANIEL and WOOD · Fitting Equations to Data
DAVID · Order Statistics
DEMING · Sample Design in Business Research
DODGE and ROMIG · Sampling Inspection Tables, *Second Edition*
DRAPER and SMITH · Applied Regression Analysis
DUNN and CLARK · Applied Statistics: Analysis of Variance and Regression
ELANDT-JOHNSON · Probability Models and Statistical Methods in Genetics
FLEISS · Statistical Methods for Rates and Proportions
GOLDBERGER · Econometric Theory
GROSS and HARRIS · Fund...
GUTTMAN, WILKS and HU... Statistics, *Second Edition*
HAHN and SHAPIRO · Statis...
HALD · Statistical Tables and...
HALD · Statistical Theory with ...ations
HARTIGAN · Clustering Algorithms

Branching Processes
with
Biological Applications

Branching Processes
with
Biological Applications

P. Jagers

Department of Mathematics
Chalmers University of Technology, Sweden

A Wiley–Interscience Publication

JOHN WILEY & SONS

London · New York · Sydney · Toronto

Library of Congress Cataloging in Publication Data:
Jagers, Peter, 1941–
   Branching processes with biological applications.

   'A Wiley-Interscience publication.'
   Includes bibliographies.
   1. Population biology—Mathematical models.
2. Branching processes. I. Title.
QH352.J33   574.5′24   74-32296

ISBN 0 471 43652 6

Typeset in Israel by Keterpress Enterprises, Jerusalem,
and printed in Great Britain by The Pitman Press, Bath, Avon.

Preface

Behind the complex growth of biological populations there hides (or, there can be imagined) a simple pattern of pure reproduction, unhampered by the hardships of limited resources, changing conditions, competition, or sex ratios, or other dependencies among individuals. This is the pattern of branching processes, of populations developing by individuals living, giving birth, and dying independently of one another. Within its framework important questions can be posed and answered. What populations die out? How do those behave who do not? Can they stabilize (i.e. in the absence of environmental checks, say dependent upon population size)? Or, if they grow beyond all limits, at what rate do they increase? How will they be composed, e.g. what will the age distribution tend to be?

For empirical populations the answers obtained are more or less relevant: asexually reproducing cell cultures in ideal laboratory conditions are much closer to the branching pattern than a system of some prey and predator in a subtle interplay. Still, the pattern in itself is fundamental enough to deserve mathematical investigation. It is also rich in implications, and can be studied in remarkable generality with concrete and easily interpreted results.

It is important that this be done, as the popular simplified models in the birth-and-death tradition require conditions which are often not satisfied. But if oversimplification is one danger, on the other hand there threatens the Charybdis of a mathematical *l'art pour l'art*. The narrow path I have tried to tread is never to hesitate before advanced methods in order to solve relevant problems, but also never to get carried away by the mathematics for its own sake.

The result is that I have treated the classical and basic questions cited generally and often by probabilistic methods. But I have shunned mathematically tempting topics like potential theory or Markovian branching processes. For such things there are fine references, K. B. Athreya's and P. Ney's Branching Processes (Springer, Berlin, 1972) and B. A. Sevast'yanov's Russian book with the same title (Nauka, Moscow, 1971). As opposed to these books I have stressed the general branching process, where individuals may give birth not only at death but at random points during life. This results in a greater mathematical complexity, but one that can be justified in view of the empirical phenomena to be studied.

v

It is the branching pattern that is the target of the analysis and therefore I have also avoided multi-type processes. After all, they concern an interaction between branching and a Markovian wandering between types. Also there is the exhaustive monograph by C. Mode, Multitype Branching Processes (American Elsevier, New York, 1971), penetrating precisely this difficult range of problems.

Though the whole book is thus concerned with an empirical pattern, it is only the last two chapters of the book that are directly applied. The former presents classical continuous time demography as part of branching process theory (as it really is). The latter treats the fascinating topic of cell proliferation, which is unusually well suited for an analysis in terms of branching processes. Mathematical readers, looking for further applied problems in that direction, can find a horn of plenty in M. Iosifescu and P. Tăutu, Stochastic Processes and Applications in Biology and Medicine II (Springer, Berlin, 1973). Biological readers will certainly deem many technical arguments difficult, or even impossible to follow. But that is unavoidable and I do hope that conclusions will arouse their interest. In particular I mention Chapter 9, where my claims sometimes do not agree with established notions.

The historical background of the subject and the contents of the book are surveyed in the first chapter. The following two chapters are devoted to the elementary Galton-Watson process. Here the time structure is the simplest possible (all individuals have the same life span and give birth only at death) and thus the branching itself is most easily understood. But this is preparatory; the main part of the book is Chapter 6, treating the general branching process. Some ramifications of it are discussed in Chapter 7, whereas Chapter 4 sketches multi-type theory and Chapter 5 gives some tools. For a basic course in branching processes I would suggest chapters 1, 2 and 6.

Literature references are given at the end of each chapter. This has obvious advantages but also the consequence that works on aspects of branching processes, omitted here are not mentioned. The order of references might seem bewildering. It is the following: references appear sectionwise, alphabetically within each section.

The mathematical prerequisites for reading this book are on the level of K. L. Chung, A Course in Probability Theory (Harcourt, Brace and World, New York, 1968) or W. Feller, An Introduction to Probability Theory and Its Applications II (Wiley, New York, 1966). However Markov chains are not needed at all and the renewal and martingale theory used is given in Chapter 5. No particular biological knowledge is required.

Upon completing the book I feel gratitude to many people: to the community of coworkers in branching processes, and especially to T. E. Harris, for his book in 1963; to my first teacher in probability theory, H. Bergström;

to K. Norrby, who introduced me to the kinetics of cell populations; to N. Keiding for several helpful observations; and to T. Lindvall, whose sensitive and intelligent comments have improved this book in many ways. I thank I. Dalemar, B. Engebrand, and E. Norrman for their good and patient typing of the manuscript and M. Härnqvist and U. Torstensson for help in reading the proofs.

Gothenburg, Sweden, 8th May, 1974.   Peter Jagers

Contents

A List of Symbols and Notation Conventions   xi

1 Introduction   1
  1.1   Historical sketch   1
  1.2   Generalities about populations   5
  1.3   A survey of results   9

2 The Galton–Watson Process   19
  2.1   Introduction   19
  2.2   Moments and the generating function   20
  2.3   The extinction probability   22
  2.4   Critical processes   24
  2.5   A simple but basic lemma   26
  2.6   Subcritical processes   28
  2.7   Supercritical processes   30
  2.8   Attaining high levels   35
  2.9   Rate of convergence results for supercritical processes   36
  2.10 Prediction in large supercritical processes and another rate
       of convergence result   37
  2.11 The total progeny of a branching process   39
  2.12 The relation between a process and its total progeny   42
  2.13 Maximum likelihood estimation of the reproduction mean
       and a Bayes example   45
  2.14 Estimation of the extinction probability   50

3 Neighbours of the Galton–Watson Process   54
  3.1   Branching processes with immigration   54
  3.2   Increasing numbers of ancestors   60
  3.3   Approximation by critical processes   63
  3.4   A diffusion approximation   67
  3.5   Galton–Watson processes in varying environments   70
  3.6   Further results for varying environments   77
  3.7   Random environments   81

4 Results for Multi-type Processes · 87
4.1 Fundamentals · 87
4.2 Analogues of classical results · 92

5 Interlude about Martingales, Renewal Theory, and Point Processes · 97
5.1 Martingales · 97
5.2 The renewal equation · 104
5.3 Refined renewal theorems · 114
5.4 Point processes · 120

6 The General Process · 123
6.1 Introduction · 123
6.2 The finiteness of the process · 126
6.3 Moments and the generating functions · 129
6.4 More about the second moments · 135
6.5 The extinction probability · 139
6.6 Critical processes · 143
6.7 The subcritical case · 156
6.8 Supercritical processes · 164
6.9 Populations counted with random characteristics · 167
6.10 Almost sure convergence in the supercritical case · 169
6.11 The stable age distribution and processes with an ancestor differing from its progeny · 176
6.12 The total progeny · 182
6.13 Integrals of branching processes and their generalizations · 183
6.14 Maximum likelihood estimation of the reproduction mean · 186

7 Neighbours of the General Process · 190
7.1 Immigration · 190
7.2 Increasing numbers of ancestors · 199
7.3 Convergence towards criticality · 200
7.4 Diffusion approximations · 202

8 Branching Processes and Demography · 207
8.1 Classical continuous time demography and age dependent birth and death processes · 207
8.2 Lotka's equation · 209
8.3 The growth of populations · 211
8.4 The age at childbearing · 214
8.5 The length of generations · 215

9 Branching Models in Cell Kinetics   224
9.1   Cell proliferation and binary splitting   224
9.2   Estimation of cell death   228
9.3   The cycle time distribution   230
9.4   The fraction labelled mitoses   232
9.5   FLM functions in binary splitting   236
9.6   Continuous labelling   241
9.7   Arrest methods   245
9.8   Synchrony   248
9.9   The composition of two-type populations: endomitosis and
       the G_0 resting phase   250

Appendix   259

Index   265

A List of Symbols and Notation Conventions

The conventions below are of two types. Some are more or less strictly adhered to, others just record the usual meaning or meanings of a symbol (example: f is more often than not the reproduction generating function of the branching process discussed, but there may be other uses of that symbol).

Symbol	Usual Meaning
a	ancestor or age
$\mathscr{B}(\mathscr{N})$	the cylinder algebra on $\mathscr{N}$, i.e.
	$\sigma(\{\{\mu \in \mathscr{N}; \mu A = k\}\, k \in Z, A \text{ Borel}\}$
$\mathscr{B}_n$	$\sigma(z_0, \ldots, z_n)$
$c_u^{a,b}(t)$	$E[z_t^a z_{t+u}^b]$
$c_u(t)$	$E[z_t z_{t+u}]$
$E[x; A]$	integration of x with respect to P over the set A, $= E[x1_A]$
f	reproduction generating function
f_n	generating function of z_n
g_n	inverse of f_n
h_n	generating function of the total progeny in the $n+1$ first generations
I_a	possible descendants of a
I	I_0
$I(n), I_a(n)$	nth generation
i.i.d	independent and identically distributed
$\inf \emptyset$	$+\infty$
$k(a)$	$\lim_{t \to \infty} e^{-\alpha t} m_t^a$
K	the stable age distribution, $K(a) = k(a)/k(\infty)$
l	mean life span
L	life distribution
m	$= \mu(\infty) = f'(1)$, reproduction mean
m_t^a	$E[z_t^a]$
m_t	$E[z_t]$
N	the positive integers
$N(a, b)$	a normal random variable mean a and variance b

xi

Symbol	Usual Meaning
$\mathcal{N}, \mathcal{N}(R_+)$	a set of integer valued measures on R_+
$O(a_n)$	$b_n = O(a_n)$ means $\lim \sup |b_n/a_n| < \infty$
$ó(a_n)$	$b_n = o(a_n)$ means $b_n/a_n \to 0$
p_k	probability of begetting k children
q	extinction probability
R	$(-\infty, \infty)$
R_+	$[0, \infty)$
s	number in $[0, 1]$
t, u, v	time
w	$\lim\limits_{n \to \infty} z_n/m^n$ or $\lim\limits_{t \to \infty} e^{-\alpha t} z_t$
x	individual
y_n	total numbers of individuals in the $n + 1$ first generations;
y_t	total progeny up to time t in a process with immigration
Z	the integers
Z_+	the non-negative integers
z_t^a	number of individuals alive and younger than a at time t
z_n	the number of individuals in the nth generation of a Galton–Watson process
z_t	z_t^∞
z_t^χ	process counted by the characteristic χ
z	$\lim\limits_{t \to \infty} e^{-\alpha t} z_t \alpha\beta/\{1 - E[e^{-\alpha t}]\}$
α	Malthusian parameter
β	average age at childbearing
ζ_n	the number of individuals in the nth generation of an imbedded Galton–Watson process
λ_x	x's life span
μ	$= E[\xi(\cdot)]$ reproduction function
$\xi(t)$	$\xi[0, t]$, the number of children begotten during the age interval $[0, t]$
ξ_x	reproduction process of x
ξ	ξ_x for some unspecified x
ξ_x	$\xi_x(\infty)$, the number of children of x
ξ	$\xi_x(\infty)$ for arbitrary x
σ_x	time of x's birth
σ^2	$\mathrm{Var}[\xi(\infty)]$, reproduction variance
$\sigma(\mathcal{A})$	the smallest σ-algebra containing the class $\mathcal{A}$ of subsets of some space, or the smallest σ-algebra with respect to which all the elements of the class $\mathcal{A}$ of real functions on the space are Borel measurable. Instead of $\sigma(\{x_1, \ldots, x_n\})$ there might be $\sigma(x_1, \ldots, x_n)$.

Symbol	**Usual Meaning**
$\sum_{1}^{0}$	$= 0$
φ	$E[\overset{\circ}{\xi}(2\alpha)]$
Φ	standardized normal distribution function
χ, χ_x	random characteristic
ϑ	$E[\overset{\circ}{\xi}{}^2(\alpha)]$
$\overset{\mathrm{d}}{=}$	distributed as
$\overset{\mathrm{d}}{\to}$	convergence in distribution
1_A	indicator function of the set A (one on A, zero on A')
$\int_0^b$	$\int_{[0,b]}$
$\int_a^b$	$\int_{(a,b]}$ for $a > 0$
$\circ$	composition of functions, $f \circ g(s) = f\{g(s)\}$
$\sim$	$a_n \sim b_n$ means that $a_n/b_n \to 1$
$*$	convolution, $g * h(t) = \int_0^\infty g(t - u)\,h(du)$
$'$	vector or matrix transpose; the complement of a set
$\wedge$	min
$\vee$	max
$\Leftrightarrow$	if and only if
$\hat{}$	Laplace–Stieltjes transform: $\hat{\mu}(s) = \int_0^\infty e^{-st}\mu(dt).$
$+$	$a^+ = a \vee 0$
$-$	$a^- = -a \wedge 0$
$a/b + c$	$(a/b) + c$
a/bc	$a/(bc)$

Chapter 1

Introduction

1.1 HISTORICAL SKETCH

The very first pages of Thomas Malthus's famous three volume Essay on the Principle of Population serve to establish the fundamental idea that a population, when unchecked, grows exponentially. In Malthus's own wording, it goes on doubling itself at regular intervals or, increases in a geometrical ratio (page 9 of Reference [4]). Towards the end of the first volume ([4], page 485) Malthus relates (rather in passing) that in the town of Berne out of 487 bourgeois families 379 became extinct in the space of two centuries, from 1583 to 1783. The theory of branching processes may be said to be born out of the recognition that such a remarkable decrease is not some odd coincidence. On the contrary, it mirrors a basic and paradoxical antipode to the rapid growth of the whole, a principle of frequent extinction of separate family lines.

This seems first to have been realized in a circle of French scholars. One of the members was the statistician L. F. Benoiston de Châteauneuf (1776–1856), who studied noble families that were founded in the tenth to twelfth centuries. He estimated their usual duration to be three hundred years [6]. It is reasonable, or at least tempting, to assume that there were connections between Benoiston and his younger Parisian colleague I. J. Bienaymé (1796–1878), who treated the problem mathematically almost simultaneously [7, 41]. Though only a verbal account of Bienaymé's work is available, it seems he was able to determine correctly the probability of a family's extinction, or anyhow the relation between this and the mean number of male children per father.

Bienaymé was known for his remarkable breadth, not only confined to the mathematical sciences. La Grande Encyclopédie (2nd ed.) claims that he knew in depth (à fond) all European languages and had studied all branches of human knowledge, except law which he abhorred. A wide range of interests also distinguished Sir Francis Galton (1822–1911) who studied the decay of the English peerage and other families of 'men of note', from his eugenic viewpoint [8]. In spite of his predilection for quantification (he made a statistical inquiry into the efficacy of prayer), Galton gave

1

social and biological explanations of the frequent family extinctions he observed. It was a third polyhistor (not abhorring jurisprudence though, he was a law doctor), the renowned Swiss botanicist Alphonse de Candolle, who turned Galton's attention to the possibility of a probabilistic interpretation (Reference [10], page 389). Neither of them seems to have known of Bienaymé's work.

Anyhow, the very same year that de Candolle's suggestion was published, Galton gave it a precise formulation appearing as problem 4001 in the Educational Times [11]: 'A large nation, of whom we will only concern ourselves with the adult males, N in number and who each bear separate surnames, colonize a district. Their law of population is such that, in each generation, a_0 per cent of the adult males have no male children who reach adult life; a_1 have one such male child, a_2 have two; and so on up to a_5 who have five. Find (1) what proportion of the surnames will have become extinct after r generations; and (2) how many instances there will be of the same surname being held by n persons'.

He received just one answer to the problem 'from a correspondent who totally failed to perceive its intricacy' [12] and turned to a friend, the clergyman and mathematician Rev. H. W. Watson. Watson transformed the problem into one of iteration of generating functions, much as it is still treated (cf. Theorem 2.3.1). If $f(x) = \sum p_j x^j$, $p_j = a_j/100$ i.e. the probability of a father begetting j male children reaching adult life, then Watson [12] defines a sequence recursively by $f_r = f_{r-1} \circ f$. He concludes that the answer to the first question is the term independent of x in $f_r(x)$ and gives the number of surnames with s representatives in the rth generation as the coefficient of x^s in $f_r(x)$ multiplied by N. Further Watson discovers, for binomial f_1 and others 'that may be compared with' the binomial form that the probability y of an ultimate extinction must satisfy $y = f(y)$, 'that is where $y = 1$', as he erroneously adds. He was well aware that this might seem to be at variance with the Malthusian principle of exponential increase of the total population and argued, subtly and rather at length, that the total population might well grow even though each specific family name will eventually get lost.

It is curious to observe that, on the mathematical side, Watson's work was almost simultaneous with Schröder's on functional iteration in Mathematische Annalen [9]. As Kendall [40] suggests, had Watson read Schröder he would have experimented with linear fractional generating functions (which are easily iterated) and he would have discovered that his last conclusion was precipitate.

But he did not, and the correct extinction probability remained unknown for another half a century. (Bienaymé's work was recovered only in 1972 [41].) In 1922 R. A. Fisher touched upon the topic in a genetic context [17], the progeny of a mutant gene, and five years later J. B. S. Haldane, the biochemist, psychologist, geneticist, biomathematician, and political publicist, sketched the correct answer [19], namely that essentially the extinction probability is one exactly when the mean $f'(1) \leq 1$.

It was a Danish actuary, J. F. Steffensen, who published the first complete analysis of the extinction probability [21]. He responded to a challenge from A. K. Erlang, given like Galton's, as a problem for solution, in Matematisk Tidskrift 1929 [20]. Evidently Erlang was not aware of the British notes around the problem but was interested in it for personal reasons, his mother belonging to a well-known but disappearing Danish family. Now the time seems to have been ripe. As evidenced by Steffensen, Erlang himself had come half way to the truth. But what is more, the small Danish language journal received two complete and one incomplete answer. Also Erlang–Steffensen's formulations are those still in use: assume that an individual and his descendants each begets n children with probability p_n. Then the probability q of his family becoming extinct is the smallest non-negative root of the equation $s = f(s)$, $f(s) = \sum_{k=1}^{\infty} p_k s^k$. Provided $p_1 < 1$, then $q < 1$ if and only if $m = f'(1) > 1$.

The relevance of this to Malthus's, Châteauneuf's, and Galton's data, is maybe the following: Assume that a population as above starts from one ancestor. The generation of his children (the first generation) will then have the expected size $m = \sum k p_k$ and generally the nth generation has an expected number of m^n members. Thus if a homogeneous population is large, say consists of r members each reproducing according to the same law $\{p_k\}$ with $m > 1$, we would expect the population to grow roughly as rm^n from generation to generation (Malthus's principle of growth in a geometric ratio). However each line of descendants from any of the r people dies out with the probability q, which is less than one. It might still be astonishingly close to one, also for large m, if f is very convex as it is if p_0 is not little but there is some chance of very many children reaching adult life. Certainly such a case should depict the situation in a plague ridden late Medieval city or among the English or French 17th century gentry fairly well.

Though of course human populations are much more complex than the simple pattern exhibited in Galton's or Erlang's problems for solution, early branching processes were thus tightly interwoven with demographic considerations. The main stream of demography, however, for obvious reasons was less concerned with family extinction than with properties of entire populations, like growth and composition. The founders of modern demography where empirical scientists. They measured frequencies of birth and death, evaluated ratios of births to marriages, and constructed the first life tables. They observed remarkable regularities and used them to estimate population sizes. One of them, of the clergy like G. H. Watson, was Johann Peter Süssmilch (1707–1767) who saw the hand of God in these regularities. He titled his work *Die göttliche Ordnung in den Veränderun-*

gen des menschlichen Geschlechts, aus der Geburt, dem Tode und der Fort-pflanzung desselben erwiesen [1]. But he was not alien to the thought that divine order might express itself in mathematical terms and on various occasions he sought the assistance of Leonhard Euler, who seems to have laid the foundations of what nowadays is called stable population theory. Euler, who shared with Süssmilch a bent for apologetics, had already formulated the principle of exponential growth in his *Introductio in analysin infinitorum* (1748) [2] and concluded from it that: 'Quam ob causam maxime ridiculae sunt eorum incredulorum hominum obiectiones, qui negant tam brevi temporis spatio ab uno homine universam terram incolis impleri potuisse'.*
Later [3] he showed that a hypothetical closed population with a given time invariant age specific mortality and fertility and a constant rate of (i.e. exponential) increase must have a fixed age distribution. Like authors in branching processes and later mathematical demographers he disregarded sex ratios, treating actually a homogeneous asexually reproducing population.

Euler's analysis of the relation between age structure and fertility-mortality was neglected for more than half a century. In 1839 the demographer Ludwig Moser [6] of Königsberg took it up, applying it to studying the populations of the United States and France. From this onwards there is an unbroken tradition via Adolphe Quételet, the great Belgian statistician, to Alfred Lotka [13, 15, 23], who shares with another actuary, L. Herbelot [14], the credit of modern stable population theory. In particular they introduced a renewal equation treatment of entities like the expected population size or the birth rate, equations like (6.3.4) and (8.2.2).

Thus, after the first quarter of the twentieth century, problems of population development had given birth to two fragments or beginnings of mathematical theories, the (Bienaymé–)Galton–Watson(–Haldane–Erlang–Steffensen) branching process and the Euler-Lotka stable population theory. The former was probabilistic and considered the fate of single family lines. The latter might be called pseudo-deterministic; it started from probabilistic assumptions but being interested only in large population behaviour contended itself with deterministic results of an expectation character. The former suffered the limitation of measuring time in generations. The coalescence of the two was gradual and occurred through the development of increasingly complex branching models. (It may still not be realized by some demographers.) Certainly this development, in its turn, was an outflow of the burst of research and results in probability theory after the thirties.

---

* For that reason are the objections of those incredulous men, who deny that in such a short space of time the whole earth could have been filled with inhabitants descending from one man, utterly ridiculous.

Briefly the story goes as follows. Continuous time processes of a birth-and-death type were introduced by McKendrick [16], Yule [18], Furry [24], Volterra [22] and Feller [25]. This development culminated in David Kendall's Stochastic processes and population growth 1949 [28, 30], where the most general populations treated satisfy:

(a) Individuals are independent of one another,
(b) An individual of age a at existing at the epoch t has a chance $\lambda(a)\,dt + o(dt)$ of producing a new individual of age zero during the subsequent time interval of length dt,
(c) An individual as above has a chance $\mu(a)\,dt + o(dt)$ of dying in the same time interval.

Meanwhile, the Galton–Watson branching process was generalized by Richard Bellman and Theodore Harris [29] into a process where independent individuals first lead a life of a random length, then give birth to a random number of children, k with probability p_k independently of the mother's life span. B. A. Sevast'yanov [32] introduced into this the possibility of dependence between life and reproduction. Harris's book [31] in 1963 tied the whole field together and furnished a basis for a rapid progress in research.

While the Kendall birth-and-death process was a rigorous probabilistic formulation of the models in vogue among demographers and actuaries, the Bellman–Harris process of individuals splitting into new individuals had its background in the physical and biological (cells, bacteria) sciences. A framework, encompassing both theories (and Sevast'yanov's) was suggested independently by Thomas Ryan [34], Kenny Crump and Charles Mode [35], and this author [36]. Here individuals are supposed to have random life spans during which births occur as a point process, time is continuous and any type of dependence between reproduction and life is allowed.

Another line of development, which we shall be little concerned with, has been that of models with individuals of different types. The first general formulations appear to be due to the Russian school, A. N. Kolmogorov, N. A. Dmitriev and B. A. Sevast'yanov [26, 27], treating discrete time or continuous time with exponential life lengths. Multi-type Bellman-Harris processes were considered by Peter Ney [32] and B. A. Sevast'yanov [33]. Into the general process, where births may occur during the mother's life, several types of individuals were introduced by Mode [37].

1.2 GENERALITIES ABOUT POPULATIONS

Actual reproducing populations consist of individuals who live, give birth to a finite number of new individuals, and die. Suppose that an enumeration of the children of the same individual has been chosen, for example that of the order of birth and of some independent random order for twins and more children born simultaneously. Then the whole population stemming from some ancestor a can be enumerated by labelling the j_nth child of the j_{n-1}th child of ... of the j_1th child of the ancestor by the vector $(a; j_1, \ldots, j_{n-1}, j_n)$.

Certainly the set of all such vectors of a and positive natural numbers is rich enough to provide labels for any sensible empiric population. In many cases it is, indeed, unnecessarily large. Consider, for instance, a cell population which multiplies by the mother splitting into two new cells. Then vectors of twos and ones would suffice.

Still the set I_a of all possible labels of descendants of a (including a itself),

$$I_a = \{a\} \cup \bigcup_{n=1}^{\infty} \{(a; x); x \in N^n\}$$

is countable. Thus, for the technical treatment it is not too large; in probability theory it is when uncountable product spaces cannot be avoided that unpleasant complications arise.

In the mathematical theory we shall identify individuals with their labels. To be precise, for any a in some unspecified set we define an *individual stemming from* a as an element of I_a. Most of the theory considers populations starting from just one ancestor. Then we take $a = 0$ and often write I for I_a and x for $(a; x)$. The *nth generation* $I_a(n)$ is the set of individuals $(a; x)$ such that $x \in N^n$. The ancestor alone belongs to the *zeroth generation*, $I_a(0)$. If $a = 0$ we write N^n for $I_a(n)$, in particular $N^0 = \{0\}$.

For a given ancestor a and $x \in N^k$, some k;

$$I_{(a;x)} = \{(a; x)\} \cup \bigcup_{n=1}^{\infty} \{(a; x, y); y \in N^n\},$$

where $(a; x, y)$ means $(a; j_1, \ldots, j_k, i_1, \ldots, i_n)$ if $x = (j_1, \ldots, j_k)$ and $y = (i_1, \ldots, i_n)$. This defines the family stemming from any individual in I_a. Obviously

$$I_a = \bigcup_{n=0}^{\infty} I_a(n), \tag{1.1}$$

$$I_a = I_a(0) \cup \bigcup_{n=1}^{\infty} I_{(a;n)} \tag{1.2}$$

a corresponding partitioning holding for any $I_{(a;x)}$.

Since our individuals can give birth we shall usually (but not consistently) refer to them by female nouns and pronouns, talking about a mother rather than a father, she rather than he or it.

Most of the topics to be discussed will be within the framework given and concerning populations where there are only individuals of one single type. But much theory and several applications deal with multi-type populations. For a mathematical description of these let T be some space, the *type space*, and τ_j elements of it. An *individual* is then a vector $(a, \tau_a; j_1, \tau_1; \ldots; j_n, \tau_n)$ to be interpreted as the j_nth type τ_n child of her mother, who ..., who was the j_1th type τ_1 child of the ancestor, who was of type τ_a.

Usually the type space is finite, $T = \{1, 2, \ldots, r\}$—then we talk of *r-type* processes. But some studies have been made of Euclidean type spaces. The type could then be viewed as energy or position in space.

In any case we define as before $I_{(a,\tau_a)}(n, \tau)$, $\tau, \tau_a \in T$, $n \in Z_+$ as the set of all individuals in the nth generation who are of type τ. If $S \subset T I_{(a,\tau_a)}(n, S)$ similarly consists of all individuals with types in S. We write $I_{(a,\tau_a)}(n) = I_{(a,\tau_a)}(n, T)$.

With each individual—in the one-type as well as multi-type case—we shall associate random entities giving her life-length and reproduction. These may be of varying complexity but those associated with different individuals should always be independent and with the same distribution for individuals of the same type except possibly the ancestors. It could well be argued that this is the defining property of all branching theory.

The simplest case concerns a one-type population with just one ancestor 0. Each $x \in I$ is associated with one Z_+-valued random variable, ξ_x, the *number of children* of x. A *realized* individual is defined inductively: the ancestor is realized, $j \in N$ is realized if $j \leq \xi_0$ and $(j_1, \ldots, j_n, j_{n+1})$ is realized if $x = (j_1, \ldots, j_n)$ is, and $j_{n+1} \leq \xi_x$. The ξ_x are assumed i.i.d. with distribution $\{p_k; k \in Z_+\}$ called the *reproduction law*. A *Galton-Watson process* is then the number of individuals realized in the different generations. Explicitly, let $z_n, n \in Z_+$, be the number of realized individuals in $I(n)$. Then $\{z_n; n \in Z_+\}$ is a Galton–Watson process.

When there are r types, the ancestor should have some type, say τ_0. We say that $(0, \tau_0)$ is realized. To each individual x we attach a random vector $\xi_x = (\xi_x^{(1)}, \ldots, \xi_x^{(r)})$, $\xi_x^{(i)}$ giving the number of x's children of type i. The individual $(0, \tau_0; j_1, \tau_1; \ldots; j_{n-1}, \tau_{n-1}; j_n, \tau_n)$ is then realized if its mother $x = (0, \tau_0; \ldots; j_{n-1}, \tau_{n-1})$ is and $j_n \leq \xi_x^{(\tau_n)}$.

Suppose that the vectors ξ_x are independent with a distribution depending only on the type of x. Let $z_n = (z_n^{(1)}, \ldots, z_n^{(r)})$ be the numbers of realized individuals in $I_{(0,\tau_0)}(n, 1), \ldots, I_{(0,\tau_0)}(n, r)$. The sequence $\{z_n; n \in Z_+\}$ is called an *r-type Galton–Watson* process.

The rest of this introductory section is concerned with just one-type processes. But the formulation of the corresponding multi-type model should make small difficulties.

We have defined the Galton–Watson process as the numbers of individuals realized in the different generations. There is an alternative interpretation. Let each individual x have a life-length of one time unit and assume that at her death she splits into ξ_x new individuals. Then $z_t = z_{[t]}$ is the number of individuals alive at time t.

This invites the obvious generalization of associating with each x two random variables, besides ξ_x a life length $\lambda_x \geq 0$ the pairs (ξ_x, λ_x) being i.i.d. Let the process start at time zero from the ancestor and assume that a realized individual is born at the death of her mother. Denote by z_t the number of realized individuals alive at time $t > 0$. Then $\{z_t, t \in R_+\}$ is traditionally called an *age-dependent branching* process. A more accurate name is, maybe, a *splitting* process. Usually it is further assumed that ξ_x and λ_x are independent. If we want to stress this we shall talk of the *Bellman–Harris model*, whereas the model allowing a real age-dependence of the reproduction ξ_x is named after *Sevast'yanov*.

The Sevast'yanov model is contained in the following generalization, where individuals beget children at randomly chosen instants during their lives and not necessarily exactly when they die: Let the ξ_x be not a random variable but a point process on R_+, that is a random number of points on R_+ with random locations, not necessarily distinct. As before the pairs (ξ_x, λ_x) for different x are assumed i.i.d., except possibly for a different distribution for ancestors, mirroring that we might prefer to discuss populations starting from not newly born ancestors. The point process ξ_x, x no ancestor, will be referred to as the *reproduction process*.

Any point process on R_+ can be translated into a point process on the interval $[t, \infty]$, $t > 0$, simply by moving the points of the process t units to the right. We write T_t for this operation. In other words, if ξ is the point process and $\xi(A)$ is the number of points placed by ξ in the set A, then $T_t\xi(A)$ equals $\xi(A - t)$, $A - t = \{u \in R; u + t \in A\}$.

With the help of this translation operator the branching process is constructed recursively: the ancestor lives time λ_0 and gives birth at the points of ξ_0 in the interval $[0, \lambda_0]$. A general individual x is born at the (random) time point σ_x defined by the λ_y, ξ_y of x's mother, grandmother, etc. (in the future we shall sometimes talk of the *predecessors* of x). She lives until time $\sigma_x + \lambda_x$, giving birth at the points of $T_{\sigma_x}\xi_x$ falling in the interval $[\sigma_x, \sigma_x + \lambda_x]$.

As before $\{z_t; t \in R_+\}$ is defined as the numbers of realized individuals alive at times $t \in R_+$. It is also interesting to study the random variables $z_t^a, t, a \in R_+$, giving only the number of realized living individuals younger than a. These processes, $\{z_t\}$ and $\{z_t^a\}$ will be called *general branching processes*. We shall further have a look at more general processes like the number of individuals in some random phase of life or having some random age-dependent property. Somewhat improperly they might be called *functionals of branching processes*. Not too much imagination should be required to see that all processes may be defined also in situations where individuals can give birth even after death.

To complete these last paragraphs we should make precise the definition of the variables σ_x and what it means for an individual $x = (j_1, \ldots, j_n)$ to be alive at some time t. The latter task is easy enough; x is *alive* at t if and only if

$$\sigma_x \leq t < \sigma_x + \lambda_x.$$

At least for age-dependent processes the variable σ_x is also easily caught. Obviously it is nothing but the sum $\lambda_0 + \lambda_{j_1} + \ldots + \lambda_{j_1 \ldots j_{n-1}}$. (We write $\lambda_{j_1 \ldots j_n}$ for $\lambda_{(j_1, \ldots, j_n)}$.) The generalized case is not much worse but still left for its proper chapter.

Background

Probability spaces of the kind indicated are known as family tree or family history spaces. They have been treated in References [42, 43] for Galton–Watson processes, in Reference [44] for so called Markov branching processes, i.e. Bellman–Harris processes with exponentially distributed lives, and in Reference [31] for the Bellman–Harris process. Modifications to suit the general process were done in References [35, 36].

1.3 A SURVEY OF RESULTS*

Consider a general branching process $\{z_t\}$ and $\{z_t^a\}$, starting for simplicity from one ancestor 0 with (ξ_0, λ_0) distributed as for the descendants. We introduce, from now on, the convention of writing ξ or λ when referring to an arbitrary ξ_x or λ_x. Also $\xi(t)$ means $\xi([0, t])$, the number of births up to and including age t. Throughout it is assumed that no children are born after their mothers died, i.e. that $P[\xi(\lambda) = \xi(\infty)] = 1$. If the process is splitting, we write $\xi = \xi(\infty)$ for the random variable giving the reproduction. A first result, more of an auxiliary nature, is

Theorem (6.2.2)

If the reproduction function $\mu(t) = E[\xi(t)]$ *satisfies* $\mu(0) < 1$ *and* $\mu(t)$ *is finite for some* $t > 0$, *then* $P[z_t$ *is finite for all* $t \geq 0] = 1$.

Introduce the *reproduction generating function* $f(s) = E[s^{\xi(\infty)}], 0 \leq s \leq 1$, and the *extinction probability* $q = P[z_t \rightarrow 0]$. A general answer to the family extinction problem is provided by

Theorem (6.5.1)

The extinction probability is the smallest non-negative root of the equation $f(s) = s$. *Provided* $P[\xi(\infty) = 1] < 1$, *it holds that* $q = 1 \Leftrightarrow m = f'(1) = \mu(\infty) \leq 1$.

---

* Theorems are given their numbers in the later text. Still the formulation here need not be exhaustive.

Elaborations of this concern the rate of the convergence $P[z_t > 0] \to 0$ and the distribution of z_t under the condition $z_t > 0$ in the so called *critical* and *subcritical* cases, $m = 1$ and $m < 1$ respectively. Here as well as in other contexts where time enters explicitly there is a technical distinction to be made, between *lattice* and *non-lattice* processes. The process is of the former type if there is a number $d > 0$ such that

$$\sum_{k=0}^{\infty} \mu(\{kd\}) = \mu(\infty)$$

and non-lattice otherwise. To simplify somewhat, limit assertions in the sequel holding without qualifications for non-lattice processes are true in the lattice case if the limit is taken over points $kd + t$, t fixed and $k \to \infty$ (cf. Theorems 5.2.6 and 5.2.7). We consider only processes with a finite *reproduction mean*, $m = \mu(\infty) = f'(1)$. A process is said to be *Malthusian* if there is a number α (called the *Malthusian parameter*) such that

$$\hat{\mu}(\alpha) = \int_0^{\infty} e^{-\alpha t} \mu(dt) = 1.$$

Besides the pathological case $\mu(0) \geq 1$ (reason out what that implies!) only subcritical processes can avoid Malthusianness and also then this is something of a mathematician's delight. For example all processes such that $\mu(0) < 1$ and $\mu(t_0) = \mu(\infty)$ for some t_0, e.g. those where no life spans exceed t_0, are Malthusian. The Malthusian parameter is positive, zero or negative according as $m >$, $=$, or < 1.

Theorem (6.7.2)

Consider a non-lattice subcritical Malthusian process such that the (*mother's*) *average age at child-bearing*

$$\beta = \int_0^{\infty} te^{-\alpha t}\mu(dt) < \infty$$

and $E[\lambda e^{-\alpha\lambda}] < \infty$. Then

$$\lim_{t \to \infty} e^{-\alpha t}P[z_t > 0]$$

exists. It is strictly positive if $E[\hat{\xi}(\alpha)\log \xi(\infty)] < \infty$.

If the process is of the splitting type, i.e. mothers give birth only at their death then $E[\hat{\xi}(\alpha)\log \xi(\infty)] = E[e^{-\alpha\lambda}\xi \log \xi]$, where $\xi = \xi(\infty)$. In the Bellman–Harris case λ is independent of ξ and since $E[e^{-\alpha\lambda}] < \infty$ by assumption the condition reduces to $E[\xi \log \xi] < \infty$. As a sequel to the theorem there follows the existence of an asymptotic distribution of the number of individuals in a process known not to be extinct.

The main result on critical processes is

Theorem (6.6.11)

Consider a critical branching process with finite reproduction variance $\sigma^2 = \text{Var}\,[\xi(\infty)]$ such that

$$\lim_{t \to \infty} t^2 \{1 - \mu(t)\} = \lim_{t \to \infty} t^2 \{1 - L(t)\} = 0.$$

Then

$$\lim_{t \to \infty} tP[z_t > 0] = 2\beta/\sigma^2$$

$(= \infty$ if $\sigma^2 = 0)$, where $0 < \beta = \int_0^\infty t\mu(\mathrm{d}t)$. If the process is further taken to be non-lattice (or the limit is reinterpreted as above) and $l = E[\lambda]$, then also

$$\lim_{t \to \infty} E[z_t/t \mid z_t > 0] = l\sigma^2/2\beta^2.$$

For Bellman–Harris processes it further holds that z_t/t, given that $z_t > 0$, is asymptotically exponential as $t \to \infty$ with parameter $2\beta/\sigma^2$.

The asymptotic exponentiality does also hold if L and μ both have finite second moments and

$$\liminf_{t \to \infty} \left| 1 - \int_0^\infty e^{itu}\mu(\mathrm{d}u) \right| > 0$$

[*Theorem* (6.6.17)].

In all these ramifications of the family extinction problem, results for the general process, treated in Chapter 6, are strong enough to encompass the special case of Galton–Watson processes (Chapter 2). When it comes to the demographic circle of problems things are somewhat more involved. Certainly questions on age distribution are pointless for the Galton–Watson process but the principle of Malthusian growth has been established slightly more generally here than for processes with a more sophisticated time structure. Indeed,

Theorem (2.7.1)

Consider a supercritical (*i.e.* $1 < m < \infty$) Galton–Watson branching process. Then $\lim_{n \to \infty} z_n/m^n$ exists a.s. and is non-zero if and only if $E[\xi \log \xi] < \infty$. If the reproduction variance $\text{Var}\,[\xi]$ is finite, the convergence holds also in mean square.

For general processes, we cannot yet avoid the second moment of the reproduction law:

Theorem (6.8.1) and Corollary (6.10.4)

For a general, non-lattice branching process with $m = \mu(\infty) > 1$ and $\mathrm{Var}\,[\xi(\infty)] < \infty$, $\lim\limits_{t \to \infty} \mathrm{e}^{-\alpha t} z_t^\alpha$ exists a.s. and in mean square. The limit can be written

$$z \int_0^a \mathrm{e}^{-\alpha t} P[\lambda > t]\,\mathrm{d}t/\beta = z\{1 - E[\mathrm{e}^{-\alpha(a \wedge \lambda)}]\}/\alpha\beta$$

where z is a random variable such that $E[z] = 1$ and $P[z = 0, z_t \not\to 0] = 0$. It follows that

$$z_t^a/z_t \to \int_0^a \mathrm{e}^{-\alpha t} P[\lambda > t]\,\mathrm{d}t \bigg/ \int_0^\infty \mathrm{e}^{-\alpha t} P[\lambda > t]\,\mathrm{d}t$$
$$= \{1 - E[\mathrm{e}^{-\alpha(a \wedge \lambda)}]\}/\{1 - E[\mathrm{e}^{-\alpha\lambda}]\}$$

a.s. on the set where $z_t \not\to 0$. As usual there is a lattice analogue. *

The theorem holds also when the ancestor differs from the progeny (provided she does not remain childless with certainty) and thus empiric population ages tend to stabilize around Euler's and Lotka's stable distribution. We make the following formal definition: The *stable age distribution K* of a branching process with Malthusian parameter α and life length distribution L is determined by

$$K(a) = \int_0^a \mathrm{e}^{-\alpha t}\{1 - L(t)\}\,\mathrm{d}t \bigg/ \int_0^\infty \mathrm{e}^{-\alpha t}\{1 - L(t)\}\,\mathrm{d}t$$

for $0 \leq a \leq \infty$, provided the denominator converges. This is always the case for $\alpha > 0$. If $\alpha = 0$ it is true exactly when the expected life span is finite and if $\alpha < 0$ only when $E[\mathrm{e}^{-\alpha\lambda}] < \infty$. If $\alpha \neq 0$

$$K(a) = \{1 - E[\mathrm{e}^{-\alpha(a \wedge \lambda)}]\}/\{1 - E[\mathrm{e}^{-\alpha\lambda}]\}.$$

A related question is that of what happens in a process starting from an ancestor with a stably distributed age. The answer is provided by

Corollary (6.11.3) (in a somewhat loose formulation).

Consider a general non-lattice branching process starting from an ancestor who has a stably distributed age. Then the expected number of individuals at time t, aged a or less, is $\mathrm{e}^{\alpha t} K(a)$, $t > 0$, $0 \leq a \leq \infty$.

With one exception I have given results from Chapter 6, since there the basic questions, that were raised during the history of branching processes and demography, are treated in the most general framework. However they meet already in the context of the generation counting Galton–Watson

---

* Added in proof: Recently K. B. Athreya and N. Kaplan obtained this for Bellman–Harris processes with $E[\xi \log \xi] < \infty$. Presumably their method works also in the general case. A related result (convergence in probability) has been given by T. Savits.

process of Chapter 2. Here the first sections serve to establish some fundamental facts like Watson's recurrence relation for the generating functions: If f_n is the generating function of the number of individuals in the nth generation, $n \in Z_+$, then $f_n = f \circ f_{n-1}$, $f_0(s) = s$, the process starting from one ancestor (Theorem 2.2.2). As a consequence of this $E[z_n] = m^n$ with a similar formula for the variance. The next section is devoted to the family extinction problem in Erlang's formulation and Section 2.4 to the Galton–Watson cases of Theorem 6.6.10, quoted above. Similarly Section 2.6 treats a generation version of Theorem 6.7.2. The chapter goes on to a martingale treatment of supercritical processes, culminating in the result already referred to as Theorem 2.7.1. The subsequent section uses this for a simple statement on the number of generations it takes for the population size to attain some prescribed high level, whereas 2.9 and 2.10 establish some normal type results on the rate of the convergence of z_n/m^n to its limit. In the next section we leave the branching process itself in order to investigate the total number of individuals that have been born in the n first generations, $y_n = z_0 + \ldots + z_n$ and its limit y as $n \to \infty$. Theorem (2.11.6) states that $P[y = k] = P[z_{n+1} = k - 1 | z_n = k]/k$ for $k \in N$. Section 2.12 then studies the pair (y_n, z_n) of random variables. The last two sections concern inference for Galton–Watson processes, the most important outcome being the maximum likelihood estimator $(y_n - 1)/y_{n-1}$ of the reproduction mean m, if observations are available of the successive generation sizes $z_0, z_1, \ldots, z_n$ (Theorem (2.13.3)).

Chapter 3 is more scattered. A Galton–Watson process with immigration (Section 3.1) is the population obtained by letting in each generation an i.i.d. number of new ancestors appear. In the supercritical case, the immigration does not affect the process so much; Malthus's increase in 'a geometrical ratio' remains (Theorem (3.1.3)). But if $m \leq 1$ new phenomena arise: Supposing $m < 1$, y_n has a limit in distribution (Theorem (3.1.1)) and if $m = 1$, $\sigma^2 < \infty$, then $y_n/n \overset{d}{\to}$ a gamma random variable (Theorem (3.1.2)). Many actual populations start from a substantial number of ancestors. Section 3.2. is devoted to processes started from r_n ancestors where $r_n \to \infty$. For suitable r_n and norming of the process a normal limit behaviour is obtained for supercritical processes (Theorem (3.2.1)), a compound Poisson for subcritical ones (Theorem (3.2.2)), and a Poisson mixture of gamma distribution (Theorem (3.2.3)) if $m = 1$. Section 3.3 considers processes where m is very close to one and in the following section z_n is studied for large n when simultaneously $m \to 1$ and the number of ancestors tends to infinity. The issue is the diffusion approximation of Theorem (3.4.1). The final three sections of the chapter expand the setting towards time inhomogeneity. In Sections 3.5 and 3.6 it is assumed that the reproduction generating function f need not be the same from generation to generation, i.e. the environment might vary. The basic result is Theorem (3.5.2), saying that the typical extinction or explosion behaviour persists: either $z_n \to 0$

or $z_n \to \infty$. Section 3.6 elaborates this, establishing classical results like Theorems (6.7.2) [(2.6.1)], (6.6.10) [(2.4.2)] and (6.8.1) [(2.7.1)] for populations in a varying environment. The chapter ends by applications of this to processes where the reproduction generating function is chosen for each generation at random.

Chapter 4 is just a brief exposé of multi-type Galton–Watson processes. Results are stated without proofs. Chapter 5 contains tools from probability theory and analysis. The martingale convergence theorem, already used in Section 2.7, is proved, the renewal theorem with some refinements is deduced, and some properties of point processes are discussed.

This takes us to what may be the main part of the book, Chapter 6, where the theory of general branching processes is described. The essence of it was sketched in the preceding. However in Section 6.11 a new concept is introduced, that of processes counted with a random characteristic. Here it is assumed that with each individual there is associated besides the life span λ and reproduction process ξ a stochastic process $\chi(t)$, $t \in R$, $\chi(t) = 0$ if $-\infty \le t < 0$. The latter is the *random characteristic*. A process z_t^χ is defined to be the sum of all $\chi_x(t - \sigma_x)$, σ_x being the birth epoch of x (infinite if x is never born). Various processes can be obtained through different choices of χ (like z_t^a if $\chi = 1_{[0,a\wedge\lambda)}$) and an analogue of Theorem (6.8.1) and Corollary (6.10.4) above holds. In Sections 6.12 and 6.13 this is applied to some special choices of random characteristics, yielding the total number of individuals born up to time t and the integral of a branching process. The last section focuses on the estimation of m for general processes.

Chapter 7 relates to Chapter 6 as does Chapter 3 to its predecessor. First processes allowing immigration are considered. Provided immigrations occur at the points of a renewal process, that the number of immigrants per renewal epoch is i.i.d., and that the processes initiated are subcritical, the existence of a limit in distribution of the total population (y_t) is established (Theorem (7.1.4)). For Bellman–Harris processes with a slightly more general type of immigration it is proved that y_t/t converges, as in the Galton–Watson case, to a gamma limit (Theorem (7.1.5)). Supercritical processes remain comparatively unaffected even by very general immigration schemes (Theorem (7.1.6)). Section 7.2 considers increasing numbers of ancestors; the counterparts of results from Section 3.2 are easily seen to hold. Problems of almost critical processes, discussed in the following section, seem more difficult. Asymptotic exponentiality is obtained only for a special class of Bellman–Harris processes. Also the diffusion approximation of Section 7.4 is deduced merely for Bellman–Harris processes.

Section 8.1 presents two topics: Lotka's model for 'self renewing aggregates' in customary demographic terminology and a special type of branching processes, the age-dependent birth and death process, which is closely connected with the demographic theory. The next section treats a relation that plays a main rôle in stable population theory, Lotka's equation. In 8.3 are discussed first Fisher's reproductive value and then the concepts of birth and death rates. It was mentioned above that the entity

$$\beta = \int_0^\infty te^{-\alpha t}\mu(dt)$$

in a Malthusian process is known (among demographers) as the average age of mothers at child-bearing. Section 8.4 applies populations counted with random characteristics to give this a basis. Section 8.5 is devoted to generations, their length, their distribution over time, and the dual problem of the generation composition of the population at a given time. To be unfairly short let us say that at time t the dominating generation tends to be $[t/\beta]$, the others contributing like a normal frequency bell (Theorem (8.5.1)). The main part of the nth generation is however born around time

$$n \int_0^\infty t\mu(dt)/\mu(\infty),$$

the distribution around this tending to be Gaussian bell-shaped as well (Theorem (8.5.5)).

Chapter 9 is exclusively concerned with cell proliferation. First I try to convery an idea of the biological background. The basic model of a binary splitting process is formulated. Relations between the expected cycle time (i.e. time from birth to division) and the doubling time are established. The mitotic index (the fraction of cells about to divide) is determined. The second section applies the statistical theory of Section 6.14 to studying the incidence of so called cell disintegration or death, i.e. the probability of cells never attaining mitotic division but dying untimely. In particular, normal and tumorous cells are compared. Section 9.3 makes use of the stable age distribution to estimate the distribution of cycle times and jointly with the obtained estimates of cell death it indicates that a more rapid proliferation of tumorous than normal cells is the effect of more rare cell disintegration rather than of shorter cycle times. Sections 9.4, 9.5, and 9.6 discuss a widely spread experimental technique in the study of cell proliferation, that of fraction labelled mitoses. The background is that cells synthesize DNA only during a particular part of their lives. This DNA can be labelled and then observed when cells enter the mitosis stage preceding division, the evolution of the fraction labelled mitoses being used for inference about the cell cycle. A critical interpretation is made in terms of the binary splitting process. Section 9.7 gives a mathematical treatment of phenomena that arise when cells are blocked at some stage, e.g. mitosis. In the next section some practical problems around the age and generation distributions are discussed. Finally a topic belonging to multi-type theory is taken up. Consider a population with two types of individuals such that one of these can beget children of its own kind only. How will the population be composed of the two types? Biological applications are given.

REFERENCES

Numbers [1–39] are in chronological order.

1. Süssmilch, J. P., *Die göttliche Ordnung in den Veränderungen des menschlichen Geschlechts, aus der Geburt, dem Tode, und der Fortpflanzung desselben erwiesen.* Berlin, 1741.

2. Euler, L., *Introductio in analysin infitorum I.* Bousquet, Lausanne, 1748. Available as: Leonhardi Euleri opera omnia, Series prima VIII, Teubner, Leipzig and Berlin, 1922.

3. Euler, L., *Recherches générales sur la mortalité et la multiplication du genre humain.* Histoire de l'Académie Royale des Sciences et Belles-Lettres, année 1760, 144–164, Berlin, 1767. Available in: *Leonhardi Euleri opera omnia,* Series prima VII, 79–100. Teubner, Leipzig and Berlin, 1923.

4. Malthus, T. R., *An Essay on the Principle of Population, as it Affects the Future Improvements of Society, with Remarks on the Speculations of Mr. Godwin, M. Condorcet, and Other Writers.* 5th ed. John Murray, London, 1817 (1st ed. 1798).

5. Bienaymé, I. J., De la loi de multiplication et de la durée des familles. *Soc. Philomath. Paris Extraits, Sér. 5,* 37–39, 1845.

6. Moser, L. F., *Die Gesetze der Lebensdauer.* Veit, Berlin, 1839.

7. Benoiston de Châteauneuf, L. F., Mémoire sur la durée des familles nobles de France. (Read Aug. 31, 1844 and Feb. 16, 1845.) *Mémoires de l'Académie royale des sciences morales et politiques de l'Institut de France* 5, 753–794, 1847.

8. Galton, F., *Hereditary Genius.* London, 1869.

9. Schröder, E., Ueber iterierte Functionen. *Math. Ann.* 3, 296–322, 1871.

10. de Candolle, A., *Histoire des sciences et des savants depuis deux siècles,* H. Georg, Geneva, 1873.

11. Galton, F., Problem 4001. *Educational Times,* 17, 1873.

12. Galton, F., and Watson, H. W., On the probability of the extinction of families. *J. Anthropol. Soc. London (Royal Anthropol. Inst. G. B. Ireland)* 4, 138–144, 1875.

13. Lotka, A., Relation between birth rates and death rates. *Science* 26, 21–22, 1907.

14. Herbelot, L., Application d'un théorème d'analyse à l'étude de la repartition par âge. *Bull. Trimestriel de l'Inst. des Actuaires Français* 19, 293–298, 1909.

15. Sharpe, F. R. and Lotka, A., A problem in age distribution. *Philos. Mag.* 21, 435–438, 1911.

16. McKendrick, A. G., Studies on the theory of continuous probabilities, with special reference to its bearing on natural phenomena of a progressive nature. *Proc. London Math. Soc. Ser. II* 13, 401–416, 1914.

17. Fisher, R. A., On the dominance ratio. *Proc. Royal Soc.* Edinburgh **42**, 321–341, 1922.
18. Yule, G. U., A mathematical theory of evolution based on the conclusions of Dr. J. C. Willis, F.R.S. *Philos. Trans. Royal Soc.* London B **213**, 21–87, 1924.
19. Haldane, J. B. S., A mathematical theory of natural and artificial selection, V: Selection and mutation. *Proc. Cambridge Philos. Soc.* **23**, 838–844, 1927.
20. Erlang, A. K., Opgave til Løsning. *Matematisk Tidsskrift B*, 36, 1929.
21. Steffensen, J. F., Om Sandsyndligheden for at Afkommet uddør. *Matematisk Tiddskrift B*, 19–23, 1930.
22. Volterra, V., *Leçons sur la théorie mathématique de la lutte pour la vie.* Gauthier-Villars, Paris, 1931.
23. Lotka, A., *Théorie analytique des associations biologiques I and II.* Hermann, Paris, 1934 and 1939.
24. Furry, W. H., On fluctuation phenomena in the passage of high energy electrons through lead. *Phys. Rev.* **52**, 569–581, 1937.
25. Feller, W., Die Grundlagen der Volterraschen Theorie des Kampfes ums Dasein in wahrscheinlichkeitstheoretischer Behandlung. *Acta Biotheoretica* **5**, 11–40, 1939.
26. Kolmogorov, A. N. and Dmitriev, N. A., Vetvyaščiesya slučaynye processy (Branching stochastic processes). *Dokl. Akad. Nauk SSSR* **56**, 7–10, 1947.
27. Kolmogorov, A. N. and Sevast'yanov, B. A., Vyčisleniye final'nyh veroyatnostey dl'a vetvyaščihs'ya slučaynyh processov (The calculation of final probabilities of branching random processes). *Dokl. Akad. Nauk SSSR* **56**, 783–786, 1947.
28. Kendall, D. G., On the generalized 'birth-and-death' process. *Ann. Math. Statist.* **19**, 1–15, 1948.
29. Bellman, R. and Harris, T. E., On the theory of age-dependent stochastic branching processes. *Proc. Nat. Acad. Sci.* **34**, 601–604, 1948.
30. Kendall. D. G., Stochastic processes and population growth. *J. Royal Statist. Soc.* B **11**, 230–264, 1949.
31. Harris, T. E., *The Theory of Branching Processes.* Springer, Berlin, 1963.
32. Ney, P. E., Generalized branching processes I and II. *Ill. J. Math.* **8**, 316–350, 1964.
33. Sevast'yanov, B. A., Vetvyaščiesya processy s prevraščeniyami, zavisyaščimi ot vozrasta častic. *Teoriya Veroyatnost. i Primenen.* **9**, 577–594, 1964. Translated as: Age-dependent branching processes. *Theory Prob. Appl.* **9**, 521–537, 1964.
34. Ryan, T. R. Jr., On age-dependent branching processes. *Ph.D. Dissertation,* Cornell University, 1968.
35. Crump, K. S. and Mode, C. J., A general age-dependent branching process I and II. *J. Math. Anal. Appl.* **24** and **25**, 494–508 and 8–17, 1968 and 1969.
36. Jagers, P., A general stochastic model for population development. *Skand. Aktuarietidskr.* **52**, 84–103, 1969.
37. Mode, C. J., *Multitype Branching Processes.* Elsevier, New York, 1971.
38. Sevast'yanov, B. A., *Vetvyaščiyesya processy (Branching Processes).* Mir, Moscow, 1971.
39. Athreya, K. B. and Ney, P. E., *Branching Processes.* Springer, Berlin, 1972.

Historical papers

40. Kendall, D. G., Branching processes since 1873. *J. London Math. Soc.* **41**, 385–406, 1966.
41. Heyde, C. C., and Seneta, E., The simple branching process, a turning point test and a fundamental identity: a historical note on I. J. Bienaymé. *Biometrika* **59**, 680–683, 1972.

See also [30, 31] above.

---

42. Everett, C. J. and Ulam, S., Multiplicative systems. *Proc. Nat. Acad. Sci.* **34**, 385–406, 1966.
43. Otter, R., The multiplicative process. *Ann. Math. Statist.* **20**, 206–224, 1949.
44. Urbanik, K., On a stochastic model of a cascade. *Studia Math.* **15**, 34–42, 1955.

Chapter 2

The Galton–Watson Process

2.1 INTRODUCTION

Let r_x be one if the individual x is realized, zero otherwise. Then the definition of Galton–Watson processes, given in Section 1.2, reduces to

$$z_n = \sum_{x \in N^n} r_x = \sum_{x \in N^{n-1}} r_x \xi_x$$

for $n \in N$. We consider only the case of one ancestor, 0. Obviously $z_0 = 1$ and for $n \geq 1$ the sum is made up of z_{n-1} i.i.d. summands ξ_x. Thus if $\{X_{nj}; n \in N, j \in N\}$ is a double array of independent random variables distributed according to the reproduction law $\{p_k; k \in Z_+\}$, then we can write

$$z_0 = 1$$

$$z_{n+1} = \sum_{j=1}^{z_n} X_{nj}, \qquad n \in N,$$

(2.1.1)

in the sense that the sequence $\{z_n\}$ given by this recursion has the same probability distribution as a Galton–Watson process. (A sum from one to zero is given the value zero.)

If we denote by $\mathscr{B}_n = \sigma(z_0, z_1, \ldots, z_n)$ the σ-algebra generated by $z_0, z_1, \ldots, z_n$, then (2.1.1) yields

$$P[z_{n+1} = k | \mathscr{B}_n] = P\left[\sum_{j=1}^{z_n} X_{nj} = k | z_n \right].$$

Hence $\{z_n\}$ is a homogeneous Markov chain with transition probabilities

$$p_{jk} = P[z_{n+1} = k | z_n = j] = \sum_{i_1 + \ldots + i_j = k} p_{i_1} \cdots p_{i_j} = p_k^{*j}, \qquad (2.1.2)$$

where the last symbol is conventional convolution notation. Conversely, any Markov chain with transition probabilities satisfying (2.1.2) has the distribution of a Galton–Watson process with reproduction law $\{p_k\}$.

A consequence of this Markovian property will be used over and over again more or less explicitly. We state it as a lemma. In this one $\mathscr{B}(Z_+^\infty)$

is the smallest σ-algebra of subsets of Z_+^∞ containing all finite dimensional sets.

Lemma (2.1.1)

Let $A \in \mathcal{B}(Z_+^\infty)$, $k \in N$ and $\{z_n^{(1)}\}$, $\{z_n^{(2)}\}$, ..., be independent Galton–Watson processes with the reproduction law of $\{z_n\}$. Then for any $r \in N$

$$P[\{z_n ; n > r\} \in A | z_r = k] = P\left[\left\{ \sum_{j=1}^{k} z_n^{(j)} ; n \geq 1 \right\} \in A \right].$$

Proof. If $A = \{r_1\} \times \ldots \times \{r_j\} \times Z_+ \times Z_+ \times \ldots$ for some $j \in N$, $r_j \in Z_+$ the assertion follows directly from (2.1.2). Since all finite dimensional sets are disjoint unions of such sets equality holds also for these. The lemma now follows from Dynkin's theorem (Appendix).   □

Example (2.1.2)

The most important branching processes in biology are those of binary splitting. *In particular, a Galton–Watson process is binary if the reproduction law satisfies $p_0 = 1 - p$, $p_2 = p$, for some $0 \leq p \leq 1$.*
Then

$$p_{jk} = \begin{cases} 0, & \text{if } k \text{ is odd}, \\ \binom{j}{k/2} p^{k/2}(1-p)^{j-k/2}, & \text{if } k \text{ is even}, \end{cases}$$

$\binom{j}{i}$ interpreted as zero if $i > j$.

Hence

$$P[z_1 = 2k_1, z_2 = 2k_2, \ldots, z_n = 2k_n]$$
$$= \prod_{i=1}^{n} \binom{2k_{i-1}}{k_i} \left(\frac{p}{1-p} \right)^{\sum_{i=1}^{n} k_i} (1-p)^{2 \sum_{i=0}^{n-1} k_i}$$

if we agree to put $2k_0 = 1$.   □

2.2 MOMENTS AND THE GENERATING FUNCTION

The conditioning applied to (2.1.1) in order to establish the Markovian character of the process is basic for the entire analysis of the Galton–Watson process. It is used to calculate the generating functions of z_n,

$$f_n(s) = E[s^{z_n}] = \sum_{k=0}^{\infty} P[z_n = k] s^k$$

for $0 \leq s \leq 1$ or, sometimes, complex $|s| \leq 1$. It is also important for the martingale arguments on which great part of the treatment of the asymptotics hinges.

Lemma (2.2.1)

Assume that $Y, X_1, X_2, \ldots$ is a sequence of independent Z_+-valued random variables. Let all X_i have the same generating function g, and Y the generating function h. Then, with

$$S = \sum_{j=1}^{Y} X_j,$$

$$E[s^S] = h \circ g(s), \qquad 0 \leq s \leq 1$$

$$E[S] = E[X_1] E[Y],$$

$$\mathrm{Var}[S] = \mathrm{Var}[X_1] E[Y] + E^2[X_1] \mathrm{Var}[Y],$$

the two last equalities holding in the sense that if one side is finite then so is the other and they are equal.

Proof. We shall only prove the first assertion. The two others follow by differentiation.

$$E[s^S] = E[E[s^S|Y]] = E[\{E[s^{X_1}]\}^Y] = E[g(s)^Y] = h \circ g(s). \qquad \square$$

Applying this lemma to (2.1.1) we obtain

$$f_n = f_{n-1} \circ f = \ldots = f \circ f_{n-1}$$

$$E[z_n] = mE[z_{n-1}]$$

$$\mathrm{Var}[z_n] = \sigma^2 E[z_{n-1}] + m^2 \mathrm{Var}[z_{n-1}].$$

Here

$$f(s) = \sum_{k=0}^{\infty} p_k s^k = E[s^{\xi_x}] = E[s^{z_1}] = f_1(s)$$

is the *reproduction generating function,*

$$m = \sum_{k=1}^{\infty} k p_k = E[\xi_x] = E[z_1] = f'(1)$$

is the *mean number of children per individual* or *reproduction mean* and

$$\sigma^2 = \sum_{k=1}^{\infty} k^2 p_k - m^2 = \mathrm{Var}[\xi_x] = \mathrm{Var}[z_1] = f''(1) + f'(1) - (f'(1))^2$$

is *the reproduction variance.* In conclusion, recursion yields:

Theorem (2.2.2)

The generating function of z_n is the composition $f \circ \dots \circ f$ (n times), its expectation is m^n, and its variance is

$$\sigma^2 m^{n-1}(m^n - 1)/(m - 1), \quad \text{if } m \neq 1,$$

$$n\sigma^2, \quad \quad \quad \quad \quad \quad \quad \text{if } m = 1.$$

2.3 THE EXTINCTION PROBABILITY

The event

$$Q = \bigcup_{n=1}^{\infty} \bigcap_{k=n}^{\infty} \{z_k = 0\} = \{z_n \to 0\}$$

is called the *extinction* of the process.

Since

$$z_k = 0 \Rightarrow z_n = 0 \quad \text{for } n \geq k$$

(prove this from the definition!)

$$Q = \bigcup_{k=1}^{\infty} \{z_k = 0\},$$

and

$$P[Q] = \lim_{n \to \infty} P\left[\bigcup_{k=1}^{n} \{z_k = 0\}\right] = \lim_{n \to \infty} P[z_n = 0] = \lim_{n \to \infty} f_n(0),$$

showing also that the last limit must exist. We shall write $q = P(Q)$ for the *extinction probability*.

Theorem (2.3.1)

The equation $f(s) = s$ has exactly one root in $[0, 1)$ if $m > 1$ and none if $m \leq 1$ and $p_1 \neq 1$. The extinction probability q is the smallest non-negative root of the equation, that is

$$m > 1 \Rightarrow q < 1,$$

$$p_1 = 1 \Rightarrow q = 0,$$

$$m = 1, \quad p_1 < 1 \Rightarrow q = 1,$$

$$m < 1 \Rightarrow q = 1.$$

Proof. We discard the trivial case $p_1 = 1$. Since $f_n(0) \uparrow q$, and f is continuous on $[0, 1]$

$$f(q) \leftarrow f \circ f_n(0) = f_{n+1}(0) \to q.$$

Hence $f(q) = q$. Assume that a is any number in $[0, 1]$ such that $f(a) = a$. Then $f_1(0) = f(0) \leq f(a)$ and $f_n(0) \leq a \Rightarrow f_{n+1}(0) = f \circ f_n(0) \leq f(a) = a$ and so, for all $n, f_n(0) \leq a$ proving that $q \leq a$. So q is the smallest root.

If $m \leq 1, p_1 \neq 1, s < 1$,

$$(f(s) - s)' = f'(s) - 1 < f'(1) - 1 \leq 0$$

and $f(s) - s$ must decrease strictly. As $f(1) = 1$ it follows that $f(s) > s$ for $0 \leq s < 1$. If on the other side $m > 1$, $f(s)$ increases quicker towards $f(1) = 1$ than does s. Hence $f(s) < s$ for s in some left neighbourhood of 1. But since $f(0) \geq 0$ there must be at least one $0 \leq s < 1$ such that $f(s) = s$. Now assume that there are two, say $f(s_1) = s_1, f(s_2) = s_2, 0 \leq s_1 < s_2 < 1$. With $\varphi(s) = f(s) - s$ in this case $\varphi(s_1) = \varphi(s_2) = \varphi(1) = 0$ and there must be $s_1 < a < s_2 < b < 1$ such that $\varphi'(a) = \varphi'(b) = 0$ and hence $f'(a) = f'(b)$ contradicting the fact that f' is strictly increasing—as it must be if $m > 1$. □

Combining this theorem with the results of the preceding section, we see that though $z_n \to 0$ a.s. when $m = 1$ (and $p_1 < 1$), $\text{Var}[z_n] \to \infty$, indicating a substantial instability. There is thus good reason for differentiating between three kinds of Galton–Watson processes: a Galton–Watson process is said to be *subcritical* if $m < 1$, *critical* if $m = 1$, and *supercritical* if $m > 1$. We shall not concern ourselves with the intricacies of the case $m = \infty$.

Before proceeding let us note that for $0 \leq s \leq q, f_n(0) \leq f_n(s) \leq f_n(q) = q$. Since $f(s) \geq s$ for $0 \leq s \leq q$ it follows by induction that $f_{n+1} \geq f_n$ on $[0, q]$ and thus $f_n \uparrow q$ here. Similarly $f_{n+1} < f_n$ on $(q, 1)$ showing that for $q \leq s < 1$, $f_n(s) \downarrow$ some number strictly less than one. Since this number must solve $f(s) = s$, actually $f_n(s) \downarrow q$ for $q \leq s < 1$. Hence, for $0 \leq s < 1$, $f_n(s) \to q$. From this and the continuity theorem (Reference [2], page 408) it follows that $\lim_{n \to \infty} P[z_n = k] = 0$ for any $k \in N$. Actually still more can be deduced with little trouble.

Theorem (2.3.2)
Provided $p_1 < 1, P[z_n \to \infty] = 1 - q$.

Proof. First note that if $q = 1$ there is nothing to prove. Otherwise $f'(q) < 1$. By induction

$$f_n'(q) = \{f'(q)\}^n, \qquad n \in N.$$

But, for any $k, n \in N$, assuming $0 < q < 1$,

$$P\{1 \le z_n \le k\} = \sum_{j=1}^{k} P[z_n = j]$$

$$\le \sum_{j=1}^{k} P[z_n = j]\, q^{j-1} j/q^k \le f'_n(q)/q^k = \{f'(q)\}^n/q^k.$$

Hence

$$\sum_{n=1}^{\infty} P[1 \le z_n \le k] < \infty$$

and the always efficient Borel–Cantelli lemma shows that z_n cannot, except for outcomes in a set of probability zero, visit the interval $[1, k]$ infinitely often.

If, finally, $q = 0$, then $p_0 = 0$, the sequence $\{z_n\}$ is non-decreasing, and the theorem follows, since $z_n \to \infty$ in probability. $\quad\square$

Example (2.3.3)

If $f(s) = 1 - p + ps^2$, then $q = 1$ if $p \le 1/2$ and $q = (1 - p)/p$ if $1/2 < p \le 1$.

Background

See Section 1.1. Processes with an infinite reproduction mean are discussed in References [1] and [3]–[5].

2.4 CRITICAL PROCESSES

The instabilities of processes with $m = 1$, $p_1 < 1$, can be grasped in many ways. We have already established that

$$P[z_n \to 0] = 1,$$

$$E[z_n] = 1,$$

$$\mathrm{Var}[z_n] \to \infty,$$

making it clear that if z_n is not zero we might expect it to be large. Still critical processes have the most easily described limit behaviour of the three cases.

Lemma (2.4.1)

Assume that $m = 1$ and $\sigma^2 < \infty$. Then,

$$\lim_{n \to \infty} \frac{1}{n} \left\{ \frac{1}{1 - f_n(s)} - \frac{1}{1 - s} \right\} = \frac{\sigma^2}{2}$$

uniformly in $0 \le s < 1$.

Proof. Let $0 \leq s < 1$. Since

$$f(s) = s + (\sigma^2/2)(1 - s)^2 + r(s)(1 - s)^2 \tag{2.4.1}$$

for some function r such that $\lim\limits_{s \uparrow 1} r(s) = 0$,

$$\frac{1}{1 - f(s)} - \frac{1}{1 - s} = \frac{f(s) - s}{(1 - f(s))(1 - s)} = \frac{(\sigma^2/2)(1 - s)^2 + r(s)(1 - s)^2}{(1 - f(s))(1 - s)}$$

$$= \frac{1 - s}{1 - f(s)} \{\sigma^2/2 + r(s)\} = \sigma^2/2 + \rho(s),$$

where again $\lim\limits_{s \uparrow 1} \rho(s) = 0$. Iterating this,

$$\frac{1}{n} \left\{ \frac{1}{1 - f_n(s)} - \frac{1}{1 - s} \right\} = \frac{1}{n} \sum_{j=0}^{n-1} \left\{ \frac{1}{1 - f \circ f_j(s)} - \frac{1}{1 - f_j(s)} \right\}$$

$$= \sigma^2/2 + \frac{1}{n} \sum_{j=0}^{n-1} \rho \circ f_j(s).$$

Since $f_n(0) \leq f_n(s) \leq 1$ and $f_n(0) \uparrow 1$ the convergence $f_n(s) \to 1$ is uniform and since ρ is bounded the lemma is proved. □

With help of this the asymptotics of critical process are easily exposed:

Theorem (2.4.2)
  If $m = 1$ and $\sigma^2 < \infty$, then

(a) $\lim\limits_{n \to \infty} nP[z_n > 0] = 2/\sigma^2$

(b) $\lim\limits_{n \to \infty} E[z_n/n | z_n > 0] = \sigma^2/2$

(c) $\lim\limits_{n \to \infty} P[z_n/n \leq u | z_n > 0] = 1 - \exp(-2u/\sigma^2), u \geq 0.$

Proof.

(a) $nP[z_n > 0] = n\{1 - f_n(0)\} = \left\{ \frac{1}{n} \left(\frac{1}{1 - f_n(0)} - 1 \right) + \frac{1}{n} \right\}^{-1} \to 2/\sigma^2$

(b) $E[z_n/n | z_n > 0] = E[z_n]/n\{1 - f_n(0)\} = 1/n\{1 - f_n(0)\} \to \sigma^2/2$

(c) Let $u > 0$.

$$E[\exp(-uz_n/n)|z_n > 0] = 1 - \frac{1 - f_n(\exp(-u/n))}{1 - f_n(0)}$$

$$= 1 - \frac{1}{n\{1 - f_n(0)\}}\left\{\frac{1}{n}\left[\frac{1}{1 - f_n(\exp(-u/n))} - \frac{1}{1 - \exp(-u/n)}\right]\right.$$

$$\left. + \frac{1}{n(1 - \exp(-u/n))}\right\}^{-1} \to 1 - (\sigma^2/2)(\sigma^2/2 + 1/u)^{-1} = 1/(1 + u\sigma^2/2) ,$$

which is the Laplace transform of the right hand side of (c). The uniformity in Lemma (2.4.1) was essential.   □

In the non-critical cases we shall obtain our results without any restrictions on the reproduction variance. Also if $m = 1$ there is some knowledge about the case $\sigma^2 = \infty$. R. S. Slack [9] has shown that if

$$f(s) = s + (1 - s)^{1+\alpha}L(1 - s)$$

where $0 < \alpha < 1$ and L is of slow variation at the origin, then for $u \geq 0$

$$\lim_{n \to \infty} E[\exp(-u[1 - f_n(0)]z_n)|z_n > 0] = 1 - u(1 + u^\alpha)^{-1/\alpha}.$$

Background

Theorem (2.4.2) is a contribution from the Russian probability school, (a) from Reference [7] and (b), (c) from Reference [10] under the condition $f'''(1) < \infty$. The present form was given in Reference [6] together with the lemma. The straightforward proof of the latter is a special case of an argument in Reference [8].

2.5 A SIMPLE BUT BASIC LEMMA

The remainder r, in the Taylor expansion of f around one,

$$f(s) = 1 - m(1 - s) + r(s)(1 - s), \qquad 0 \leq s \leq 1 \qquad (2.5.1)$$

plays a crucial role in the analysis of subcritical as well as supercritical processes. Obviously,

$$r(s) = m - \{1 - f(s)\}/(1 - s),$$

$$r(0) = m - (1 - p_0) \geq 0,$$

$$r(q) = m - 1 > 0, \qquad \text{if } q < 1,$$

$$r(1-) = 0, \qquad \text{and}$$

$$r'(s) \leq 0, \qquad 0 \leq s < 1.$$

Hence, r is a decreasing function from $[0, 1)$ into $[0, m]$.

Lemma (2.5.1)

For any δ, $0 < \delta < 1$,

$$\sum_{k=1}^{\infty} r(1 - \delta^k) < \infty \Leftrightarrow \sum_{k=1}^{\infty} p_k k \log k < \infty.$$

Proof.

$$r(s) = m - \sum_{k=0}^{\infty} s^k \left(1 - \sum_{k=0}^{\infty} p_k s^k \right)$$

$$= m - \sum_{k=0}^{\infty} s^k + \sum_{n=0}^{\infty} \left(\sum_{k=0}^{n} p_k \right) s^n = m - \sum_{n=0}^{\infty} a_n s^n,$$

where

$$a_n = \sum_{k>n} p_k.$$

Note that $\sum a_n = m$, write $\alpha = -\log \delta$, and let $j \in N$, $j > 1$. Then

$$r(1 - \delta) + \int_1^j r(1 - e^{-\alpha x}) \, dx \geq \sum_{k=1}^{j} r(1 - \delta^k) \geq \int_1^j r(1 - e^{-\alpha x}) \, dx$$

$$= \alpha^{-1} \int_{1-\delta}^{1-\delta^j} \frac{r(s) \, ds}{1 - s},$$

after the substitution $s = 1 - e^{-\alpha x}$. Hence

$$\sum r(1 - \delta^k) < \infty \Leftrightarrow \int_0^1 \frac{r(s) \, ds}{1 - s} < \infty.$$

But

$$0 \leq \int_0^1 \frac{r(s) \, ds}{1 - s} = \int_0^1 \sum s^k \left(m - \sum a_n s^n \right) ds$$

$$= \int_0^1 \sum_k \left(\sum_{n>k} a_n \right) s^k ds = \sum_{n=1}^{\infty} a_n \sum_{k=0}^{n-1} 1/(k+1)$$

$$= \sum_{n=1}^{\infty} a_n [\log n + O(1)] < \infty \Leftrightarrow$$

$$\sum_{n=1}^{\infty} a_n \log n = \sum_{n=1}^{\infty} \left(\sum_{k>n} p_k \right) \log n$$

$$= \sum_{k=2}^{\infty} p_k \left(\sum_{n=1}^{k-1} \log n \right) = \sum_{k=2}^{\infty} p_k \log(k-1)!$$

$$= \sum_{k=2}^{\infty} p_k [(k-1) \log(k-1) + o(k \log k)] < \infty \Leftrightarrow \sum_{k=1}^{\infty} p_k k \log k < \infty. \qquad \square$$

Though we shall have use mostly for the lemma as stated we shall also need the closely allied relation

$$\sum_{k=1}^{\infty} \{1 - f(1 - \delta^k)\} < \infty \Leftrightarrow \sum_{k=1}^{\infty} p_k \log k < \infty, \qquad (2.5.2)$$

valid for any probability generating function and $0 < \delta < 1$. It is proved much as Lemma (2.5.1) and indeed implies it.

Background

This is from Reference [11].

2.6 SUBCRITICAL PROCESSES

If we replace s by $f_k(s)$ in Relation (2.5.1), we obtain

$$\frac{1 - f_{k+1}(s)}{1 - f_k(s)} = m\{1 - r \circ f_k(s)/m\} \qquad (2.6.1)$$

The product of these equalities for $0 \leq k < n$ is

$$\{1 - f_n(s)\}/(1 - s) = m^n \prod_{k=0}^{n-1} \{1 - r \circ f_k(s)/m\} \qquad (2.6.2)$$

Since $0 \leq r/m \leq 1$, $m^{-n}\{1 - f_n(s)\}/(1 - s)$ decreases as $n \to \infty$ to some limit $\varphi(s) \geq 0$. In particular

$$P[z_n > 0] = 1 - f_n(0) \sim m^n \varphi(0).$$

But by the well-known relation between convergence of sums and of products, $\varphi(0) > 0$ if and only if

$$\sum r \circ f_k(0) < \infty.$$

Now $1 - f(s) \leq m(1 - s)$ and by induction it follows that

$$1 - f_k(s) \leq m^k(1 - s)$$

for any k. Similarly, if $s \geq s_0$,

$$1 - f_k(s) \geq (f'(s_0))^k (1 - s);$$

and with $s_0 = p_0$ (being positive in the subcritical case) we obtain, writing $a = f'(p_0) > 0$,

$$1 - m^k \leq f_k(0) = f_{k-1}(p_0) \leq 1 - a^{k-1}(1 - p_0) \leq 1 - b^k \qquad (2.6.3)$$

where $b = a \wedge (1 - p_0)$. Therefore Lemma (2.5.1) shows that

$$\sum r \circ f_k(0) < \infty \Leftrightarrow \sum p_k k \log k < \infty$$

and the following has been proved:

Theorem (2.6.1)

In a subcritical process,

$$\lim_{n \to \infty} m^{-n} P[z_n > 0] = \begin{cases} 0 & \text{if } \sum p_k k \log k = \infty \quad \text{or} \quad p_0 = 1, \\ \\ \varphi(0) > 0 & \text{otherwise.} \end{cases}$$

The same method applies to prove what is known as Yaglom's theorem:

Theorem (2.6.2)

In the subcritical case (with $p_0 < 1$)

$$\lim_{n \to \infty} P[z_n = k | z_n > 0] = b_k$$

exists for $k \in N$,

$$\sum_{k=1}^{\infty} b_k = 1,$$

$$\sum_{k=1}^{\infty} k b_k = 1/\varphi(0) < \infty \Leftrightarrow \sum_{k=1}^{\infty} p_k k \log k < \infty$$

and if $g(s) = \sum b_k s^k$, then

$$g \circ f = mg + 1 - m \tag{2.6.5}$$

Proof. Put

$$g_n(s) = E[s^{z_n} | z_n > 0] = \frac{f_n(s) - f_n(0)}{1 - f_n(0)} = 1 - \frac{1 - f_n(s)}{1 - f_n(0)} =$$

$$= 1 - (1 - s) \prod_{k=0}^{n-1} \{1 - r \circ f_k(s)/m\} \{1 - r \circ f_k(0)/m\}^{-1}.$$

Since $f_k(s) \geq f_k(0)$ and r does not increase, the terms in the product are greater than one and there is a function g such that $g_n \downarrow g$. Since the g_n are probability generating, g can be written

$$g(s) = \sum_{k=1}^{\infty} b_k s^k.$$

Evidently $g(0) = 0$. Now, as $n \to \infty$,

$$g_n \circ f_k(0) = 1 - \frac{1 - f_k \circ f_n(0)}{1 - f_n(0)} \to 1 - m^k,$$

which can be made arbitrarily close to one by a large choice of k.

Hence $g(1-) = 1$. Further,

$$\sum k b_k = g'(1-) = \lim_{k \to \infty} \frac{1 - g \circ f_k(0)}{1 - f_k(0)}$$

$$= \lim_{k \to \infty} m^k / (1 - f_k(0)) = 1/\varphi(0),$$

and

$$g_n \circ f = 1 - \frac{1 - f_{n+1}}{1 - f_{n+1}(0)} \cdot \frac{1 - f \circ f_n(0)}{f - f_n(0)} \to 1 - (1 - g)\, m. \qquad \square$$

Though Equation (2.6.5) seems explicit and actually determines g (provided $g'(1) < \infty$ at least) it is not too helpful in computing g. For example in the binary case it reduces to

$$g(1 - p + ps^2) = 2pg(s) + 1 - 2p,$$

which the reader might try and solve.

Later we shall need the variance of $\{b_k\}$. As opposed to the expectation this can be obtained by differentiation of (2.6.5). Indeed, $c = \varphi(0)$,

$$g''(1) = f''(1)/c(m - m^2), \tag{2.6.6}$$

and $g''(1)$ is finite together with σ^2.

Background

Yaglom's theorem is from 1947 [10] and Theorem (2.6.1) is due to Kolmogorov [7], both assuming that $f'''(1) < \infty$. The final formulation was given in [11].

2.7 SUPERCRITICAL PROCESSES*

The generating function f is strictly increasing. It has an inverse g, whose nth iterate we shall denote by g_n, $n \in N$. We set $g_0(s) = s$. The function g is increasing, concave, differentiable and maps the interval $[q, 1]$ onto itself. For any s in this interval and $n \in N$ we define

$$x_n(s) = g_n(s)^{z_n}. \tag{2.7.1}$$

This random variable is measurable with respect to $\mathscr{B}_n = \sigma(z_1, \ldots, z_n)$ and

$$E[x_{n+1}(s)|\mathscr{B}_n] = E[g_{n+1}(s)^{z_{n+1}}|z_n] = E[g_{n+1}(s)^{z_1}]^{z_n}$$

$$= \{f \circ g_{n+1}(s)\}^{z_n} = g_n(s)^{z_n} = x_n(s).$$

Hence $\{x_n(s)\}$ is a non-negative martingale and by Theorem (5.1.2) $\lim_{n \to \infty} x_n(s) = x_\infty(s)$ exists almost surely. Since $0 \le x_n(s) \le 1$, so is $x_\infty(s)$,

* If you are not acquainted with martingales, you may feel like consulting Section 5.1.

and by dominated convergence

$$E[x_\infty(s)] = E[x_1(s)] = s.$$ (2.7.2)

Since $E[x_{n+1}^2(s)|\mathscr{B}_n] \geq E^2[x_{n+1}(s)|\mathscr{B}_n]$, $\{x_n^2(s)\}$ is a submartingale, again between zero and one. Therefore, for $s < 1$,

$$E[x_\infty^2(s)] \geq E[x_1^2(s)] > E^2[x_1(s)]$$

provided we assume that z_1 is not degenerate. It follows that $x_\infty(s)$ is itself a random variable with positive variance. Defining $c_n(s) = -\log g_n(s)$, $y(s) = -\log x_\infty(s)$ (we shall see in a moment that $x_\infty(s) > 0$), we have proved that, as $n \to \infty$

$$c_n(s) z_n \to y(s) \qquad \text{a.s.}$$

where $y(s)$ is a non-degenerate random variable. This is the first part of:

Theorem (2.7.1)

For any supercritical process with non-degenerate reproduction law there exists a sequence $\{k_n\}$ of positive numbers such that $k_n z_n$ converges a.s., as $n \to \infty$, to a non-degenerate, finite, and non-negative random variable. The sequence can always be chosen as $\{c_n(s)\}$ above for $q < s < 1$. For $a < m^{-1}$, $a^n z_n \to 0$ a.s. whereas, for $a > m^{-1}$, $a^n z_n \to 0$, if $z_n \to 0$, and $a^n z_n \to \infty$ otherwise. If

$$\sum_{k=1}^\infty p_k k \log k < \infty$$

$\{k_n\}$ can be chosen $\{m^{-n}\}$, whereas $m^{-n} z_n \to 0$ a.s. if this sum diverges.

Proof. First observe that since $f(s) \leq s$ for $q \leq s \leq 1$ $g(s)$ must be $\geq s$ and therefore $g_n \uparrow$ some limit g_∞. Since

$$s = f_n \circ g_n(s) \leq f_n \circ g_\infty(s) \to q,$$

if $g_\infty(s) < 1$, $g_\infty(s) = 1$ for $s > q$.
Next turn to the Taylor expansion (2.5.1), $1 - f(s) = \{m - r(s)\}(1 - s)$. Replace s by $g(s)$ for $q < s < 1$ to obtain

$$\{1 - g(s)\}/(1 - s) = m^{-1}/\{1 - r \circ g(s)/m\}.$$

As usual, repeat and take the product:

$$m^n\{1 - g_n(s)\} = (1 - s)/\prod_{k=1}^n \{1 - r \circ g_k(s)/m\}.$$ (2.7.3)

This representation says something about $c_n(s)$ since $-\log x \sim 1 - x$ as $x \downarrow 1$. In particular

$$\lim_{n \to \infty} c_n/c_{n-1} = \lim_{n \to \infty} (1 - g_n)/(1 - g_{n-1}) = 1/m$$ (2.7.4)

a fact that helps in proving the a.s. finiteness of

$$y(s) = \lim_{n \to \infty} c_n(s)\, z_n.$$

Indeed,

$$P[y(s) < \infty] = E\big[P\big[\lim_{n \to \infty} c_n z_n < \infty | z_1\big]\big]$$

$$= E\left[P\left[\lim_{n \to \infty} (c_n/c_{n-1}) \cdot c_{n-1} z_{n-1} < \infty \right]^{z_1} \right]$$

$$= f(P[y(s) < \infty]).$$

In the same manner

$$P[y(s) = 0] = f(P[y(s) = 0]).$$

Hence these probabilities can only be q or one. Since $y(s)$ is not degenerate,

$$P[y(s) = 0] = q.$$

As to the finiteness, the arguments preceding the theorem told us that

$$s = E[x_\infty(s)] = E[e^{-y(s)}] \le P[y(s) < \infty].$$

Hence $P[y(s) < \infty] = 1$ if $s > q$.

These results show that the suggested norming is the proper one; the limit of $c_n(s)\, z_n$ is almost always finite and zero (almost) only if $\{z_n\}$ itself becomes extinct. Its drawback is, of course, that it is complicated. However Equation (2.7.3) makes it clear that

$$c_n(s) = m^n \not\to \infty \Leftrightarrow \prod_{n=1}^{\infty} \{1 - r \circ g_n(s)/m\} > 0 \Leftrightarrow \sum_{n=1}^{\infty} r \circ g_n(s) < \infty.$$

Choose $q < s_0 < 1$ such that $m_0 = f'(s_0) > 1$ and k such that $g_k(s) \ge s_0$. From

$$m_0^n(1 - s) \le 1 - f_n(s) \le m^n(1 - s)$$

valid for $s_0 \le s \le 1$, we deduce

$$1 - m_0^{-(n-k)}\{1 - g_k(s)\} \le g_n(s) \le 1 - m^{-n}(1 - s).$$

Lemma (2.5.1) therefore shows that for $s_0 \le s < 1$

$$\lim_{n \to \infty} c_n(s)\, m^n \begin{cases} < \infty & \text{if } \sum p_k k \log k < \infty, \\[2mm] = \infty & \text{if } \sum p_k k \log k = \infty. \end{cases}$$

It remains to realize that $a^n z_n$ cannot have a non-zero limit if $\sum p_k k \log k$ diverges (the other case being trivial). For $a \leq m^{-1}$ this is already clear. For $a > m^{-1}$

$$\{1 - g_n(s)\}/a^n = (1 - s)/\prod_{k=1}^{n} a\{m - r \circ g_k(s)\},$$

which must tend to zero for $q < s < 1$ since

$$1 = m - r(q) \leq m - r \circ g_k(s) \to m.$$

Thus, (recall that $z_n \to 0$ means that z_n actually turns zero)

$$a^n z_n = \{a^n/c_n(s)\}\, c_n(s)\, z_n \to \begin{cases} 0, & \text{if } z_n \to 0, \\ \infty, & \text{if } z_n \to \infty. \end{cases} \qquad \square$$

Theorem (2.7.2)

The moment generating function φ *of* $y(s) = \lim\limits_{n \to \infty} c_n(s) z_n$, $q < s < 1$, $\varphi(u) = E[e^{-uy(s)}]$ *satisfies*

$$\varphi(mu) = f \circ \varphi(u), \qquad u \geq 0. \qquad (2.7.5)$$

$E[y(s)] = -\varphi'(0)$ *is finite if and only if*

$$\sum p_k k \log k < \infty$$

and in this case there is only one solution of Equation (2.7.5), whose derivative at zero exists and has a given value.

Proof. According to Equation (2.7.4) $c_{n+1}/c_n \to 1/m$. Therefore, as $n \to \infty$,

$$\varphi(mu) \leftarrow E[\exp(-muc_{n+1}(s) z_{n+1})] = E[E[\exp(-muc_{n+1}(s) z_{n+1})|z_1]]$$

$$= E[\{E[\exp(-um\{c_{n+1}(s)/c_n(s)\}\, c_n(s)\, z_n)]\}^{z_1}] \to f \circ \varphi(u),$$

by a repeated use of the dominated convergence theorem. Further, the relation $\varphi(mu) = f \circ \varphi(u)$ can be written $\varphi(u/m) = g \circ \varphi(u)$. Since $\varphi(u) \geq \geq \varphi(\infty) = q$, this can be iterated to

$$1 - \varphi(u/m^n) = 1 - g_n \circ \varphi(u) = O(m^{-n}) \Leftrightarrow \sum p_k k \log k < \infty$$

showing the second assertion. Next, if φ and ψ are two solutions with $\varphi'(0) = \psi'(0)$ finite, then for any $u \geq 0$

$$|\varphi(u) - \psi(u)| = |f \circ \varphi(u/m) - f \circ \psi(u/m)|$$

$$\leq m|\varphi(u/m) - \psi(u/m)| \leq \ldots \leq m^n|\varphi(u/m^n) - \psi(u/m^n)| \to 0. \qquad \square$$

Actually the distribution of w is absolutely continuous on $(0, \infty)$ and its density never vanishes [12, 17]. The proofs of this are long and analytic; Equation (2.7.5) is used to establish the integrability of w's characteristic function via the Fourier inversion formula.

Define $w_n = z_n/m^n$. We know that $\lim_{n \to \infty} w_n = w$ exists a.s. and that $w = 0$ a.s. if and only if $\sum p_k k \log k = \infty$. Since $E w_n = 1$, $\{w_n\}$ cannot, in this case, be uniformly integrable.

Theorem (2.7.3)

The sequence w_n is uniformly integrable if and only if

$$\sum p_k k \log k < \infty.$$

In other words,

$$E[|w_n - w|] \to 0 \Leftrightarrow \sum p_k k \log k < \infty.$$

Proof. It is already clear that the left hand side implies the convergence at right. Instead of the converse we shall prove the stronger assertion that

$$\sum p_k k \log k < \infty \Rightarrow E\left[\sup_n w_n\right] < \infty.$$

Let $w^{(j)}$ be independent replicae of w and note that

$$P[w > at] \geq P\left[w > at, \sup_n w_n > t\right]$$

$$= \sum_{n=0}^{\infty} P[w > at, w_n > t, w_k \leq t, 0 \leq k < n]$$

$$= \sum_{n=0}^{\infty} P[w > at \,|\, z_n > tm^n, w_k \leq t, 0 \leq k < n]$$

$$\cdot P[w_n > t, w_k \leq t, 0 \leq k < n]$$

$$\geq \sum_{n=0}^{\infty} P\left[\sum_{j=1}^{tm^n} w^{(j)}/tm^n > a\right] P[w_n > t, w_k \leq t, 0 \leq k < n]$$

$$\geq bP[\sup_n w_n > t]$$

for some $b > 0$ if $t \geq 1$ and some positive $a < E[w]$ (which is positive) by the law of large numbers. Hence

$$E\left[\sup_n w_n\right] = \int_0^{\infty} P\left[\sup_n w_n > t\right] dt$$

$$\leq 1 + \int_0^{\infty} P[w > at]\, dt/b = 1 + E[w]/ab < \infty.$$

Finally, all this section has been concerned with the case $1 < m < \infty$. May be it is worthwhile pointing out that if $m = \infty$ there is no norming such that $c_n z_n$ has a finite non-degenerate limit even in probability [Theorem (4.4.) of Reference [3]].

Background

Convergence theorems for w_n have been known since the forties. Some names to be mentioned are Hawkins and Ulam [14], Yaglom [10], establishing convergence in distribution under moment restrictions. Harris proved L^2-convergence [13] and J. L. Doob seems first to have applied martingale arguments to w_n. The martingales $x_n(s)$ used here are trite modifications of those in References [15] and [17]. The log-condition for $w_n \to w$ was found by Stigum and Kesten [16].

2.8 ATTAINING HIGH LEVELS

The almost sure convergence theorem also subsumes some easily exposed information on the hitting of high levels. For a more profound analysis in a different vein the reader should have a look at Reference [20].
Define v_k, $k \in N$, by

$$v_k = \inf\{n; z_n \geq k\},$$

interpreting, as usual, the infimum of the empty set as $+\infty$. Let $^m\log$ denote the logarithm with base m.

Theorem (2.8.1)

In a supercritical process with

$$\sum p_k k \log k < \infty$$

it holds almost surely that

$$\limsup_{k \to \infty} (v_k - {}^m\log k) \leq 1 - {}^m\log w$$

and

$$\liminf_{k \to \infty} (v_k - {}^m\log k) \geq - {}^m\log w.$$

Proof. It is clear that $v_k \to \infty$ a.s. as $k \to \infty$. Therefore Theorem (2.7.1) applies to show that

$$z_{v_k}/m^{v_k} \to w,$$

almost surely again. However, with probability one,

$$w/m \leftarrow z_{v_{k-1}}/m^{v_k} < k/m^{v_k} \leq z_{v_k}/m^{v_k} \to w.$$

The assertion to be proved follows upon logarithmization.   □

2.9 RATE OF CONVERGENCE RESULTS FOR SUPERCRITICAL PROCESSES

This section is concerned with the speed of the convergence $w_n \to w$ in the case $m > 1, \sigma^2 < \infty$. Further results may be found in the list of references.

Lemma (2.9.1)

If $m > 1$, $\sigma^2 < \infty$, then $z_n/m^n \to w$ in mean square and $\mathrm{Var}[w] = \sigma^2/(m^2 - m)$.

This is Harris's 1948 result for the supercritical case [13]. We give the traditional proof, using the completeness of L^2. It does, by the way, also yield an a.s. theorem (Reference [21], page 13). Recall that the finiteness of m did not ensure L^1-convergence [Theorem (2.7.3)].

Proof. By Theorem (2.2.1)

$$E[w_n^2] = 1 + \sigma^2(1 - m^{-n})/(m^2 - m).$$

Since

$$E[w_{n+k}w_n] = E[w_n E[w_{n+k}|w_n]] = E[w_n^2],$$

$$E[(w_{n+k} - w_n)^2] = \sigma^2 m^{-n}(1 - m^{-k})/(m^2 - m)$$

which can be made small uniformly in $k \in Z_+$ by choosing n large. Hence there exists a square integrable random variable, which is of course w, such that $E[(w_n - w)^2] \to 0$. Also

$$\mathrm{Var}[w] = \lim_{n \to \infty} \mathrm{Var}[w_n] = \sigma^2/(m^2 - m). \qquad \square$$

Consider the difference

$$z_n - m^n w = z_n - \lim_{k \to \infty} z_{n+k}/m^k \overset{\mathrm{d}}{=} \lim_{k \to \infty} \sum_{j=1}^{z_n} (1 - z_k^{(j)}/m^k),$$

where the $\{z_k^{(j)}\}$, $j \in N$, and $\{z_n\}$ are all i.i.d. Write $w^{(j)} = \lim_{k \to \infty} z_k^{(j)}/m^k$. Evidently

$$(w_n - w) m^n/\sqrt{z_n} \overset{\mathrm{d}}{=} \sum_{j=1}^{z_n} (1 - w^{(j)})/\sqrt{z_n}. \qquad (2.9.1)$$

By the variance computed in Lemma (2.9.1) and Theorem A.2 in the Appendix we have, conditionally upon $z_n \not\to 0$, that

$$(w_n - w) m^n/\sqrt{z_n} \overset{\mathrm{d}}{\to} N(0, \sigma^2/(m^2 - m)). \qquad (2.9.2)$$

To obtain a non-random norming note that

$$m^{n/2}(w_n - w) \to 0$$

on the set where so does z_n, whereas on the complement

$$m^{n/2}(w_n - w) = \{(w_n - w)m^n/\sqrt{z_n}\}\sqrt{z_n/m^n} \xrightarrow{d} N(0, \sigma^2/(m^2 - m))\sqrt{w}. \quad (2.9.3)$$

Since the $w^{(j)}$ in (2.9.1) are independent of z_n, the normal and w random variables in the limit are independent as well. We have proved:

Theorem (2.9.2)

Assume that $m > 1$ and $\sigma^2 < \infty$. Conditionally upon $z_n \nrightarrow 0$ then Equation (2.9.2) holds. Further (since $w = 0 \Leftrightarrow z_n \to 0$)

$$m^{n/2}(w_n - w) \xrightarrow{d} N(0, \sigma^2/(m^2 - m)) \cdot \sqrt{w},$$

the limit product factors being independent.

Background

Results of this type are recent and all due to C. C. Heyde with collaborators. They have also obtained extimates of the rate of convergence [26] and an iterated logarithm law [24]. The proof given here is different from those published and also due to Heyde.

2.10 PREDICTION IN LARGE SUPERCRITICAL PROCESSES AND ANOTHER RATE OF CONVERGENCE RESULT

The following problem is closely related to that of the preceding section: As $n \to \infty$, what can be said about

$$y_n(j) = \frac{z_{n+j} - m^j z_n}{a_j \sqrt{z_n}},$$

where a_j is some suitable norming? Its solution is fairly direct—with the help of the central limit theorem.

To make things easy let us first choose

$$a_j^2 = \operatorname{Var}[z_j] = \sigma^2 m^{j-1}(m^j - 1)/(m - 1).$$

Then, as before

$$y_n(j) = \frac{1}{\sqrt{z_n}} \sum_{i=1}^{z_n} (z_j^{(i)} - m^j)/a_j$$

where the i.i.d. processes $\{z_j^{(i)}; j \in Z_+\}$ are initiated by the different members of the nth generation. The terms in the sum have expectation zero and variance a_j^2 and for $j < k$ the covariances of terms in $y_n(j)$ and $y_n(k)$ multiplied by $\sqrt{z_n}$ are independent of n:

$$E[(z_j^{(i)} - m^j)(z_k^{(i)} - m^k)]/(a_j a_k) = \{m^k(m^j - 1)/(m^j(m^k - 1))\}^{1/2}.$$

Hence again Theorem A.2 of the Appendix implies that the distribution of any vector $(y_n(1), \ldots, y_n(r))$, $r \in N$, given that $z_n > 0$, converges as $n \to \infty$ to the k-dimensional normal distribution with means zero, and covariances $[m^k(m^j - 1)/m^j(m^k - 1)]^{1/2}, 0 < j \le r$.

We get a somewhat more elegant limit by using the norming suggested in Reference [27] and noting that convergence in distribution of a sequence of sequences of random variables is nothing but the convergence of all finite dimensional vectors:

Theorem (2.10.1)
Define

$$y_n(j) = \frac{z_{n+j} - m^j z_n}{\sigma m^{j-1} \sqrt{z_n}}$$

for $j \in N$, $\{z_n\}$ a supercritical process with $\sigma^2 < \infty$. Then, conditionally upon $z_n > 0$, $\{y_n(j); j \in Z_+\}$ tends in distribution as $n \to \infty$ to $\{y(j); j \in Z_+\}$. Here $y(0) = 0$, the increments $y(j) - y(j-1)$ are independent and normal with mean zero and variance m^{1-j}, $1 \le j \le k$.

The proof amounts to a computation of covariances and is gladly left for the reader. As some compensation, here is a multidimensional version of Theorem (2.9.2):

Theorem (2.10.2)
In a supercritical process with reproduction mean m and variance $\sigma^2 < \infty$ define for $j \in Z_+$

$$y_n(j) = \begin{cases} m^n(w - w_{n+j})/\sqrt{z_n} & \text{if } z_n > 0 \\ 0 & \text{if } z_n = 0 \end{cases}$$

Then $\{y_n(j); j \in Z_+\} \overset{d}{\to} \{y(j); j \in Z_+\}$, where all $y(j) = 0$ with probability q and with probability $1 - q$ the $y(j)$ are normal with zero mean and

$$\text{Cov}(y(j), y(j+k)) = \sigma^2/m^{k+1}(m - 1).$$

The proof of this can be patterned after the preceding one.

2.11 THE TOTAL PROGENY OF A BRANCHING PROCESS

The equation

$$y_\infty = \sum_{n=0}^{\infty} z_n \qquad (2.11.1)$$

defines y_∞ as an integer or infinite valued random variable. Obviously

$$P[y_\infty < \infty] = q$$

and

$$y_n = \sum_{k=0}^{n} z_k \uparrow y_\infty.$$

The generating functions h_n of y_n are recursively related:

$$h_{n+1}(s) = sE[E[s^{z_1 + \cdots + z_{n+1}}|z_1]] = sE[\{h_n(s)\}^{z_1}] = sf \circ h_n(s). \qquad (2.11.2)$$

It follows that h_∞,

$$h_\infty(s) = E[s^{y_\infty}],$$

satisfies

$$h_\infty(s) = sf \circ h_\infty(s). \qquad (2.11.3)$$

If g is some other function between zero and one solving Equation (2.11.3) then

$$|h_\infty(s) - g(s)| \le sm|h_\infty(s) - g(s)|$$

and $h_\infty(s) = g(s)$ for $0 \le s < m^{-1}$. Thus (2.11.3) can only have one solution, which is analytic on the unit interval.

In the case $m \le 1$ the function taking s into $s/f(s)$ is strictly increasing and its inverse is easily checked to be nothing but h_∞. In the supercritical case with $q > 0$ consider a new process with reproduction generating function $f(qs)/q$. Since $f'(q) < 1$ this process is subcritical. Denote the generating function of the total progeny here by h. Then

$$h(s) = sf\{qh(s)\}/q$$

and qh must satisfy Equation (2.11.3). Therefore h_∞ is q multiplied by the inverse of $qs/f(qs)$.

It is a matter of straightforward computations to verify that

$$E[y_n] = \begin{cases} (1 - m^{n+1})/(1 - m), & \text{if } m \ne 1 \\ n + 1, & \text{if } m = 1 \end{cases} \qquad (2.11.4)$$

$$\text{Var}[y_n] = \begin{cases} \sigma^2\{(1 - m^{2n+1})/(1 - m) - (2n + 1)m^n\}/(1 - m), & \text{if } m \ne 1 \\ (2n + 1)(n + 1)n\sigma^2/6, & \text{if } m = 1. \end{cases}$$
$$(2.11.5)$$

Example (2.11.1)

For binary splitting Equation (2.11.3) reduces to

$$h_\infty(s) = s(1 - p) + sp\{h_\infty(s)\}^2$$

with the solutions

$$h_\infty(s) = \{1 \pm \sqrt{1 - 4s^2p(1 - p)}\}/2ps$$

Since $h_\infty(s)$ must be ≤ 1 and $2ps \leq 1$ *if* $p \leq 1/2$ we see in this case that

$$h_\infty(s) = \{1 - \sqrt{1 - 4s^2p(1 - p)}\}/2ps$$
$$= \sum_{k=1}^{\infty} \frac{(2k - 3)!!}{k!} \{2p(1 - p)\}^{k-1}(1 - p)s^{2k-1}. \quad (2.11.6) \qquad \square$$

For the main result of this section we shall make use of a simple consequence of the so-called Ballot Theorem: Let $X_1, X_2, \ldots\ldots$ be i.i.d. with values in Z_+ and let S_k be their partial sums,

$$S_k = \sum_{j=1}^{k} X_j.$$

Then, if $E[X_1] < 1$,

$$P[S_k < k \text{ for all } k] = 1 - E[X_1].$$

This is proved as Corollary (5.1.9) in the martingale section.

Theorem (2.11.2)

Write $p_{jk} = P[z_{n+1} = k|z_n = j]$. *Then*

$$P[y_\infty = k] = (1/k)\, p_{k,k-1}, \qquad k \in N.$$

More generally, if $z_0 = j \in N$, *then*

$$P[y_\infty = k] = (j/k)\, p_{k,k-j}, \qquad k \geq j.$$

Proof. For $0 \leq s \leq 1$ let $X, X_1, \ldots$ be independent with distribution $\{p_k\}$ and let $X(s), X_1(s), X_2(s), \ldots$ be i.i.d. with

$$P[X(s) = k] = p_k s^k/f(s).$$

Define the partial sums $S_n(s)$ as above and $S_0(s) = 0$. Suppose to begin with that $m \leq 1$. Then

$$E[X(s)] = sf'(s)/f(s) < 1$$

for $0 \leq s < 1$ and

$$E[X(1)] = m.$$

Further

$$P[S_n(s) < n, \forall n \in N] = 1 - sf'(s)/f(s).$$

The law of large numbers says that

$$\lim_{n \to \infty} S_n(s)/n = sf'(s)/f(s) < 1$$

for $0 \leq s < 1$. Therefore for any $k \in N$ $S_n(s)$ can be $\geq n - k$ only for a finite number of indices n. Hence, since $S_0(s) = 0$,

$$1 = \sum_{n=k}^{\infty} P[S_n(s) = n - k, S_{n+j}(s) < n - k + j, \forall j \in N]$$

$$= P[S_j(s) < j, \forall j \in N] \sum_{n=k}^{\infty} P[S_n(s) = n - k]$$

$$= \{1 - sf'(s)/f(s)\} \sum_{n=k}^{\infty} P[S_n = n - k] s^{n-k}/\{f(s)\}^n. \qquad (2.11.7)$$

Here the relation

$$P[S_n(s) = k] = P[S_n = k] s^k/\{f(s)\}^n$$

was used. Now let $u = h_\infty^{-1}(s) = s/f(s)$. Then

$$sf'(s)/f(s) = h_\infty(u) f' \circ h_\infty(u)/f \circ h_\infty(u) = f' \circ h_\infty(u) u.$$

However differentiation of Equation (2.11.3) at u yields

$$h_\infty'(u) = f \circ h_\infty(u) + uf' \circ h_\infty(u) h_\infty'(u) = f(s) + \{sf'(s)/f(s)\} h_\infty'(u)$$

or

$$1 - sf'(s)/f(s) = f(s)/h_\infty'(u) = f \circ h_\infty(u)/h_\infty'(u) = h_\infty(u)/uh_\infty'(u)$$

by Equation (2.11.3) again. Insertion of this, $s/f(s) = u$, and $s = h_\infty(u)$, into Equation (2.11.7) results in

$$\{h_\infty(u)\}^{k-1} h_\infty'(u) = \sum_{n=k}^{\infty} P[S_{n-1} = n - k] u^{n-1}.$$

Integrate from zero to s to obtain for $k \geq 1$

$$\{h_\infty(s)\}^{k-1} = \sum_{n=k}^{\infty} P[S_{n-1} = n - k] s^{n-1}(k-1)/(n-1)$$

since $h_\infty(0) = 0$.

This concludes the proof for $m \leq 1$. To a supercritical process with

$q > 0$ the argument preceding Example (2.11.1) applies; since

$$\{f(qs)/q\}^j = \sum_{k=0}^{\infty} p_{jk} q^{k-j} s^k,$$

$$\{h_\infty(s)\}^{k-1} = q^{k-1} \{h(s)\}^{k-1}$$

$$= q^{k-1} \sum_{n=k}^{\infty} p_{n-1,n-k} q^{1-k} s^{n-1} (k-1)/(n-1)$$

$$= \sum_{n=k}^{\infty} p_{n-1,n-k} s^{n-1} (k-1)/(n-1). \qquad \square$$

Example (2.11.3)

Invoking Example (2.1.2) we see that for binary splitting

$$P[y_\infty = k] = \begin{cases} \dbinom{k}{(k-1)/2} p^{(k-1)/2} (1-p)^{(k-1)/2}/k, & \text{if } k \text{ is odd} \\ \\ 0 & \text{otherwise,} \end{cases}$$

which is (fortunately) the result of Example (2.11.1). $\qquad \square$

Background

Theorem (2.11.2) is from Reference [28].

2.12 THE RELATION BETWEEN A PROCESS AND ITS TOTAL PROGENY

Recursive equations like Equation (2.11.2) hold also for the two-dimensional generating functions

$$H_n(s, t) = E[s^{y_n} t^{z_n}]:$$

$$\begin{cases} H_{n+1}(s, t) = sf \circ H_n(s, t) & n \in Z_+ \\ H_0(s, t) = st \end{cases} \qquad (2.12.1)$$

From this it is a matter of computation to verify that

$$\text{Cov}[y_n, z_n] = \begin{cases} \sigma^2 \{nm^{n-1} - m^n(1-m^n)/(1-m)\}/(1-m) \\ n(n+1)\sigma^2/2, & \text{if } m = 1. \end{cases} \qquad (2.12.2)$$

So if ρ_n is the correlation coefficient of y_n and z_n Equation (2.11.5) yields

$$\lim_{n \to \infty} \rho_n = \begin{cases} 0, & \text{if } m < 1, \\ (\sqrt{3})/2, & \text{if } m = 1, \\ 1, & \text{if } m > 1. \end{cases} \qquad (2.12.3)$$

The strong relationship between the process and its progeny in the super-critical case is further exposed by the following theorem:

Theorem (2.12.1)

In a supercritical process let $c_n(s)$ be the constants from Theorem (2.7.1). Then, for $q < s < 1$,

$$c_n(s)\, y_n \to y(s)\, m/(m-1) \qquad a.s.$$

where $y(s) = \lim_{n \to \infty} c_n(s)\, z_n$. Similarly, with $w = \lim_{n \to \infty} z_n/m^n$,

$$y_n/m^n \to wm/(m-1).$$

Proof. Deleting the dependence on s we see from Equation (2.7.3) that $c_{n-j}/c_n \to m^j$, as $n \to \infty$, and therefore for any $k \le n$

$$c_n y_n \ge c_n \sum_{j=0}^{k} z_{n-j} = \sum_{j=0}^{k} (c_n/c_{n-j})\, c_{n-j} z_{n-j} \to y \sum_{j=0}^{k} m^{-j} \qquad a.s.$$

Hence, $\liminf c_n y_n \ge ym/(m-1)$. On the other hand it is directly checked from Equation (2.7.3) that

$$(1 - g_n)/(1 - g_k) = m^{k-n} \prod_{j=k+1}^{n} (1 - r \circ g_j/m) \le m^{k-n}$$

Let $\varepsilon > 0$. There is a (random) v such that $(1 - g_n) z_n < y + \varepsilon$ for $n \ge v$ (since $c_n/(1 - g_n) \to 1$).
Hence

$$c_n y_n \le c_n y_v + \{c_n/(1 - g_n)\}\, (y + \varepsilon) \sum_{k=v}^{n} (1 - g_n)/(1 - g_k)$$

$$\le c_n y_v + \{c_n/(1 - g_n)\}\, (y + \varepsilon) \sum_{k=0}^{n} m^{-k} \to (y + \varepsilon)\, m/(m-1).$$

The case with norming m^{-n} follows simply. $\qquad \square$

We turn to the somewhat more complex subcritical case.

Theorem (2.12.2)

In a subcritical process with finite reproduction variance σ^2

$$\lim_{n \to \infty} P[y_n/n \le u | z_n > 0] = \begin{cases} 0, & \text{if } u < \dfrac{\sigma^2}{m - m^2} + 1 \\[2mm] 1, & \text{if } u \ge \dfrac{\sigma^2}{m - m^2} + 1 \end{cases}$$

Proof. With

$$h_n(s) = E[s^{y_n}]$$

$$g_n(s) = E[s^{y_n}; z_n = 0]$$

we wish to prove that

$$E[s^{y_n/n} \mid z_n > 0] = \{h_n(s^{1/n}) - g_n(s^{1/n})\}/\{1 - f_n(0)\} \to s^{1 + \sigma^2/(m - m^2)}.$$

First note that

$$g_0(s) = 0,$$

$$g_{n+1}(s) = sf \circ g_n(s),$$

$$g_{n+1} \geq g_n, \qquad g_n \uparrow h_\infty.$$

Next with $\vartheta(s) = sf' \circ h_\infty(s)$, $\varphi_n = (h_n - h_\infty)/\vartheta^n$, $\psi_n = (h_\infty - g_n)/\vartheta^n$ we have (at least for positive arguments)

$$(h_n - g_n)/\{1 - f_n(0)\} = (\varphi_n + \psi_n)\,\vartheta^n/\{1 - f_n(0)\}$$

Repeated use of the mean value theorem yields, with $h_\infty(s) < \eta_k < h_k(s)$,

$$\varphi_n(s) = \{s - h_\infty(s)\} \prod_{k=1}^{n} f'(\eta_k)/f' \circ h_\infty(s). \tag{2.12.4}$$

Since

$$0 \leq h_k - h_\infty = f \circ h_{k-1} - f \circ h_\infty \leq m(h_{k-1} - h_\infty) \leq \ldots \leq m^k$$

and

$$f'(\eta_k) - f' \circ h_\infty(s) \leq f' \circ h_k(s) - f' \circ h_\infty(s) \leq f''(1)\, m^k,$$

the series

$$\sum \{f'(\eta_k) - f' \circ h_\infty(s)\}$$

converges uniformly for $0 \leq s \leq 1$. Hence so does the product in Equation (2.12.4) and φ_n increases to some continuous function φ_∞. By Dini's theorem the convergence is uniform and

$$\lim_{n \to \infty} \varphi_n(s^{1/n}) = \varphi_\infty(1) = \lim_{n \to \infty} \varphi_n(1) = 0.$$

Similarly ψ_n decreases uniformly to a continuous limit ψ_∞ and

$$\lim_{n \to \infty} \psi_n(s^{1/n}) = \psi_\infty(1) = \lim_{n \to \infty} \{1 - f_n(0)\}/m^n = c$$

where c is the positive $\lim P[z_n > 0]/m^n$ of Theorem (2.6.1).

However with $a = f''(1)/(m - m^2)$ and ρ a well behaved rest, $\rho(s) \to 0$

as $s \uparrow 1$,

$$f' \circ h_\infty(s) = m\{1 - (a - \rho(s))(1 - s)\},$$

and

$$\{\vartheta(s^{1/n})\}^n = sm^n\{1 - (a - \rho(s^{1/n}))(1 - s^{1/n})\}^n \sim sm^n \exp(a \log s)$$

$$= m^n s^{a+1}, \quad \text{as } n \to \infty.$$

Hence

$$\{\vartheta(s^{1/n})\}^n/\{1 - f_n(0)\} \to 1/c. \qquad \square$$

There is also a result for critical branching processes. We state it without proof [see 29].

Theorem (2.12.3)

Let $m = 1$ and $a = \sigma^2/2 < \infty$. Then for $0 \le u < \infty$

$$\lim_{n \to \infty} P[y_n/n^2 \le u | z_n > 0] = F(u)$$

where, $s > 0$

$$\int_0^\infty e^{-su} F(du) = 2\sqrt{as} \, \text{cosech}(2\sqrt{as}).$$

Background

All results in this section are due to Pakes [29]. Pakes also has a central limit analogue to the 'law of large numbers' of Theorem (2.12.2).

2.13 MAXIMUM LIKELIHOOD ESTIMATION OF THE REPRODUCTION MEAN AND A BAYES EXAMPLE

As a prelude consider binary splitting:

Example (2.13.1)

Assume that the process started from z_0 ancestors and that $z_1, z_2, \ldots, z_n$ individuals were observed in the n first generations. Except z_0 these numbers are all even, $z_j = 2x_j$ and their likelihood is

$$l(p, z_0, \ldots, z_n) =$$

$$= \binom{z_0}{x_1}\binom{z_1}{x_2}\cdots\binom{z_{n-1}}{x_n} p^{(y_n - z_0)/2}(1 - p)^{y_{n-1} - (y_n - z_0)/2}$$

$$= c(z_0, \ldots, z_n) \, p^{(y_n - z_0)/2}(1 - p)^{y_{n-1} - (y_n - z_0)/2}. \qquad (2.13.1)$$

This is maximized by

$$\hat{p}_n = (y_n - z_0)/2y_{n-1}$$

and a maximum likelihood estimator of m is thus $(y_n - z_0)/y_{n-1}$. $\qquad \square$

In the general case the same estimator maximizes the likelihood. We shall arrive at this by roundabout means—assuming that we can observe more than the Galton–Watson process and then apply the following lemma. It is certainly known among statisticians but seldom stated explicitly.

Lemma (2.13.2)

Let $\{p_{\vartheta,\sigma}, \vartheta \in \Theta, \sigma \in \Sigma\}$ be a class of densities with respect to some σ-finite measure μ on $(\mathscr{X}, \mathscr{A})$. Let T be some statistic on this space (i.e. simply a function on $\mathscr{X}$). If there is a maximum likelihood estimator $\hat\vartheta$ of ϑ based on observations in $\mathscr{X}$ and $\hat\vartheta = g \circ T$ for some g, then g is a maximum likelihood estimator of ϑ based on observation of $T(x)$, $x \in \mathscr{X}$, the density of T being with respect to μT^{-1}.

Proof. Write $P_{\vartheta,\sigma}$ for the probability measure on $\mathscr{X}$ with μ-density $p_{\vartheta,\sigma}$. By the Radon–Nikodym theorem the measure $P_{\vartheta,\sigma}T^{-1}$ on the range space of T has some density $p_{\vartheta,\sigma}^T$ with respect to μT^{-1}. For any B in the range space let $A = T^{-1}B$. Then

$$\int_B p_{\vartheta,\sigma}^T(y)\, \mu T^{-1}(\mathrm{d}y) = \int_{T^{-1}(B)} p_{\vartheta,\sigma}(x)\, \mu(\mathrm{d}x)$$

$$\leq \int_{T^{-1}(B)} p_{g\circ T(x),\sigma}(x)\, \mu(\mathrm{d}x).$$

If you agree that this last integral equals

$$\int_B p_{g(y),\sigma}^T(y)\, \mu T^{-1}(\mathrm{d}y)$$

the proof is complete since the resulting inequality for all subsets B of T's range space implies that $p_{g,\sigma}^T \geq p_{\vartheta,\sigma}^T$. If you do not, here is an argument:

Let $h: T(\mathscr{X}) \times \mathscr{X} \to R_+$ be measurable enough and define measures H and H_ϑ, $\vartheta \in T(\mathscr{X})$, by

$$H(A) = \int_A h(T(x), x)\, \mu(\mathrm{d}x)$$

$$H_\vartheta(A) = \int_A h(\vartheta, x)\, \mu(\mathrm{d}x).$$

Write h^T for the Radon–Nikodym derivative $\mathrm{d}HT^{-1}/\mathrm{d}\mu T^{-1}$ and h_ϑ^T for $\mathrm{d}H_\vartheta T^{-1}/\mathrm{d}\mu T^{-1}$ regarded as functions on $T(\mathscr{X})$. Let $\mathscr{H}$ be the class of functions h such that $h^T(y, y) = h_y^T(y)$. Start from functions h of the form $f \circ T(x) \cdot g(x)$, f an indicator. Use monotonicity and linearity to see that $\mathscr{H}$ contains all non-negative integrable functions h. □

To apply this lemma consider the random variables z_{nk} = the number of realized individuals in generation n with exactly k children.

Theorem (2.13.3)

In a Galton–Watson process with z_0 ancestors maximum likelihood estimators of p_k and m, based on observation of $\{z_{jk}; j \leq n\}$ are, respectively,

$$\hat{p}_k = \sum_{j=0}^{n} z_{jk}/y_n$$

$$\hat{m}_n = (y_{n+1} - z_0)/y_n.$$

From Lemma (2.13.2) the wanted consequence is immediate:

Corollary (2.13.4)

The maximum likelihood estimator of m based on $\{z_0, z_1, \ldots, z_n\}$ is $\hat{m}_n = (y_n - z_0)/y_{n-1}$.

Proof of the theorem. For given z_j we have

$$P[z_{j0} = i_0, z_{j1} = i_1, \ldots | z_j] = z_j ! \prod_{k=0}^{\infty} p_k^{i_k} / \prod_{k=0}^{\infty} (i_k !) \tag{2.13.2}$$

provided

$$\sum_{k=0}^{\infty} i_k = z_j.$$

Otherwise the probability is zero. Since

$$z_j = \sum_{k=0}^{\infty} k z_{j-1k}, \tag{2.13.3}$$

the likelihood of an array $\{z_{jk}, j \leq n\}$ is

$$\frac{z_0 !}{\prod z_{0k} !} \prod p_k^{z_{0k}} \frac{(\sum k z_{0k}) !}{\prod z_{1k} !} \prod p_k^{z_{1k}} \cdots \frac{(\sum k z_{n-1k}) !}{\prod z_{nk} !} \prod p_k^{z_{nk}}. \tag{2.13.4}$$

Choosing a distribution $\{p_k\}$ to maximize this is equivalent to maximization of the logarithm, which except a term independent of the reproduction law is

$$\sum_{k=0}^{\infty} \left(\sum_{j=0}^{n} z_{jk} \right) \log p_k. \tag{2.13.5}$$

However, if $\{a_k\}_0^{\infty}$ is any probability distribution, and all sums involved converge,

$$\sum_0^{\infty} a_k \log p_k - \sum_0^{\infty} a_k \log a_k = \sum_0^{\infty} a_k \log p_k/a_k \leq \log \sum_0^{\infty} p_k = 0$$

according to Jensen's inequality. Hence Expression (2.13.4) is maximized by the choice

$$\hat{p}_k = \sum_{j=0}^{n} z_{jk} / \sum_{k=0}^{\infty} \sum_{j=0}^{n} z_{jk} = \sum_{j=0}^{n} z_{jk}/y_n.$$

Recall the general principle that if $\hat{\vartheta}$ is a maximum likelihood estimator of ϑ, then so is $f(\hat{\vartheta})$ of $f(\vartheta)$. We obtain, using Equation (2.13.3)

$$\hat{m} = \sum_{k=0}^{\infty} k\hat{p}_k = (y_{n+1} - z_0)/y_n. \qquad \square$$

As always with maximum likelihood estimators it is important to study the large sample properties. Obviously if $m \le 1, (y_{n+1} - z_0)/y_n \to 1 - z_0/y_\infty$ almost surely. Even in the supercritical case this occurs with probability q. Theorem (2.11.2) tells us that this limit satisfies (with $z_0 = 1, 0 \le u < 1$),

$$P[1 - 1/y_\infty \le u] = \sum_{k \le 1/(1-u)} p_{k,k-1}/k.$$

However, if $z_n \to \infty$, and we consider for simplicity the case $\sum p_k k \log k < \infty$, $z_0 = 1$, then

$$y_n/m^n \to wm/(m - 1)$$

almost surely and

$$P[(y_{n+1} - 1)/y_n \to m | z_n \to \infty] = 1.$$

In other words the estimator is strongly consistent, conditionally upon non-extinction. What about the rate of convergence? Actually

$$(y_{n+1} - z_0)/y_n - m = (y_{n+1} - z_0 - my_n)/y_n = \sum_{j=0}^{n} (z_{j+1} - mz_j)/y_n.$$

Any term $z_{j+1} - mz_j$ is the sum of z_j independent random variables each with the reproduction distribution but translated to zero mean. Hence $y_{n+1} - z_0 - my_n$ has the distribution of a sum

$$\sum_{j=1}^{y_n} \eta_{nj}$$

where the η_{nj} are i.i.d. with expectation zero. The random variable y_n is of course depending upon these in some intricate manner. Still a central limit theorem is applicable and from Theorem A.2 in the Appendix we obtain

Theorem (2.13.5)

Consider a supercritical branching process with the finite reproduction variance σ^2. As $n \to \infty$, the distribution of

$$\{(y_{n+1} - z_0)/y_n - m\} \sqrt{y_n},$$

given that $z_n > 0$, is asymptotically normal with mean zero and variance σ^2.

We conclude this section by some words on Bayesian inference in the binary case. The almost binomial form (2.13.1) of the likelihood makes a beta prior distribution of p most attractive (the principle of choosing a 'conjugate prior'—the ethics of assuming a prior just to guarantee neat computations being of course debatable. But the beta densities,

$$B_{a,b}(p) = \frac{\Gamma(a + b)}{\Gamma(a)\,\Gamma(b)} p^{a-1}(1 - p)^{b-1},$$

$a, b > 0$, $0 \le p \le 1$, can assume many varying forms and, in particular, include the case of p uniformly distributed on the unit interval; choose $a = b = 1$.)

The beta assumption yields after routine computations that the posterior density of p given $z_0, \ldots, z_n$ is again of beta form, $B_{(y_n - z_0)/2 + a, y_n - 1 - (y_n - z_0)/2 + b}$. The posterior expectation—which is the Bayes estimator for a quadratic loss function—is

$$E[p|z_0, \ldots, z_n] = \frac{y_n - z_0 + 2a}{2(y_{n-1} + a + b - 2)}$$

and, of course,

$$E[m|z_0, \ldots, z_n] = \frac{y_n - z_0 + 2a}{y_{n-1} + a + b - 2}$$

which is never the same as the maximum likelihood estimator. In particular, the latter is smaller than the estimator for a uniform prior. Furthermore, in applications to cell proliferation there are often good reasons to assume that $p \ge 0.75$. This means that we should choose a beta frequency with a large and b little. Then the maximum likelihood estimator might be substantially smaller and result in severe underestimation in cases where the process becomes extinct.

Background

Theorem (2.13.3) is old, from Reference [13]. For the corollary and Theorem (2.13.4) priority belongs to Dion [30], who has also constructed estimators of the reproduction variance [31]. Nagayev [32] estimated m by z_{n+1}/z_n.

2.14 ESTIMATION OF THE EXTINCTION PROBABILITY

Changing the setup of the last section slightly, we assume that we have observed the reproduction of n individuals. This gives a sample $x_1, x_2, \ldots, x_n$ of independent random variables each with distribution $\{p_k\}$. From Theorem (2.13.3) it is clear that the maximum likelihood estimator of p_k is

$$\hat{p}_k(n) = (\text{the number of } x_i = k)/n.$$

By the law of large numbers $\hat{p}_k(n)$ is strongly consistent. It is obviously unbiased. The same applies to the empiric generating function and its derivatives: with probability one

$$\hat{f}_n^{(j)}(s) = \sum_{k=j}^{\infty} k! \, \hat{p}_k(n) s^{k-j}/(k-j)! \to f^{(j)}(s)$$

uniformly for $0 \le s \le 1$ if $f^{(j)}(1-) < \infty$ and uniformly on any closed subset of $[0, 1)$ otherwise. Also

$$E[\hat{f}_n^{(j)}(s)] = f^{(j)}(s).$$

To obtain a normal convergence we make the preceding argument a little preciser. Write e_{ik} for the indicator function of the event $\{x_i = k\}$. For any $0 \le s \le 1$

$$\hat{f}_n(s) = \sum_{i=1}^{n} \left\{ \sum_{k=0}^{\infty} e_{ik} s^k \right\} /n.$$

The random variables

$$\sum_{k=0}^{\infty} e_{ik} s^k$$

$1 \le i \le n$ are i.i.d with expectation $f(s)$ and variance

$$E\left[\sum_{k=0}^{\infty} e_{ik} s^k \right]^2 - f^2(s) = \sum_{k=0}^{\infty} E[e_{ik}] s^{2k} - f^2(s) = f(s^2) - f^2(s)$$

since $e_{ij} e_{ik}$ is zero for $j \ne k$. This variance is zero if and only if the reproduction law is degenerate. Thus, if this is not the case

$$\sqrt{n} \{ \hat{f}_n(s) - f(s) \} / \sqrt{f(s^2) - f^2(s)}$$

is asymptotically, as $n \to \infty$, $N(0, 1)$. In the same way normal convergence holds for the derivatives.

From what we have seen about $\hat{f}_n$ it follows that $\hat{q}_n$,

$$\hat{q}_n = \inf\{s \ge 0; \hat{f}_n(s) = s\}$$

is a consistent maximum likelihood estimator of q. Further,

Theorem (2.14.1)

If $m > 1$ and $p_0 > 0$, then $(\hat{q}_n - q) \sqrt{n}$ is asymptotically normal with mean zero and variance $\{f(q^2) - q^2\}/\{1 - f'(q)\}^2$. If $m = 1$ and $0 < \sigma^2 < \infty$, then

$$\lim_{n \to \infty} P[(1 - \hat{q}_n) \sqrt{n} \leq u] = \begin{cases} \Phi(u/2), & \text{if } u \geq 0 \\ 0, & \text{if } u < 0. \end{cases}$$

If $m < 1$, then with probability one $\hat{q}_n$ actually equals one from some index on.

Proof. Since $\hat{f}_n(\hat{q}_n) = \hat{q}_n$,

$$\hat{f}_n(q) = \hat{q}_n - (\hat{q}_n - q)\hat{f}_n'(s_n)$$

for some s_n between $\hat{q}_n$ and q. Hence

$$\{\hat{f}_n(q) - f(q)\} \sqrt{n} = (\hat{q}_n - q) \sqrt{n}\{1 - \hat{f}_n'(s_n)\}.$$

But the left hand side was seen to be asymptotically normal with variance $f(q^2) - f^2(q)$, $s_n \to q$ a.s. and with probability one $\hat{f}_n' \to f'$ uniformly. This proves the first assertion.

The second one is a consequence of similar arguments, and convexity properties of $\hat{f}_n$. For $u \geq 0$

$$P[(1 - \hat{q}_n) \sqrt{n} \leq u] = P[1 - u/\sqrt{n} \leq \hat{q}_n] =$$

$$= P[\hat{f}_n(1 - u/\sqrt{n}) \geq 1 - u/\sqrt{n}] =$$

$$= P[1 - \hat{f}_n'(1) u/\sqrt{n} + \hat{f}_n''(s_n) u^2/2n \geq 1 - u/\sqrt{n}] =$$

$$= P[\hat{f}_n''(s_n) u \geq 2\{\hat{f}_n'(1) - 1\}/\sqrt{n}],$$

where $1 - u/\sqrt{n} < s_n < 1$. Hence $s_n \to 1$ and $\hat{f}_n''(s_n) \to f''(1) = \sigma^2$ a.s. But we know that $\{\hat{f}_n'(1) - 1\} \sqrt{n} \xrightarrow{d} N(0, \sigma^2)$, since

$$\hat{f}_n'(1) = \sum k\hat{p}_k(n) = \sum_1^n x_i/n.$$

Finally, if $m < 1$, then by the law of large numbers

$$\hat{f}_n'(1) = \sum_1^n x_j/n$$

is a.s. strictly less than one from some index on. From that index onwards $\hat{q}_n = 1$. $\qquad \square$

The results can be used in the customary way to construct approximate confidence intervals.

Background

This is from Stigler [33]. He also has a paper about estimating the age of a Galton–Watson process (i.e. the number of a generation whose size has been observed) provided the reproduction law is known. This is in *Biometrika* **57**, 505–512, 1971. Further investigations into that problem have been made by Crump [34].

REFERENCES

1. Darling, D. A., The Galton–Watson process with infinite mean. *J. Appl. Prob.* **7**, 455–456, 1970.
2. Feller, W., *An Introduction to Probability Theory and Its Applications II.* Wiley, New York, 1966.
3. Seneta, E., Functional equations and the Galton–Watson process. *Adv. Appl. Prob.* **1**, 1–42, 1969.
4. Seneta, E., The simple branching process with infinite mean. I. *J. Appl. Prob.* **10**, 206–212, 1973.
5. Seneta, E., Regularly varying functions in the theory of simple branching processes. *Adv. Appl. Prob.* **6**, 106–113, 1974.
6. Kesten, H., Ney, P., and Spitzer, F., The Galton–Watson process with mean one and finite variance. *Teor. Veroyatnost. i Primenen.* **11**, 579–611, 1966.
7. Kolmogorov, A. N., Zur Lösung einer biologischen Aufgabe. K rešeniyu odnoy biologičeskoy zadači (To the solution of one biological problem). *Izv. NII matem. i mek. Tomskogo un.* **2**. 1–6 (German version), 7–12 (Russian), 1938.
8. Lindvall, T., Limit theorems for some functionals of certain Galton–Watson branching processes. *J. Appl. Prob.* **11**, 320–327, 1974.
9. Slack, R. S., A branching process with mean one and possibly infinite variance. *Z. Wahrscheinlichkeitstheorie verw. Geb.* **9**, 139–145, 1968.
10. Yaglom, A. M., Nekotorye predelnye teoremy teorii vetvyaščihsya slučaynyh processov (Some limit theorems from the theory of branching stochastic processes). *Dokl. Akad. Nauk SSSR* **56**, 795–798, 1947.
11. Heathcote, C. R., Seneta, E., and Vere-Jones, D., A refinement of two theorems in the theory of branching processes. *Teor. Veroyatnost. i Primenen.* **12**, 342–346, 1967.
12. Dubuc, S., La fonction de Green d'un processus de Galton–Watson. *Stud. Math.* **XXXIV**, 69–87, 1970.
13. Harris, T. E., Branching processes. *Ann. Math. Statist.* **19**, 474–494, 1948.
14. Hawkins, D. and Ulam, S., Theory of multiplicative processes I. *Los Alamos Scientific Laboratory*, LADC-265, 1944.
15. Heyde, C. C., Extension of a result of Seneta for the supercritical Galton–Watson process. *Ann. Math. Statist.* **41**, 739–742, 1970.
16. Kesten, H. and Stigum, B., A limit theorem for multidimensional Galton–Watson processes. *Ann. Math. Statist.* **37**, 1211–1223, 1966.
17. Seneta, E., On recent theorems concerning the supercritical Galton–Watson process. *Ann. Math. Statist.* **39**, 2098–2012, 1968.

18. Stigum, B. P., A theorem on the Galton–Watson process. *Ann. Math. Statist.* **37**, 695–698, 1966.

19. Badalbayev, I. S., Ob odnom svoystve nadkritičeskogo vetvyaščegosya slučaynogo processa s nepreryvnym vremenem (On a certain property of the supercritical branching random process with continuous time). *Izv. Akad. Nauk Uzb. SSR Ser Fiz.–Mat. Nauk* **17**, 9–13, 1973.

20. Nagayev, A. V., Odna predelnaya teorema dlya nadkritičeskogo vetvyaščegosya processa (One limit theorem for the supercritical branching process). *Mat. Zametki* **9**, 585–592, 1971.

21. Harris, T. E., *The Theory of Branching Processes*. Springer, Berlin–Göttingen–Heidelberg, 1963.

22. Heyde, C. C., A rate of convergence result for the supercritical Galton–Watson process. *J. Appl. Prob.* **7**, 451–454, 1970.

23. Heyde, C. C., Some central limit analogues for supercritical Galton–Watson processes. *J. Appl. Prob.* **8**, 52–59, 1971.

24. Heyde, C. C., Some almost sure convergence theorems for branching processes. *Z. Wahrscheinlichkeitstheorie verw. Geb.* **20**, 189–192, 1971.

25. Heyde, C. C. and Brown, B. M., An invariance principle and some convergence rate results for branching processes. *Z. Wahrscheinlichkeitstheorie verw. Geb.* **20**, 271–278, 1971.

26. Heyde, C. C. and Leslie, J. R., Improved classical limit analogues for Galton–Watson processes with or without immigration. *Bull. Austr. Math. Soc.* **5**, 145–156, 1971.

27. Bühler, W. J., Ein zentraler Grenzwertsatz für Verzweigungsprozesse. *Z. Wahrscheinlichkeitstheorie verw. Geb.* **11**, 139–141, 1969.

28. Dwass, M., The total progeny in a branching process and a related random walk. *J. Appl. Prob.* **6**, 682–686, 1969.

29. Pakes, A. G., Some limit theorems for the total progeny of a branching process. *Adv. Appl. Prob.* **3**, 176–192, 1971.

30. Dion, J.–P., Estimation of the mean and the initial probabilities of a branching process. *J. Appl. Prob.* **11**, 1974.

31. Dion, J.–P., Estimation of the variance of a branching process. *Tech. Report. Département de Mathématiques,* Université du Quebec à Montréal, 1974.

32. Nagayev, A. V., Ob ocenke srednego čisla neposredstvennyh potomkov časticy v vetvyăščemsya slučaynom processe. *Teoriya Veroyatnost. i Primenen* **12**, 363–369, 1967. Translated as: On estimating the mean number of direct descendants of a particle in a branching process. *Theory Prob. Appl.* **12**, 314–320.

33. Stigler, S. M., The estimation of the probability of extinction and other parameters associated with branching processes, *Biometrika* **58**, 499–508, 1971.

34. Crump, K. S. and Howe, R. B., Nonparametric estimation of the age of a Galton–Watson branching process, *Biometrika* **59**, 535–538, 1972.

Chapter 3

Neighbours of the Galton–Watson Process

3.1 BRANCHING PROCESSES WITH IMMIGRATION

As we have seen repeatedly, Galton–Watson populations either expire or tend to grow without limits. Immigration into subcritical processes can be viewed as an artifice, aimed at stabilizing population sizes. But immigration is a natural object of study in its own right and not only in the subcritical case.

Our framework will be the following. At times $n \in Z_+$ k i.i.d. Galton–Watson processes are initiated with probability a_k, $k \in Z_+$, $a_0 < 1$,

$$h(s) = \sum_{k=0}^{\infty} a_k s^k.$$

Processes started at different times are independent. The interpretation is that k immigrants each become ancestors of a process, and that the immigration is steady but random. The numbers of immigrants at different times should also be independent.

Denote by y_n the number of individuals at time n and by $z_n(j)$ the number of individuals then stemming from an immigrant at time j. Obviously— or by definition, if you prefer—

$$y_n = z_n(0) + \ldots + z_n(n). \tag{3.1.1}$$

The $z_n(j)$ are independent with the distribution of k independent replicae of z_{n-j} with probability a_k. If $\mathscr{B}_n$ is the σ-algebra generated by events occurring at times $0, 1, \ldots, n$, then branching and independence properties imply that

$$E\left[s^{y_{n+1}} \middle| \mathscr{B}_n\right] = \{f(s)\}^{y_n} h(s)$$

and hence, with

$$h_n(s) = E\left[s^{y_n}\right],$$
$$h_{n+1} = h \cdot h_n \circ f \tag{3.1.2}$$

54

or

$$h_n = \prod_{j=0}^{n} h \circ f_j. \tag{3.1.3}$$

Thus h_n decreases to some limit h_∞ satisfying

$$h_\infty = h \cdot h_\infty \circ f. \tag{3.1.4}$$

The question arises if h_∞ is ever a proper generating function.

Theorem (3.1.1)

Consider a subcritical branching process with immigration, the latter with generating function h. Then the number of individuals y_n at time n tends, as $n \to \infty$, in distribution to a random variable y_∞, whose generating function satisfies Equation (3.1.4) and is determined by it. Also

$$h_\infty(1) = 1 \Leftrightarrow \sum_{k=2}^{\infty} a_k \log k < \infty,$$

Proof. The convergence has already been demonstrated and

$$h_\infty = \prod_{j=0}^{\infty} h \circ f_j.$$

The product is uniformly (for $0 \le s \le 1$) convergent together with

$$\sum_{j=0}^{\infty} \{1 - h \circ f_j\}$$

that is, if and only if

$$\sum_{j=0}^{\infty} \{1 - h \circ f_j(0)\} < \infty,$$

A reference to Relations (2.6.3) and (2.5.2) completes the proof of the theorem, except the asserted unicity of the solution of Equation (3.1.2). This one is easily checked though. Assume that φ and ψ are two solutions of the equation. Then

$$|\varphi - \psi| = |h| \, |\varphi \circ f - \psi \circ f| \le |\varphi \circ f - \psi \circ f| \le \dots$$
$$\dots \le |\varphi \circ f_n - \psi \circ f_n| \to |\varphi(1-) - \psi(1-)| = 0,$$

f $\varphi(1-) = \psi(1-)$. □

There is an intriguing parallel with subcritical processes conditioned upon non-extinction as in Section 2.6. Equation (2.6.5) is nothing but a special case of Equation (3.1.4). Write

$$h_\infty(s) = \{1 - g(s)\}/(1 - s)$$

and

$$h(s) = \{1 - f(s)\}/m(1 - s)$$

to see that $g \circ f = mg + 1 - m$ reduces to Equation (3.1.4). Here h_∞ is not usually a generating function. But if $g'(1) < \infty$ this can obtained through consideration of $h_\infty/g'(1)$.

Like (2.6.5), Equation (3.1.4) provides limited information about the form of h. The reader might feel like trying his/her efforts (in vain) with the simple case of binary splitting and Poisson distributed immigration. But in the critical case much more explicit results can be reached.

Theorem (3.1.2)

With h as before the generating function of the number of immigrants per generation assume that $a = h'(1) < \infty$, $m = 1$, and $f''(1) = \sigma^2 < \infty$. Then y_n/n converges, as $n \to \infty$, in distribution to a random variable with the gamma density

$$(1/\Gamma(\alpha)\,\beta^\alpha)\,u^{\alpha-1}e^{-u/\beta}, \qquad u \in R_+$$

where $\alpha = 2a/\sigma^2$, $\beta = \sigma^2/2$.

Note. *Here is an interpretation: At each time point on the average a individuals immigrate. By Theorem (2.4.2) roughly $a/\beta n$ of the processes initiated at one immigration remain n time units later. Therefore we should expect $\alpha = a/\beta$ processes to survive in the long run. But each one of these has an exponentially distributed size with mean β.*

Proof. Taking Laplace transforms, we see that it is enough to prove the convergence

$$-\log h_n(e^{-t/n}) \to \alpha \log(1 + t\beta)$$

for fixed $t > 0$, h_n still the generating function of y_n.

The relations to be used are a logarithmized version of Equation (3.1.3),

$$-\log h_n(e^{-t/n}) = -\sum_{j=0}^{n} \log h \circ f_j(e^{-t/n}),$$

a Taylor expansion of the function $-\log h$ around 1,

$$-\log h(s) = \{a + \rho(s)\}\{1 - s\},$$

$$\lim_{s \uparrow 1} \rho(s) = 0,$$

and a somewhat manipulated Lemma (2.4.1),

$$1 - f_j(s) = (1 - s)/\{1 + j(\beta + r_j(s))(1 - s)\}, \qquad (3.1.5)$$

where $r_j(s) \to 0$ uniformly for $0 \leq s < 1$. Thus we have

$$-\log h_n(e^{-t/n}) = \sum_{j=0}^{n} \frac{\{a + \rho \circ f_j(e^{-t/n})\}\{1 - e^{-t/n}\}}{1 + j\{\beta + r_j(e^{-t/n})\}\{1 - e^{-t/n}\}}.$$

Let $0 < \varepsilon < \beta$ be given. Since $f_j \geq f$ there is an n_0 such that $|\rho \circ f_j(e^{-t/n})| < \varepsilon$ for $n \geq n_0$. This n_0 can also be so chosen that $|r_j(e^{-t/n})| < \varepsilon$ for all n and $j \geq n_0$. Obviously

$$\lim_{n \to \infty} \sum_{j=0}^{n_0 - 1} \frac{\{a + \rho \circ f_j(e^{-t/n})\}\{1 - e^{-t/n}\}}{1 + j\{\beta + r_j(e^{-t/n})\}\{1 - e^{-t/n}\}} = 0$$

and for $n \geq n_0$

$$\sum_{j=n_0}^{n} \frac{(a - \varepsilon)(1 - e^{-t/n})}{1 + j(\beta + \varepsilon)(1 - e^{-t/n})} \leq \sum_{j=n_0}^{n} \frac{\{a + \rho \circ f_j(e^{-t/n})\}\{1 - e^{-t/n}\}}{1 + j\{\beta + r_j(e^{-t/n})\}\{1 - e^{-t/n}\}} = S(n)$$

$$\leq \sum_{j=n_0}^{n} \frac{(a + \varepsilon)(1 - e^{-t/n})}{1 + j(\beta - \varepsilon)(1 - e^{-t/n})}.$$

However, by the usual inequalities between sums and integrals it holds for any $A, B > 0$ that

$$(A/B)\log \frac{1 + B(n - 1)}{1 + Bn_0} \leq \sum_{j=n_0}^{n} \frac{A}{1 + Bj} \leq (A/B)\log \frac{1 + Bn}{1 + B(n_0 - 1)}.$$

Thus,

$$\frac{a - \varepsilon}{\beta + \varepsilon} \log \frac{1 + (\beta + \varepsilon)(1 - e^{-t/n})(n - 1)}{1 + (\beta + \varepsilon)(1 - e^{-t/n})n_0} \leq S(n)$$

$$\leq \frac{a + \varepsilon}{\beta - \varepsilon} \log \frac{1 + (\beta - \varepsilon)(1 - e^{-t/n})n}{1 + (\beta - \varepsilon)(1 - e^{-t/n})(n_0 - 1)}.$$

From this

$$\{(a - \varepsilon)/(\beta + \varepsilon)\} \log \{1 + (\beta + \varepsilon)t\} \leq \liminf_{n \to \infty} S(n)$$

$$\leq \limsup_{n \to \infty} S(n) \leq \{(a + \varepsilon)/(\beta - \varepsilon)\} \log \{1 + (\beta - \varepsilon)t\}.$$

But ε was arbitrary between zero and β. $\qquad \square$

As might be expected, the rapid proliferation of supercritical processes renders the immigration aspect less prominent in this last case. The functions c_n of Section 2.7 provide a proper norming for y_n as well as z_n. Indeed, using the notation and conventions of Section 2.7 (but retaining the meaning of $\mathscr{B}_n$) it holds for $q < s < 1$ that

$$E\left[g_{n+1}(s)^{y_{n+1}} \mid \mathscr{B}_n\right] = \{f \circ g_{n+1}(s)\}^{y_n} h \circ g_{n+1}(s)$$

$$= g_n(s)^{y_n} h \circ g_{n+1}(s) \le g_n(s)^{y_n}.$$

Since $0 \le g_n^{y_n} \le 1$, the sequence $\{g_n(s)^{y_n}\}$ is a bounded non-negative super-martingale and it converges almost surely [Theorem (5.1.2)] as well as in any $L^p, p > 0$. Deleting the s, we write

$$g_n^{y_n} \to \bar{x}$$

$$c_n y_n \to \bar{y} = -\log \bar{x}.$$

For the further analysis we use a sister of Equation (3.1.2), the relation

$$h_n = h \circ f_n \cdot h_{n-1}. \tag{3.1.6}$$

It is a direct sequel to Equation (3.1.3). Let ψ be the Laplace transform function of $\bar{y}$. On one hand then

$$E\left[e^{-uc_n y_n}\right] \to \psi(u).$$

On the other hand, Equation (3.1.6) means that

$$E\left[e^{-uc_n y_n}\right] = E\left[g_n^{u y_n}\right] = h_n \circ g_n^u = h\{E\left[g_n^{u z_n}\right]\} \cdot E\left[g_n^{u y_{n-1}}\right].$$

From Theorem (2.7.2) it follows that

$$h\{E\left[g_n^{u z_n}\right]\} = h\{E\left[e^{-uc_n z_n}\right]\} \to h \circ \varphi(u)$$

and

$$E\left[g_n^{u y_{n-1}}\right] = E\left[e^{-uc_{n-1} y_{n-1} c_n/c_{n-1}}\right] \to \psi(u/m)$$

by the first sentence in the proof of that theorem. Hence

$$\psi(u) = h \circ \varphi(u) \cdot \psi(u/m).$$

Repetition yields

$$\psi(u) = \psi(0) \prod_{j=0}^{\infty} h \circ \varphi(u/m^j).$$

According to Equation (2.7.5) this can be written

$$\psi(u) = \psi(0) \prod_{j=0}^{\infty} h \circ g_j \circ \varphi(u).$$

But the particular case $u = 1$ can be calculated directly:

$$\psi(1) = E[\bar{x}] = \lim_{n \to \infty} E[g_n^{y_n}] = \lim_{n \to \infty} \prod_{j=0}^{n} h \circ f_j \circ g_n$$

$$= \lim_{n \to \infty} \prod_{j=0}^{n} h \circ g_{n-j} = \prod_{j=0}^{\infty} h \circ g_j.$$

Furthermore, by Equation (2.7.2)

$$\varphi(1) = E[x] = s.$$

The argument used in the proof of Theorems (2.7.1) and (3.1.1) and Relation (2.5.2) show that the infinite product is positive, provided

$$\sum_{1}^{\infty} a_k \log k < \infty.$$

Hence, in this case $\psi(0) = 1$. If on the other side $\sum a_k \log k$ diverges, obviously $\psi = 0$. Hence

$$P[\bar{y} < \infty] = 1 \quad \text{if} \quad \sum_{1}^{\infty} a_k \log k < \infty$$

$$P[\bar{y} = \infty] = 1 \quad \text{if} \quad \sum_{1}^{\infty} a_k \log k = \infty.$$

In the first case the Laplace transform function of $\bar{y}$ is given explicitly by

$$\psi(u) = \prod_{j=0}^{\infty} h \circ \varphi(u/m^j). \tag{3.1.7}$$

It follows from Theorem (2.7.1) that $\bar{y} \geq 0$ is non-degenerate and finite and, by the remark following Theorem (2.7.2), that $\bar{y}$ is a continuous random variable. By differentiation finally, $a = h'(1)$,

$$E[\bar{y}] = E[y] \, am/(m - 1). \tag{3.1.8}$$

This needs to be summarized:

Theorem (3.1.3)

Let $\{y_n\}$ be a supercritical branching process with immigration. Let $c_n(s)$ and φ be as in Section 2.7 and $h(s) = \sum a_k s^k$ the generating function of the number of immigrants. Then, if

$$\sum_{k=1}^{\infty} a_k \log k = \infty,$$

$$P\left[\lim_{n \to \infty} c_n(s) y_n = \infty \right] = 1, \quad q < s < 1,$$

If, on the contrary $\sum a_k \log k$ *converges, the limit* $\lim_{n \to \infty} c_n(s) y_n = \bar{y}$ *exists a.s. for any* $q < s < 1$. *The random variable* $\bar{y}$ *is positive, continuous and not degenerate. Its Laplace transform* ψ *satisfies Equation* (3.1.7) *and its expectation* (3.1.8). *If furthermore*

$$\sum p_k k \log k < \infty,$$

$c_n(s)$ *can be replaced by* m^n.

The last assertion is a simple corollary to Theorem (2.7.1).

Background

Branching processes with immigration is a favourite topic of the anti-podean school of probability. The first results concerned the subcritical case and were due to Heathcote. The critical and supercritical cases were solved by Seneta, who also has an approximation theorem along the lines of Section 3.3, [6]. Theorem (3.1.2) was also obtained in Reference [1] and again in Reference [2], where not necessarily identically distributed immigrations are treated. The latter paper also gives a necessary and sufficient condition for $y_n \overset{d}{\to}$ some proper random variable in the case $m \leq 1$, namely that

$$\int_0^1 \{1 - h(s)\}/\{1 - f(s)\} \, ds < \infty.$$

3.2. INCREASING NUMBERS OF ANCESTORS

Hitherto it has been assumed that $z_0 = 1$ or at least fixed. It might be interesting to investigate sequences of processes where $z_0 = r_k \to \infty$, as generations pass.

Theorem (3.2.1)

Consider the supercritical case with finite reproduction variance. Let r_k *be integers* $\to \infty$ *and* $\{z_n(r_k); n \in Z_+\}$ *the process started from* r_k *ancestors. Then*

$$\lim_{n \to \infty} P[\{z_n(r_n) - r_n m^n\}/b_n \leq u] = \Phi(u) \tag{3.2.1}$$

provided

$$b_n \sim m^n \sigma \sqrt{r_n/(m^2 - m)}. \tag{3.2.2}$$

Φ *is, as usual, the standardized normal distribution.*

Compare this with Theorems (2.9.2) and (2.10.2).

Theorem (3.2.2)

If $m < 1$, $\sum p_k k \log k < \infty$ and $r_n \sim am^{-n}$, than the limit law of $z_n(r_n)$ is compound Poisson:

$$\lim_{n \to \infty} E\left[s^{z_n(r_n)}\right] = \exp ac\{g(s) - 1\}, \tag{3.2.3}$$

where g is the generating function from Theorem (2.6.2) and $c = \varphi(0)$ the limit of $P[z_n > 0]/m^n$. If instead $r_n m^n \to \infty$, and $\sigma^2 < \infty$, Equation (3.2.1) holds with b_n as there. Except degenerate laws and trivial changes in the norming these are the only possible limit distributions.

Theorem (3.2.3)

Consider the critical case with $\sigma^2 < \infty$. If $r_n \sim an$, then for $u > 0$

$$\lim_{n \to \infty} E\left[e^{-uz_n(r_n)/n}\right] = \exp\{-2au/(2 + u\sigma^2)\},$$

which is the Laplace transform of a Poisson mixture of gamma-distributions. If $r_n/n \to \infty$, then Equation (3.2.1) holds with $b_n \sim \sigma\sqrt{nr_n}$. Again these limits are essentially unique.

Proof of Theorem (3.2.1). This is nothing but a fairly simple consequence of the central limit theorem. Write

$$\sigma_n^2 = \operatorname{Var} z_n = \sigma^2 m^{n-1}(m^n - 1)/(m - 1)$$

and let z_{nj}, $j \in N$, for each n be independent random variables with the distribution of z_n. Then

$$\{z_n(r_n) - r_n m^n\}/\sigma_n\sqrt{r_n}$$

has the distribution of

$$\sum_{j=1}^{r_n} (z_{nj} - m^n)/\sigma_n\sqrt{r_n}.$$

By the classical Lindeberg condition (Reference 27) this random variable tends, in law as $n \to \infty$ to the standardized normal one if and only if for any $\delta > 0$

$$E\left[(z_n - m^n)^2/\sigma_n^2; \quad |z_n - m^n| > \delta\sigma_n\sqrt{r_n}\right] \to 0.$$

However this is evident since (w as before $= \lim\limits_{n \to \infty} z_n/m^n$)

$$x_n = (z_n - m^n)/\sigma_n \to (w - 1)(m^2 - m)^{\frac{1}{2}}/\sigma = x$$

in mean square and a.s. Here is a more detailed argument: Let $\varepsilon > 0$ be given. Choose r such that $E[x^2; |x| > r] < \varepsilon/2$ and $P[x = r] = 0$. By dominated convergence $E[x_n^2; |x_n| \leq r] \to E[x^2; |x| \leq r]$ and $E[x_n^2; |x_n| > r] = E[x_n^2] - E[x_n^2; |x_n| \leq r] \to E[x^2] - E[x^2; |x| \leq r]$, implying that $E[x_n^2; |x_n| > r] < \varepsilon$ for n large enough. $\square$

The first part of Theorem (3.2.2) is proved analogously by an application of convergence to the Poisson law:

Proof of Theorem (3.2.2). Write $y_{nj} = z_{nj} \wedge 1$ for z_{nj} as before independent replicae of z_n. By Theorem (2.6.1) $P[z_n > 0]/m^n \to c$ and if $r_n \sim am^{-n}$,

$$r_n P[y_{nj} = 1] \sim am^{-n}P[z_n > 0] \to ac.$$

This means that

$$P\left[\sum_{j=1}^{r_n} y_{nj} = k\right] \to e^{-ac}(ac)^k/k \,!$$

and, by virtue of Theorem (2.6.2), with

$$g_n(s) = E[s^{z_n}|z_n > 0],$$

$$E[s^{z_n(r_n)}] = \sum_{k=0}^{r_n} P\left[\sum_{j=1}^{r_n} y_{nj} = k\right]\{g_n(s)\}^k \to \exp\{ac(g(s) - 1)\}.$$

The convergence is dominated, since $E[s^{z_n}|z_n > 0] \leq s$.

The normal convergence part follows like Theorem (3.2.1): Let $a_n = [1/m^n]$ and $b_n = [r_n/a_n]$. Then $z_n(r_n)$ converges in distribution as above. But $E[z_n(a_n)] \to 1$ and $\text{Var}[z_n(a_n)] \to \sigma^2/(m - m^2)$. By Equation (2.6.5) the limit law of $z_n(a_n)$ has these same first and second moments and therefore the argument concluding the proof of Theorem (3.2.1) applies to show that for any $\delta > 0$

$$E[\{z_n(a_n) - a_n m^n\}^2 ; \quad |z_n(a_n) - a_n m^n| > \delta a_n] \to 0.$$

Next let $z_{nj}(a_n)$ for $j \in N$ be the nth generation of independent processes with a_n ancestors. Note that $a_n b_n/r_n \to 1$. In distribution the following equality holds with the z_{nj} as before and also independent of the $z_{nj}(a_n)$:

$$\{z_n(r_n) - r_n m^n\}/\sigma_n\sqrt{r_n} = \sqrt{a_n b_n/r_n}\sum_{j=1}^{b_n}\{z_{nj}(a_n) - a_n m^n\}/\sigma_n\sqrt{a_n b_n}$$

$$+ \sum_{j=1}^{r_n - a_n b_n}(z_{nj} - m^n)/\sigma_n\sqrt{r_n}.$$

The first sum is asymptotically normal with mean zero and variance one. The second sum tends to zero in mean square.

It is easy to see that these are the only possible limit laws. If $r_n m^n \to \infty$ the limit must be normal and the norming the proposed one. If $r_n m^n \to 0$, $z_n(r_n)$ must tend to zero in probability. And if neither of these relations hold there must be a subsequence for which $r_n m^n \to$ some positive constant, ensuring that the only possible limit is Equation (3.2.3).

Finally the

Proof of Theorem (3.2.3). This is very close to the preceding and we give only the first half: with y_{nj} as before and $r_n \sim an$

$$r_n P[y_{nj} = 1] \to 2a/\sigma^2.$$

Hence for $u > 0$

$$E[e^{-uz_n(r_n)/n}] = \sum_{k=0}^{\infty} P\left[\sum_{j=1}^{r_n} y_{nj} = k\right] \{E[e^{-uz_n/n}|z_n > 0]\}^k$$

$$\to \sum_{k=0}^{\infty} \frac{(2a/\sigma^2)^k}{k!} e^{-2a/\sigma^2} \left(\frac{2}{2 + u\sigma^2}\right)^k = \exp\{-2au/(2 + u\sigma^2)\}. \quad \square$$

Background

These facts were found by John Lamperti [10], who has some remarks on the case with infinite reproduction variance as well.

3.3 APPROXIMATION BY CRITICAL PROCESSES

The investigations up to this point have disclosed radical differences between asymptotics of the three types of processes. Only in the critical case did an explicit and simple limit distribution appear. Therefore it is tempting to ask how nearly critical processes behave in the long run. The essence of the following theorem is that for very rich classes of Galton–Watson processes z_n is asymptotically exponential as $n \to \infty$ and $m \to 1$. This holds even uniformly.

Theorem (3.3.1)

For any $\alpha > 0$ let $\mathscr{K}_\alpha$ be a class of Galton–Watson processes with reproduction variances not less than α and uniformly convergent second reproduction moments (i.e. for each $\varepsilon > 0$ it should be possible to choose k such that

$$\sum_{k > k_\varepsilon} k^2 p_k < \varepsilon$$

for all reproduction laws of processes in the class). Suppose that the number 1 belongs to the closure of the set of reproduction means of processes in $\mathscr{K}_\alpha$. Write $a = f''(1)/2$ and interpret $(m^n - 1)/(m - 1)$ as n for $m = 1$. Then,

(a) $\qquad \{a(1 - m^{-n})/(m - 1)\} P[z_n > 0] \to 1$

(b) $\qquad E[z_n(m - 1)/a(1 - m^{-n})|z_n > 0] \to 1$

(c) $\qquad P[z_n(m - 1)/a(1 - m^{-n}) \le u|z_n > 0] \to 1 - e^{-u}$

uniformly for all processes in $\mathscr{K}_\alpha$, as $n \to \infty$, $m \to 1$. Here the norming of course means that the a and m of the corresponding processes should be used.

Note. *The theorem can be given a more transparent form by consideration of special classes $\mathcal{K}_\alpha$. One such class is for given $\alpha, \beta > 0$ the set of branching processes with $\sigma^2 > \alpha$ and*

$$\sum k^2 p_k \log k \le \beta$$

or even

$$\sum k^2 p_k a_k \le \beta$$

for any $a_k \to \infty$. It can also be viewed the following way: Let $\{z_{jn}\} j = 1, 2, \ldots$ be a sequence of branching processes with reproduction means $m_j \to 1$ and variances $\sigma_j^2 \to \sigma^2 > 0$. Then, provided the reproduction variances are uniformly convergent

$$\lim_{\substack{n \to \infty \\ j \to \infty}} (1 - m_j^{-n})(m_j - 1) P[z_{jn} > 0] = 2/\sigma^2,$$

$$\lim_{\substack{n \to \infty \\ j \to \infty}} E[z_{jn}(m_j - 1)/(1 - m_j^{-n})|z_{jn} > 0] = \sigma^2/2,$$

$$\lim_{\substack{n \to \infty \\ j \to \infty}} P[z_{jn}(m_j - 1)/(1 - m_j^{-n}) \le u|z_{jn} > 0] = 1 - e^{-2u/\sigma^2}. \qquad \square$$

The proof starts from the following generalization of the expansions in Lemma (2.4.1):

$$\frac{m}{1 - f(s)} - \frac{1}{1 - s} = \frac{m(1 - s) - 1 + f(s)}{\{1 - f(s)\}(1 - s)} = \frac{1 - s}{1 - f(s)}\{a + r(s)\}$$

$$= a/m + \rho(s), \qquad (3.3.1)$$

where r is the usual remainder and

$$\rho(s) = r(s)/m + \{a + r(s)\}\left\{\frac{1 - s}{1 - f(s)} - 1/m\right\}. \qquad (3.3.2)$$

For $m \ne 1$ this leads to, $0 \le s < 1$,

$$\frac{m^n}{1 - f_n(s)} - \frac{1}{1 - s} = \sum_{j=0}^{n-1} m^j \left\{\frac{m}{1 - f \circ f_j(s)} - \frac{1}{1 - f_j(s)}\right\}$$

$$= \sum_{j=0}^{n-1} m^j \{a/m + \rho \circ f_j(s)\} = \frac{a(m^n - 1)}{m^2 - m} + \sum_{0}^{n-1} m^j \rho \circ f_j(s). \qquad (3.3.3)$$

For $m = 1$ the equation is true with n instead of $(m^n - 1)/(m^2 - m)$, but since that case was treated in Section 2.4 we use the given form here. Equation (3.3.3) yields

$$
\frac{1}{1 - f_n(s)} - \frac{m^{-n}}{1 - s} = \frac{a(1 - m^{-n})}{m^2 - m} \left\{ 1 + \frac{m}{a} \frac{\sum\limits_0^{n-1} m^j \rho \circ f_j(s)}{\sum\limits_0^{n-1} m^j} \right\}
$$

$$
= \frac{a(1 - m^{-n})}{m - 1} \{ 1 + R_n(s) \}, \tag{3.3.4}
$$

giving some hope for the following analogue of Lemma (2.4.1).

Lemma (3.3.2)

Let $\alpha > 0$ and $\mathscr{K}_\alpha$ be as in Theorem (3.3.1). Then for $0 \leq s < 1$

$$
\frac{1}{1 - f_n(s)} = \frac{m^{-n}}{1 - s} + \frac{a(1 - m^{-n})}{m - 1} \{ 1 + R_n(s) \},
$$

where, as $n \to \infty, m \to 1, R_n(s) \to 0$ uniformly in $0 \leq s < 1$ and for all processes in $\mathscr{K}_\alpha$.

The proof of Theorem (3.3.1) from this lemma follows the path of proving Theorem (2.4.2). It is left for the reader. The lemma itself is the consequence of a sequel of simpler assertions:

Lemma (3.3.3)

For any $\mathscr{K}_\alpha$ as in Theorem (3.3.1) there is a $\delta > 0$ such that, for processes in $\mathscr{K}_\alpha$ with m sufficiently close to one, $p_0 \geq \delta$.

Proof. For m close to one and j large enough

$$
\sum_{k=2}^{j} k(k - 1) p_k = \sigma^2 + m^2 - m - \sum_{k>j} k(k - 1) p_k
$$

$$
\geq \alpha + m(m - 1) - \alpha/2 \geq \alpha/3 > 0
$$

Hence, for such m,

$$
p_0 = 1 - m + \sum_{k=2}^{\infty} (k - 1) p_k \geq 1 - m + \sum_{k=2}^{j} k(k - 1) p_k/j
$$

$$
\geq 1 - m + \alpha/3j
$$

provided j was chosen large enough. It follows that $p_0 \geq \alpha/6j$ provided m also $\leq 1 + \alpha/6j$.

Lemma (3.3.4)

As $n \to \infty$ and $m \to 1, f_n \to 1$ uniformly in $0 \leq s \leq 1$ and any fixed $\mathscr{K}_\alpha$.

Proof. Since $(1 - s)/\{1 - f(s)\} \geq 1/m$, r increases and $a + r(s) \geq f''(a)/2$, we have from Equations (3.3.2) and (3.3.3) that

$$1/(1 - f_n) \geq \sum_{j=0}^{n-1} m^{j-1-n}(a + r \circ f_j) \geq \{a + r(p_0)\} \sum_{j=1}^{n-1} m^{j-1-n}$$

$$\geq f''(p_0)(1 - m^{-n-1})/2m^2(m - 1).$$

By Lemma (3.3.3) therefore

$$1/(1 - f_n) \geq f''(\delta)(1 - m^{-n-1})/2m^2(m - 1)$$

if only m is not too far away from one. However with j as in Lemma (3.3.3)

$$f''(\delta) \geq \sum_{k=1}^{j} k(k - 1)\delta^k P_k \geq \delta^j \sum_{k=1}^{j} k(k - 1)p_k \geq \delta^k \alpha/3$$

Hence $1/(1 - f_n) \geq \delta^k \alpha(1 - m^{-n-1})/6m^2(m - 1) \to \infty$ as $m \to 1$ and $n \to \infty$. $\square$

These results will now be used to estimate first r in Equation (3.3.1), then ρ, and finally R_n.

Lemma (3.3.5)
   In any $\mathscr{K}_\alpha$ (even with $\alpha = 0$) it holds uniformly that $\lim_{s \uparrow 1} r(s) = 0$.

Proof. Since $f''(s)/2 \leq a + r(s) \leq a = f''(1)/2$

$$|r(s)| \leq f''(1) - f''(s) \to 0$$

uniformly in any class of $\{p_k\}$ such that $\sum k(k - 1)p_k$ is uniformly convergent. $\square$

Lemma (3.3.6)
   As $n \to \infty$ and $m \to 1$, $r \circ f_n \to 0$ uniformly for all processes in any $\mathscr{K}_\alpha$, $\alpha > 0$.

Proof. Combine the two preceding lemmata. $\square$

Lemma (3.3.7) ·
   As $s \to 1$, $(1 - s)/\{1 - f(s)\} - 1/m \to 0$ uniformly in the class of all processes with $f''(1) \leq 2a < \infty$ and $m \geq b > 0$.

Proof. By the boundedness of second reproduction moments, $\sum kp_k$ is uniformly convergent and therefore $f'(s) \geq b/2$ for s big enough. Hence

$$0 \le \frac{1-s}{1-f(s)} - \frac{1}{m} = \frac{m(1-s) - 1 + f(s)}{m(1-f(s))} \le \frac{a(1-s)^2}{m(1-f(s))} \le \frac{a(1-s)}{mf'(s)}$$

$$\le 2a(1-s)/b^2$$

as $s \to 1$.

Lemma (3.3.8)

As $n \to \infty$ and $m \to 1$, $\rho \circ f_n \to 0$ uniformly in $0 \le s \le 1$ and for processes in any class $\mathscr{K}_\alpha$, $\alpha > 0$.

This is, of course, an immediate consequence of Lemmata (3.3.4), (3.3.5), and (3.3.7). The reader can also check that $|\rho|$ is bounded by some constant K for processes in a given $\mathscr{K}_\alpha$.

To prove Lemma (3.3.2) let $\varepsilon > 0$ and some $\mathscr{K}_\alpha$, $\alpha > 0$, be given. First choose $k \ge 1$ such that $|\rho \circ f_j| < \varepsilon$ for $j \ge k$ and all processes in $\mathscr{K}_\alpha$. Next observe that

$$\left| \sum_{j=0}^{k-1} m^j \rho \circ f_j \right| \Big/ \sum_{j=0}^{n-1} m^j \le k(1 \vee m^{k-1})(m-1)/(m^n - 1) \to 0$$

as $m \to 1, n \to \infty$. Finally

$$\left| \sum_{j=k}^{n-1} m^j \rho \circ f_j \right| \Big/ \sum_{j=0}^{n-1} m^j < \varepsilon,$$

Background

Theorem (3.3.1) stems from Nagayev and Muhamedhanova [13] with the third reproduction moment required not to exceed some constant. A further discussion was given in Reference [11]. The Expansion (3.3.1), which the given treatment relies heavily upon, is due to Lindvall [12].

3.4 A DIFFUSION APPROXIMATION

Population developments rather different from the ones in the preceding two sections can be obtained if both the number of ancestors and the mean reproduction are varying. Results in this direction were first outlined by Feller [14], wanting to reconcile branching and diffusion models of population development. They were made rigorous by Jiřina [17] and generalized by Lindvall [12]. A final formulation is due to Grimvall [16].

First some heuristics. Consider a Galton–Watson process with a very large population. Therefore measure it in units of r individuals. In this new unit each individual has weight $1/r$ and therefore $m - 1$, being the expected increment of the population due to one individual, should be of the same order of magnitude. Accordingly put $m = 1 + \alpha/r$. With time unit also r, tr generations pass during time t. Start the process from r ancestors, i.e., in the new unit the population has size one at time zero. At time t the expected size and variance are, by Theorem (2.2.2),

$$E[z_{tr}(r)/r] = (1 + \alpha/r)^{tr} \approx e^{\alpha t},$$

$$\mathrm{Var}\,[z_{tr}(r)/r] = \sigma^2 (1 + \alpha/r)^{tr-1} \{(1 + \alpha/r)^{tr} - 1\}/\alpha$$

$$\approx \sigma^2 e^{\alpha t}(e^{\alpha t} - 1)/\alpha, \tag{3.4.1}$$

for $\alpha \neq 0$, and $t\sigma^2$ for $\alpha = 0$. The approximations are actually nothing but the same two moments of a diffusion process describing population growth: Consider a diffusion x_t with infinitesimal drift and variance proportional to the process.

$$E[x_{t+u} - x_t | x_t] = \alpha x_t u + o(u), \qquad x_t > 0,$$
$$\mathrm{Var}\,[x_{t+u} - x_t | x_t] = (\sigma^2/2)\, x_t u + o(u), \quad x_t > 0, \tag{3.4.2}$$

as $u \to 0$, both entities being zero if $x_t = 0$. The proportionality to x_t is analogous to the branching assumption of individuals reproducing independently. The density $p(t, x)$ of x_t satisfies for $x > 0$ the forward equation

$$\frac{\partial}{\partial t} p(t, x) = (\sigma^2/2) \frac{\partial^2}{\partial x^2} \{xp(t, x)\} - \alpha \frac{\partial}{\partial x} \{xp(t, x)\} \tag{3.4.3}$$

(for this and more about diffusions see pages 320–331 of Reference [15]). Since $p(t, x) = 0$ for $x < 0$, the Laplace transform

$$\varphi(t, u) = E[e^{-ux_t}]$$

is well defined for $u > 0$ and satisfies ($x_0 = 1$)

$$\frac{\partial}{\partial t} \varphi(t, u) = u\{\alpha - (\sigma^2/2) u\} \frac{\partial}{\partial u} \varphi(t, u) \tag{3.4.4}$$

$$\varphi(0, u) = u$$

The reader can verify that the solution of this is

$$\varphi(t, u) = \begin{cases} \exp\{-ue^{\alpha t}/(1 + u\sigma^2(e^{\alpha t} - 1)/2\alpha)\}, & \text{if } \alpha \neq 0, \\ \exp\{-u/(1 + u\sigma^2/2)\}, & \text{if } \alpha = 0. \end{cases} \tag{3.4.5}$$

Differentiation (or, directly, multiplication of Equation (3.4.3) by x or x^2 and integration by parts) yields

$$E[x_t] = e^{\alpha t}$$

$$\mathrm{Var}\,[x_t] = \sigma^2 e^{\alpha t}(e^{\alpha t} - 1)/\alpha.$$

All this makes the following plausible.

Theorem (3.4.1)

For each j let $\{z_{jn}; n \in Z_+\}$ be a Galton–Watson process with reproduction mean $1 + \alpha_j/j$ and variance σ_j^2. Assume that

$$\alpha_j \to \alpha, \quad \sigma_j^2 \to \sigma^2 > 0 \tag{3.4.6}$$

$$\sum_{k=0}^{\infty} k^2 p_{jk} \text{ converges uniformly in } j. \tag{3.4.7}$$

Here $\{p_{jk}; k \in Z_+\}$ is the reproduction distribution of $\{z_{jn}\}$. Then for $t \geq 0$ and $r_n \to \infty$ $z_{r_n, r_n t}(r_n)/r_n \overset{d}{\to} x_t$, where $\{x_t; t \in R_+\}$ is the diffusion starting from one with infinitesimal drift and variance αx and $\sigma^2 x/2$ respectively and having zero as an absorbing barrier. (We write z_t for $z_{[t]}$ etc.)

Proof. Suppose $\alpha \neq 0$ and n so large that no divisions by zero occur in the sequel. For simplicity take $r_n = n$. Write $f_{jn}(s) = E[s^{z_{jn}}]$, $a_j = f_{j1}''(1)/2$, and R_{jn} for the R_n corresponding to $\{z_{jn}\}$. Expansion (3.3.4) at the point $e^{-u/n}$, $u \geq 0$, yields for $t > 0$.

$$1/\{1 - f_{n,nt}(e^{-u/n})\} = (1 + \alpha_n/n)^{-nt}/(1 - e^{-u/n}) + a_n\{1 - (1 + \alpha_n/n)^{-nt}\}$$
$$\{1 + R_{n,nt}(e^{-u/n})\} n/\alpha_n.$$

By Relations (3.4.6) and (3.4.7) and the note following Theorem (3.3.1), Lemma (3.3.2) implies that

$$\lim_{n \to \infty} 1/\{n(1 - f_{n,nt}(e^{-u/n}))\} = e^{-\alpha t}/u + (\sigma^2/2\alpha)(1 - e^{-\alpha t}).$$

Hence

$$\lim_{n \to \infty} \log \{1 - f_{n,nt}(e^{-u/n})\}^n = -ue^{\alpha t}/\{1 + (\sigma^2/2\alpha)(e^{\alpha t} - 1)\}.$$

The case $\alpha = 0$ is left for the reader. $\square$

Note. *The requirement (3.4.7) above can be relaxed by a shade to: For all $t > 0$*

$$\lim_{j \to \infty} \sum_{k > jt} k^2 p_{jk} = 0. \tag{3.4.8}$$

Indeed, if that is the case and $f^{(j)}$, $\rho^{(j)}$ are the remainders in Equation (3.3.1) corresponding to $\{p_{jk}; k \in Z_+\}$, then for any $\varepsilon > 0$

$$\limsup_{n \to \infty} |r^{(n)}(e^{-u/n})| \leq \limsup_{n \to \infty} \{f_{n1}''(1) - f_{n1}''(e^{-u/n})\}$$

$$\leq \limsup_{n \to \infty} \left\{ u \sum_{k=0}^{\varepsilon n} k^3 p_{nk}/n + \sum_{k > \varepsilon n} k^2 p_{nk} \right\} \leq \varepsilon \sigma^2 u.$$

Hence $r^{(n)}(e^{-u/n}) \to 0$. So does $\rho^{(n)}(e^{-u/n})$ and it follows that $\lim_{n \to \infty} R_{n,nt}(e^{-u/n}) = 0$. Actually Equation (3.4.8) is also necessary [16].

Background

We have not touched the topic of weak convergence in a function space sense. Such questions are discussed in References [12] and [16].

3.5 GALTON-WATSON PROCESSES IN VARYING ENVIRONMENTS

In the preceding sections we have seen that provided there is a steady immigration or provided individuals stemming from a constantly increasing number of ancestors are counted, then the processes can result in stable population sizes. These are, of course, fairly artificial devices and the question arises whether stabilizing populations could be arrived at if the reproduction law is no longer the same from generation to generation. For example it might seem conceivable that a stationary population could be obtained by letting supercritical processes approach criticality as time passes. The purpose of this section is to answer such questions—essentially in the negative.

It is sometimes said that this explosion or extinction behaviour of branching processes is in sharp contrast to that of biological populations, which indeed tend to stabilize with time. My feeling is rather that branching processes illustrate a basic property of populations under reproduction: left for themselves all populations explode or die out; stability is a result of competition between species and subtle environmental checks.

Consider now a Galton-Watson process but assume that individuals in the nth generation multiply according to a reproduction law $\{p_{nk}\}$ with probability generating function φ_n. Keeping our old notation, we wish to investigate

    a. if there is a random variable z_∞ (possibly infinite) such that $z_n \to z_\infty$ in a suitable sense,

    b. when $P[z_\infty = 0]$, $P[z_\infty = \infty]$, or their sum is one.

Here are the answers.

Theorem (3.5.1)

    There is a random variable z_∞ taking values in $Z_+ \bigcup \{\infty\}$ and such that $z_n \to z_\infty$ almost surely.

Theorem (3.5.2)

    Provided all $p_{n0} < 1$, this z_∞ satisfies $P[z_\infty = 0] + P[z_\infty = \infty] < 1$ if and only if

$$\sum_{n=0}^{\infty} \{1 - p_{n1}\} < \infty.$$

Theorem (3.5.3)

Write μ_j for the reproduction means,

$$\mu_j = \sum_{k=1}^{\infty} kp_{jk}, \qquad m_n = \prod_{j=0}^{n-1} \mu_j$$

It holds that

$$E[z_\infty] \leq \liminf_{n \to \infty} m_n.$$

There is an immediate but informative

Corollary (3.5.4)

Suppose that

$$\sum_{n=0}^{\infty} \{1 - p_{n1}\} = \infty$$

and

$$\liminf_{n \to \infty} m_n < \infty.$$

Then, the extinction probability $P[z_\infty = 0] = 1$.

Before the proofs a comment: By "essentially in the negative" I meant Theorem (3.5.2), saying that a limit behaviour different from the classical Galton–Watson case can occur only if $p_{n1} \to 1$ rapidly, i.e. the population size is fairly constant. Indeed, in that case the a.s. convergence of the z_n, being integer valued, implies that when $z_n \not\to \infty$, the population will not change at all from some finite generation on. For investigations into processes where the reproduction law of individuals in a generation depends on its size see References [23] and [24].

Example (3.5.5)

Consider a binary splitting process with the splitting probability varying from generation to generation,

$$p_{n0} = 1 - p_n$$

$$p_{n2} = p_n.$$

Since $p_{n1} = 0$, z_n can only become extinct or grow indefinitely, $P[z_n \to 0] = 1$ if

$$\liminf_{n \to \infty} \left\{ n \log 2 + \sum_{k=0}^{n-1} \log p_k \right\} < +\infty.$$

This shows that there are processes which are supercritical in all generations but still bound to extinction. Indeed if all $p_n > 1/2$ but $\sum (p_n - 1/2)$ converges, so does $\sum \log 2p_n$. □

Now to the proofs. First two lemmata from Reference [19] are needed to prove the difficult part of the first theorem (concerning the situation where the range of z_∞ has more than two points). Write $f_n = \varphi_0 \circ \dots \circ \varphi_{n-1}$ for the generating function of z_n.

Lemma (3.5.6)

If $g = \lim\limits_{n \to \infty} f_n$ exists and is strictly increasing, then $\varphi_n \to \varepsilon$, uniformly. Here $\varepsilon(s) = s$.

Proof. Fix $0 \le s < 1$ and let $\delta > 0$ be arbitrary but such that $s + \delta < 1$. Assume that there were a subsequence $\{n_k\}$ such that $\varphi_{n_k}(s) \ge s + \delta$. Then

$$f_{n_k+1}(s) = f_{n_k} \circ \varphi_{n_k}(s) \ge f_{n_k}(s + \delta)$$

and (letting $k \to \infty$)

$$g(s) \ge g(s + \delta)$$

would be a necessary but impossible consequence. Hence

$$\limsup_{n \to \infty} \varphi_n(s) \le s.$$

In a similar way

$$\liminf_{n \to \infty} \varphi_n(s) \ge s.$$

Further $\varphi_n(1) = 1 = \varepsilon(1)$ and convergence of probability generating functions to such a function (like ε) is always uniform. $\quad\square$

Lemma (3.5.7)

Let $\{g_n\}$ and $\{h_n\}$ be two sequences of generating functions. Suppose that $g_n \circ h_n \to \varepsilon$. Then, both $g_n \to \varepsilon$ and $h_n \to \varepsilon$.

Proof. Let X_{nk}, Y_n n, $k \in N$ be independent random variables with generating functions h_n and g_n respectively. Then

$$\sum_{k=1}^{Y_n} X_{nk}$$

has the generating function $g_n \circ h_n$.

The convergence $g_n \circ h_n \to \varepsilon$ means that

$$\sum_{k=1}^{Y_n} X_{nk} \xrightarrow{\text{d}} 1.$$

Since Y_n and X_{nk} are non-negative and integer valued, this requires that both $Y_n \xrightarrow{\text{d}} 1$ and $X_{nk} \xrightarrow{\text{d}} 1$ as $n \to \infty$. $\quad\square$

Lemma (3.5.8)

Theorem (3.5.1) *is true for convergence in distribution.*

Proof. First $f_n(0) = \varphi_0 \circ \ldots \circ \varphi_{n-1}(0)$ does not decrease and

$$q = P[z_n \to 0] = \lim_{n \to \infty} f_n(0).$$

If for all $0 < s < 1$

$$\limsup_{n \to \infty} f_n(s) = q,$$

obviously $f_n \to q$ on $[0, 1)$. Let us therefore assume that there is an $0 < s < 1$ such that

$$\limsup_{n \to \infty} f_n(s) = a > q.$$

Then

$$f_{n_k}(s) \to a$$

for some subsequence $\{n_k\}$ and by the Helly selection theorem (page 261 of Reference [15]); this one can be so chosen that $f_{n_k} \to$ some g uniformly on compact subsets of $[0, 1)$. Since the f_{n_k} are generating functions so is g (though possibly defective). And as $g(s) = a > q = g(0)$, g must be strictly increasing.

    Next

$$f_{n_k} = (\varphi_0 \circ \ldots \circ \varphi_{n_1 - 1}) \circ \ldots \circ (\varphi_{n_{k-1}} \circ \ldots \circ \varphi_{n_k - 1})$$

and by Lemma (3.5.6)

$$\lim_{k \to \infty} \varphi_{n_{k-1}} \circ \ldots \circ \varphi_{n_k - 1} = \varepsilon.$$

For any $n \in N$ define $k(n)$ by

$$n_{k(n)} \leq n < n_{k(n)+1}.$$

Obviously then

$$f_n = f_{n_{k(n)}} \circ (\varphi_{n_{k(n)}} \circ \ldots \circ \varphi_{n-1})$$

But

$$(\varphi_{n_{k(n)}} \circ \ldots \circ \varphi_{n-1}) \circ (\varphi_n \circ \ldots \circ \varphi_{n_{k(n)+1} n - 1} \to \varepsilon$$

implies by Lemma (3.5.7) that

$$\varphi_{n_{k(n)}} \circ \ldots \circ \varphi_{n-1} \to \varepsilon.$$

Hence the *uniform* convergence

$$f_{n_{k(n)}} \to g$$

on closed subsets of $[0, 1)$ shows that $f_n \to g$.   ☐

Postponing the a.s. convergence, we turn to the

Proof of Theorem (3.5.2). Let us first show that

$$\sum_{n=0}^{\infty} \{1 - p_{n1}\} < \infty \Rightarrow P[0 < z_{\infty} < \infty] > 0, \qquad (3.5.1)$$

if only all $p_{n0} < 1$. If the left hand side holds, then there are positive integers i, j such that $P[z_i = j]$ and $p_{n1} > 0$ for all $n > i$. Therefore

$$P[z_{\infty} = j] \geq P[z_i = j] \prod_{n > i} (p_{n1})^j > 0$$

and Relation (3.5.1) follows.

For the converse, we shall work analytically, comparing

$$\prod_{k=n}^{\infty} \varphi_k'(0)$$

for large n with the product

$$\prod_{k=n}^{\infty} \varphi_k' \circ g_{k+1}(0).$$

Here

$$g_k = \lim_{n \to \infty} \varphi_k \circ \ldots \circ \varphi_n$$

exists by Lemma (3.5.8). Assume that $P[0 < z_{\infty} < \infty]$ is strictly positive. Then g is strictly increasing. Also $g = f_k \circ g_k$. The argument proving Lemma (3.5.6) applies to show that $g_k \to \varepsilon$. Hence

$$\prod_{k=n}^{\infty} \varphi_k' \circ g_{k+1}(0) = g_n'(0) \to 1$$

as $n \to \infty$. Further

$$\prod_{k=n}^{\infty} \varphi_k'(0) \Bigg/ \prod_{k=n}^{\infty} \varphi_k' \circ g_{k+1}(0) = \exp - \sum_{k=n}^{\infty} \log \{\varphi_k' \circ g_{k+1}(0)/\varphi_k'(0)\}$$

$$\geq \exp - \sum_{k=n}^{\infty} \{\varphi_k' \circ g_{k+1}(0) - \varphi_k'(0)\}/\varphi_k'(0)$$

$$\geq \exp - \sum_{k=n}^{\infty} \varphi_k'' \circ g_{k+1}(0) \, g_{k+1}(0)/\varphi_k'(0).$$

Since $\varphi_k'(0) \to 1$ we can choose n so that $\varphi_k'(0) \geq 1/2$ for $k \geq n$. For such n

$$\prod_{k=n}^{\infty} \varphi_k'(0) \geq \prod_{k=n}^{\infty} \varphi_k' \circ g_{k+1}(0) \exp\left\{ -2 \sum_{k=n}^{\infty} \varphi_k'' \circ g_{k+1}(0) \right\}.$$

However since $\varphi'_k \circ g_{k+1}(0) \to 1$, $g'_n(0) \to 1$, and $\varphi''_k \geq 0$ it holds for large n that

$$0 \leq \sum_{k=n}^{\infty} \varphi''_k \circ g_{k+1}(0) \leq 2g'_n(0) \sum_{k=n}^{\infty} \varphi''_k \circ g_{k+1}(0) \, g'_{k+1}(0)/\varphi'_k \circ g_{k+1}(0)$$

$$= 2g''_n(0) \to 0,$$

as $n \to \infty$. Hence,

$$\prod_{k=n}^{\infty} \varphi'_k(0) > 0$$

for large n and since $\varphi'_n(0) = p_{n1} \leq 1$,

$$\sum_n (1 - p_{n1}) < \infty. \qquad \Box$$

Before completing the proof of Theorem (3.5.1), we note that Theorem (3.5.3) is just a consequence of Fatou's lemma:

$$E[z_\infty] \leq \liminf_{n \to \infty} E[z_n] = \liminf_{n \to \infty} m_n.$$

Proof of Theorem (3.5.1).

A. The case $\sum (1 - p_{n1}) < \infty$. Clearly

$$|\varphi_n(s) - s| = \left| \sum_{k \neq 1} p_{nk}s^k - (1 - p_{n1})s \right| \leq \sum_{k \neq 1} p_{nk} + 1 - p_{n1} = 2(1 - p_{n1}).$$

Hence,

$$\sum_{n > j} \sup_{0 < s < 1} |\varphi_n(s) - s| \leq 2 \sum_{n \geq j} (1 - p_{n1}) < 1/4$$

for j large enough. Define $a_j = 1/2$ and recursively, $n > j$, $a_{n+1} = \varphi_n^{-1}(a_n)$. Since $1 - p_{n1} < 1/8$ for $n > j$, the φ_n are strictly increasing and have inverses. Further $|a_{n+1} - a_n| = |a_{n+1} - \varphi_n(a_n)| \leq 2(1 - p_{n1})$ and $|a_{n+k} - a_n| \leq 2 \sum_{i \geq n}(1 - p_{i1}) < 1/4$, showing that $\lim_{n \to \infty} a_n = a$ must exist. Choosing $n = j$ and letting $k \to \infty$ above we see that $|a - a_j| \leq 1/4$ or $1/4 \leq a \leq 3/4$. As usual let $\mathscr{B}_n = \sigma(z_1, \ldots, z_n)$ and note that

$$E[a_{n+1}^{z_{n+1}} | \mathscr{B}_n] = E[a_{n+1}^{z_{n+1}} | z_n] = \{\varphi_n(a_{n+1})\}^{z_n} = a_n^{z_n}.$$

Thus the random variables $0 \leq a_n^{z_n} \leq 1$ constitute a martingale and by Theorem (5.1.2) $a_n^{z_n} \to$ some x a.s. Since $a_n \to a$, it follows that $z_n \to \to z_\infty \log x/\log a$ a.s.

B. The case $\sum (1 - p_{n1}) = \infty$. Whereas the preceding case was proved directly, this one needs Theorem (3.5.2) and Lemma (3.5.8). If $q = = \lim_{n \to \infty} f_n(0) = P[z_n \to 0] = 1$, there is nothing to prove. Thus assume $q < 1$. $E[s^{z_\infty}] = g(s) = q$ for $0 \le s < 1$ and

$$\sup_{k \ge n} \sup_{0 \le s < 1/2} |f_k(s) - q| = t_n \downarrow 0$$

by Lemma (3.5.8). Take j to be large enough to ensure that $q + 2t_n \le 1$ for $n \ge j$. For such n define $a_n = f_n^{-1}(q + 2t_n)$. Since $f_n(a_n) - q = 2t_n$, a_n must be $> 1/2$ if only $t_n > 0$. But $t_n = 0$ only if some f_k is constant, i.e. some $\varphi_k = 1$. In that case $z_n = 0$ for $n \ge k$ and so we can restrict ourselves to $t_n > 0$ for all n. Further

$$\varphi_n(a_{n+1}) = \varphi_n \circ f_{n+1}^{-1}(q + 2t_{n+1}) = f_n^{-1}(q + 2t_{n+1}) \le f_n^{-1}(q + 2t_n) = a_n,$$

which we shall use to show that $x_n = a_n^{z_n} - 1_{\{0\}}(z_n)$, $n \ge k$, yields a non-negative supermartingale with respect to $\{\mathscr{B}_n\}$. Indeed,

$$E[x_{n+1}|\mathscr{B}_n] = E[a_{n+1}^{z_{n+1}} - 1_{\{0\}}(z_{n+1})|z_n]$$
$$= \{\varphi_n(a_{n+1})\}^{z_n} - \{\varphi_n(0)\}^{z_n} \le a_n^{z_n} - \{\varphi_n(0)\}^{z_n} \le a_n^{z_n} - 1_{\{0\}}(z_n) = x_n.$$

Again supermartingale convergence yields $x_n \to$ some x a.s. But since

$$E[x_n] = f_n(a_n) - f_n(0) = q + 2t_n - f_n(0) \le 3t_n \downarrow 0,$$
$$E[x] \le \lim_{n \to \infty} E[x_n] = 0$$

by Fatou's lemma, and $x_n \to 0$ a.s. If $z_n \to 0$, then clearly $x_n = a_n^{z_n} - 1_{\{0\}}(z_n) \to 1 - 1 = 0$. But on the set where $z_n \not\to 0$ we must have a.s. that $a_n^{z_n} \to 0$. As $a_n \ge 1/2$, this requires that $z_n \to \infty$. $\qquad \square$

Background

The idea of branching processes in varying environments seems first to have been formulated by I. J. Good in the discussion following the representation of D. G. Kendall's Stochastic processes and population growth. The reference is J. Roy. Statist. Soc. Ser. B. 11, 271–272, 1949. The a.s. convergence in Theorem (3.5.1) is from Reference [25], whereas the other results of this section [except the simple Theorem (3.5.1)] are discoveries of J. D. Church [21] (see also [19]).

Another early inquiry is Reference [22]. Agresti [18] has obtained an elegant necessary and sufficient condition for $P[z_n \to 0] = 1$, namely that $\sum 1/m_j = \infty$ (plus some regularity requirements).

3.6 FURTHER RESULTS FOR VARYING ENVIRONMENTS

In analogy with the supercritical Galton–Watson process define $w_n = z_n/m_n$. (We assume throughout that all $p_{n0} < 1$). The sequence $\{w_n\}$ still constitutes a non-negative martingale with an almost sure limit $w \geq 0$. The further analysis will be made under the simplifying assumption $\varphi_k''(1) = 2a_k < \infty$.

Write

$$m_{kn} = \prod_{j=k+1}^{n-1} \mu_j, \qquad 0 \leq k < n-1, \qquad m_{n-1\,n} = 1.$$

Any reader, patient enough, can convince him/herself that

$$f_n''(1) = 2 \sum_{k=0}^{n-1} a_k m_k m_{kn}^2. \tag{3.6.1}$$

By martingale properties

$$E[(w_{n+r} - w_n)^2] = E[w_{n+r}^2] - E[w_n^2] = \{f_{n+r}''(1) + m_{n+r}\}/m_{n+r}^2$$

$$- \{f_n''(1) + m_n\}/m_n^2 = 2 \sum_{k=n}^{n+r-1} a_k/m_k\mu_k^2 + 1/m_{n+r} - 1/m_n. \tag{3.6.2}$$

Invoking L^2-completeness we conclude:

Theorem (3.6.1)
Let $w_n = z_n/m_n$. Then

$$w = \lim_{n \to \infty} w_n$$

exists a.s. Provided $m_n \to m$, $0 < m \leq \infty$, the convergence $w_n \to w$ holds in mean square if and only if

$$\sum_{n=0}^{\infty} a_n/m_n\mu_n^2 < \infty$$

Then

$$\mathrm{Var}[w] = 2 \sum_{n=0}^{\infty} a_n/m_n\mu_n^2 + (1-m)/m, \quad (1 - \infty)/\infty \quad \text{interpreted as} \quad -1. \tag{3.6.3}$$

As stated this theorem covers two quite different situations:
  (a) $m_n \to \infty$ rapidly and a_n remains bounded,
  (b) $m_n \to m$, $0 < m < \infty$ but $\sum a_n$ converges.
The latter case is an L^2-variant of the results in Section 3.5 (with non-degenerate $z_\infty = mw$). The former yields a close affinity to supercriticality.

The counterpart of critical processes turns to be processes satisfying

(c) $0 < \lim\limits_{n \to \infty} m_n < \infty$ and $\sum\limits_{n} a_n/m_n \mu_n^2 = \infty$

and a process is "subcritical" if

(d) $m_n \to 0$.

The critical case resembles the setup in Section 3.3 and will be treated under similar regularity conditions: In obvious notation Equation (3.3.1) reads

$$\mu_k/(1 - \varphi_k) - 1/(1 - \varepsilon) = a_k/\mu_k + \rho_k. \tag{3.6.4}$$

Writing $f_{kn} = \varphi_{k+1} \circ \ldots \circ \varphi_{n-1}$ and proceeding as in Section 3.3, we obtain

$$m_n/(1 - f_n) - 1/(1 - \varepsilon) = \sum_{k=0}^{n-1} m_{kn}\{\mu_k/(1 - \varphi_k \circ f_{kn}) - 1/(1 - f_{kn})\}$$

$$= \sum_{k=0}^{n-1} m_{kn}(a_k/\mu_k + \rho_k \circ f_{kn}). \tag{3.6.5}$$

With

$$1/B_n = \sum_{k=0}^{n-1} a_k/m_k \mu_k^2,$$

$$R_n = \sum_{k=0}^{n-1} \rho_k \circ f_{kn}/m_{k+1},$$

$$1/(1 - f_n) = 1/m_n(1 - \varepsilon) + 1/B_n + R_n. \tag{3.6.6}$$

Provided B_n/m_n and $B_n R_n \to 0$ uniformly on $[0, 1)$, this leads to, $u \geq 0$,

$$E[e^{-uz_n B_n/m_n}|z_n > 0] = 1 - \{1 - f_n(e^{-uB_n/m_n})\}/\{1 - f_n(0)\}$$

$$= 1 - \frac{1 + B_n/m_n + B_n R_n(0)}{1 + B_n/m_n(1 - e^{-uB_n/m_n}) + B_n R_n(e^{-uB_n/m_n})} \to 1/(1 + u),$$

as $n \to \infty$. Hence under suitable conditions the exponential behaviour, typical of critical processes, persists. Let us find these conditions.

Lemma (3.6.2)

Suppose that

$$\sum_{j} p_{kj} j^2 \quad \text{converges uniformly in } k, \quad \inf \mu_k > 0. \tag{3.6.8}$$

Then the functions ρ_k of Equation (3.6.4) satisfy $\lim\limits_{s \to 1} \rho_k(s) = 1$ uniformly in k.

Proof. This just combines Lemmata (3.3.5) and (3.3.7). □

Lemma (3.6.3)
Assume that Condition (3.6.8) holds and further that

$$\inf_n a_n > 0, \qquad 0 < \lim_{n \to \infty} m_n < \infty. \qquad (3.6.9)$$

$$\lim_{n \to \infty} f_{kn}(s) = 1$$

uniformly in s and in $n - k$. (This means that f_{kn} is close to one for all k whenever $n - k$ is large.)

Proof. Obviously $\mu_n \to 1$. Hence by Lemma (3.3.3), there is an r such that

$$p = \inf_{j \geq r} \varphi_j(0) > 0 \quad \text{and} \quad \sup_{j \geq r} \mu_j \leq 2.$$

Then [argue as in Lemma (3.3.4)]

$$c = \inf_{j \geq r} \varphi_j''(p) > 0.$$

Further from Equation (3.3.1),

$$a_j/\mu_j + \rho_j \geq \varphi_j''/2\mu_j \geq c/4$$

for $s \geq p$ and therefore Equation (3.6.5) yields for these s

$$1/(1 - f_n) \geq (c/4) \sum_{j=r}^{n-1} 1/m_j \to \infty.$$

Generally, $f_{kn+1} - f_{kn+1}(0) = f_{kn} \circ \varphi_n(0) \geq f_{kn}(p)$ and for $s \geq p$

$$1/(1 - f_{kn}) \geq (c/4) m_k \sum_{j=r \vee (k+1)}^{n-1} 1/m_j$$

$$\geq (c/4) \{n - 1 - r \vee (k+1)\} \inf_j m_j / \sup_j m_j. \qquad □$$

Lemma (3.6.4)
Suppose that Conditions (3.6.8) and (3.6.9) are satisfied. Then $B_n \to 0$ and $B_n R_n \to 0$ uniformly.

Proof. Let $\varepsilon > 0$ be given. Choose j such that $|\rho_k \circ f_{kn}| < \varepsilon$ for $n - k \geq j$. Then,

$$|B_n R_n| \leq \left| \sum_{k=1}^{n-j} \rho_k \circ f_{kn}/m_{k+1} \right| \Big/ \sum_{k=1}^{n} a_k/m_k \mu_k^2$$

$$+ \left| \sum_{k=n-j+1}^{n} \rho_k \circ f_{kn}/m_{k+1} \right| \Big/ \sum_{k=1}^{n} a_k/m_k \mu_k^2 \leq \varepsilon/(\inf a_k/\mu_k) + o(n),$$

since $\sum a_k/m_k \mu_k^2$ diverges. That $B_n \to 0$ is trivial. □

Theorem (3.6.5)

Let Conditions (3.6.8) and (3.6.9) be satisfied. Then, as $n \to \infty$,

(a) $P[z_n > 0]/B_n \to 1$

(b) $P[B_n z_n/m_n > u | z_n > 0] \to e^{-u}$

for $u \geq 0$.

The proof is direct from the preceding.

We turn, finally, to what emerges as subcritical processes in the case of a not necessarily constant environment.

Theorem (3.6.6)

Suppose that all $p_{n0} < 1$, and that

$$\limsup_{n \to \infty} \sum_{k=0}^{n} a_k m_n/m_{k+1} < \infty. \tag{3.6.10}$$

Then, $\lim_{n \to \infty} P[z_n > 0]/m_n$ *is a finite strictly positive number. So (though possibly zero) is*

$$\lim_{n \to \infty} P[z_n = k | z_n > 0] = b_k,$$

where

$$\sum_{k=1}^{\infty} b_k = 1$$

and

$$\sum_{k=1}^{\infty} k b_k = \lim_{n \to \infty} m_n/P[z_n > 0].$$

Proof. The Expansion (2.6.1) still renders faithful service:

$$1 - \varphi_k = \mu_k(1 - r_k/\mu_k)(1 - \varepsilon),$$

$$1 - f_{k-1\,n} = 1 - \varphi_k \circ f_{kn} = \mu_k(1 - r_k \circ f_{kn}/\mu_k)(1 - f_{kn}),$$

and, since $f_{-1\,n} = f_n$, $f_{n-1\,n} = \varepsilon$,

$$1 - f_n = (1 - \varepsilon) \prod_{k=0}^{n-1} \mu_k(1 - r_k \circ f_{kn}/\mu_k). \tag{3.6.11}$$

Therefore

$$(1 - f_n)/m_n = (1 - \varepsilon) \prod_{k=0}^{n-1} (1 - r_k \circ f_{kn}/\mu_k),$$

which decreases to some limit $(1 - \varepsilon)\,\varphi$.

The function φ is strictly positive by Inequality (3.6.10) and as $r_k < \mu_k$ and

$$0 \leq r_k \circ f_{kn} \leq a_k(1 - f_{kn}) \leq a_k m_{kn},$$

the convergence $(1 - f_n)/m_n(1 - \varepsilon) \to \varphi$ actually is uniform (if the left hand is defined as one in that point).

Proceeding as in the classical case, we conclude that

$$P[z_n > 0]/m_n \to \varphi(0)$$

and

$$E[s^{z_n}|z_n > 0] = 1 - \{1 - f_n(s)\}/\{1 - f_n(0)\}$$

$$= 1 - (1 - s) \prod_{k=1}^{n} \{1 - r_k \circ f_{kn}(s)/\mu_k\}/\{1 - r_k \circ f_{kn}(0)/\mu_k\} \downarrow \text{ some } g(s).$$

Under Condition (3.6.10) $g(s) = 1 - (1 - s)\varphi(s)/\varphi(0)$ and g is probability generating. Also as $s \to 1$,

$$g'(1)g'(1) \leftarrow \{1 - g(s)\}/(1 - s) = \varphi(s)/\varphi(0) \to 1/\varphi(0) < \infty. \qquad \square$$

Background

Theorem (3.6.1) is due to Fearn [22], who also gave part of Theorem (3.6.5). An account of all the results of this section was given in Reference [26].

3.7 RANDOM ENVIRONMENTS

The two preceding sections have an obvious bearing on branching processes where the reproduction law of each generation is chosen randomly from the class of all reproduction laws (or some suitable subclass). Such processes have been said to be in random environments.

To be more formal, let Ω_r denote the class of all probability distributions $p = \{p_k; k \in Z_+\}$ on Z_+ such that $\sum kp_k < \infty$. Endow Ω_r with the usual product algebra, that is the one generated by all sets

$$\{p \in \Omega_r; [p_0, ..., p_k] \in A\}$$

for $k \in Z_+$, A a measurable subset of R_+^k.

Next suppose that $(\Omega, \mathscr{S}, P)$ is a probability space rich enough for all the following functions to be defined on it:

(a) a sequence $\pi = \{\pi_n; n \in Z_+\}$ of measurable maps $\Omega \to \Omega_r$, $\pi_n = \{\pi_{nk}; k \in Z_+\}$ the *environmental process*,

(b) for each $p \in \Omega_r$ a double array of i.i.d. random variables X_{nj}^p with law p.

Define $\{z_n\}$, a *Galton–Watson process in random environment*, by

$$z_0 = 1$$

$$z_{n+1} = \sum_{j=1}^{z_n} X_{nj}^{\pi_n}.$$

Then

$$E[s^{z_n}|\pi] = f_n^\pi(s) = \varphi_0 \circ \ldots \circ \varphi_{n-1}(s)$$

if

$$\varphi_n(s) = \sum_{k=0}^{\infty} \pi_{nk} s^k.$$

In other words, conditional upon π a branching process in a random environment reduces to a process in a varying environment. This makes the following assertions rather easy consequences of the preceding.

The limit $z_\infty = \lim_{n \to \infty} z_n$ exists almost surely and by Theorem (3.5.2)

$$P[z_\infty = 0 \text{ or } \infty] = 1 \Leftrightarrow P\left[\exists \pi_{n0} = 1 \text{ or } \sum_{n=1}^{\infty} (1 - \pi_{n1}) = \infty \right] = 1$$

From the ergodic theorem (page 421 of Reference [32]) we see that the process is of the extinction or explosion type (i.e. z_∞ is either zero or infinite) if the sequence $\{\pi_{n1}\}$ is stationary with $P[\pi_{n1} < 1] > 0$. If the π_{n1}, $n \in N$ are independent, the Kolmogorov three series theorem applies to show that the process cannot stabilize without becoming extinct if either one of the series

$$\sum (1 - E[\pi_{n1}])$$

or

$$\sum \text{Var}[\pi_{n1}]$$

diverges. For the case of a general environmental process criteria like 'there is a $c > 0$ such that for all $n \in N$ $P[\pi_{n1} \leq 1 - c/n] = 1$' guarantee the same instability.

Theorem (3.5.3) has several direct consequences as well. With

$$\mu_n = \sum_k k \pi_{nk}$$

$$m_n = \prod_0^{n-1} \mu_k$$

it yields first

$$E[z_\infty|\pi] \leq \liminf_{n \to \infty} m_n$$

and by Fatou's lemma

$$E[z_\infty] \le \liminf_{n \to \infty} E[m_n].$$

Evidently

$$\liminf m_n < \infty \Leftrightarrow \liminf \sum_1^n \log \mu_k < \infty$$

and z_∞ is a.s. finite provided

$$P\left[\liminf \sum_1^n \log \mu_k < \infty\right] = 1.$$

When the sequence $\{\mu_n\}$ is stationary and ergodic this is the case if and only if

$$E[\log \mu_1] \le 0.$$

If the reproduction means are independent the three series criterion is again applicable: The convergences of

$$\sum E[(\log \mu_k)^+]$$

and

$$\sum \mathrm{Var}[(\log \mu_k)^+]$$

imply that z_∞ is finite. For a third illustration, let

$$\mathscr{A}_n = \sigma(\mu_0, \dots, \mu_n)$$

and assume that

$$E[\mu_{n+1}|\mathscr{A}_n] \le 1$$

Then $\{m_n\}$ is a non-negative supermartingale adapted to $\{\mathscr{A}_n\}$ and $m_\infty = \lim_{n \to \infty} m_n$ exists with

$$E[z_\infty] \le E[m_\infty] \le E[\mu_1].$$

If the criteria in this paragraph, ensuring that z_∞ is finite, are combined with those of the preceding, guaranteeing that $\{z_n\}$ can only explode or become extinct, simple conditions for

$$P[z_n \to 0] = 1$$

result.

In the same vein the arguments proving Theorem (3.6.1) can be applied to processes in random environments, showing that

$$\sum_{n=1}^\infty E[a_n/m_n\mu_n^2] < \infty \Rightarrow z_n/m_n \to \text{some } w \text{ in } L^2.$$

Here $a_n = \sum_k k(k-1)\pi_{nk}/2$.

If the environmental process has i.i.d. first and second moments, the convergence at the left takes place, if and only if $E[a_n/\mu_n^2] < \infty$ and $E[1/\mu_n] < 1$. More generally a martingale argument works to prove a.s. convergence of z_n/m_n; Let

$$\mathscr{B}_n = \sigma(\pi_0, \ldots, \pi_{n-1}, z_1, \ldots, z_n)$$

$$\mathscr{C}_n = \sigma(\pi_0, \ldots, \pi_n, z_1, \ldots, z_{n-1})$$

Obviously

$$E[z_{n+1}/m_{n+1}|\mathscr{B}_n] = E[E[z_{n+1}/m_{n+1}|\mathscr{C}_n]|\mathscr{B}_n] = E[z_n/m_n|\mathscr{B}_n] = z_n/m_n.$$

Critical random environments seem to be rare. Logarithmization turns Condition (3.6.9) for Theorem (3.6.5) into the convergence of

$$\sum_1^\infty \log \mu_k.$$

If the μ_n are i.i.d. this can only occur if (Theorem (8.2.5) of Reference [27]) $P[\mu_n = 1] = 1$.

The subcritical case, finally, is left to the reader.

Background

The idea of branching processes in random environments stems from Reference [32] where the environmental process is assumed to consist of i.i.d. random elements. The other References [19, 20, 28–33] (except Loève) concern stationary, ergodic environmental processes, though in Reference [20] less (exchangeability) is assumed. For an intriguing biological application, consult Reference [31].

REFERENCES

1. Foster, J., *Branching processes allowing immigration*. Thesis, University of Wisconsin, 1969.
2. Foster, J. and Williamson, L. A., *Limit theorems for the Galton–Watson process with time-dependent immigration*. Report, University of Colorado, 1970.
3. Heathcote, C. R., A branching process allowing immigration, *J. Roy. Statist. Soc.* **B27**, 138–143, 1965. Corrections and comments in the same journal, **B28**, 213–217, 1966.
4. Heyde, C. C. and Seneta E., Analogues of classical limit theorems for the supercritical Galton–Watson process with immigration, *Math. Biosci.* **11**, 249–259, 1971.
5. Pakes, A. G., An asymptotic result for a subcritical branching process with immigration. *Bull. Austr. Math. Soc.* **2**, 223–228, 1970.
6. Quine, M. P. and Seneta, E., A limit theorem for the Galton–Watson process with immigration. *Austr. J. Statist.* **11**, 166–173, 1969.

7. Seneta, E., An explicit limit theorem for the critical Galton–Watson process with immigration. *J. Roy. Statist. Soc.* **32**, 149–152, 1970.

8. Seneta, E., A note on the supercritical Galton–Watson process with immigration, *Math. Biosci.* **6**, 305–312, 1970.

9. Seneta, E., On the supercritical Galton–Watson process with immigration, *Math. Biosci.* **7**, 9–14, 1970.

10. Lamperti, J., Limiting distributions for branching processes. *Proc. Fifth Berkely Symp. Math. Statist. Prob.*, 225–241, Univ. or California Press, 1967.

11. Fahady, K. S., Quine, M. P. and Vere–Jones, D., Heavy traffic approximations for the Galton–Watson process. *Adv. Appl. Prob.* **3**, 282–300, 1971.

12. Lindvall, T., Limit theorems for some functionals of certain Galton–Watson branching processes. *Adv. Appl. Prob.* **2**, 309–321, 1974.

13. Nagayev, S. V. and Muhamedhanova, R., *Perehodnyie yavleniya v vetvyaščihsya slučaynyh processah s diskretnym vremenem.* (Limit phenomena in branching stochastic processes with discrete time). In Siraždinov, S. H. (ed.), Predelnyie teoremy i statističeskiye vyvody, 83–89. Fan, Tashkent, 1966. Corrections to this in 1968, same publication.

14. Feller, W., Diffusion processes in genetics. *Proc. Second Berkeley Symp. Math. Statist. Prob.* 227–246. Univ. of California Press, Berkeley, 1951.

15. Feller, W., *An Introduction to Probability Theory and its Applications II.* John Wiley and Sons, Inc., New York, 1966.

16. Grimvall, A., On the convergence of sequences of branching processes. *Ann. Prob.* **2**, 1974.

17. Jiřina, M., On Feller's branching diffusion processes. *Časopis Pěst. Mat.* **94**, 84–90, 1969.

18. Agresti, A., On the extinction times of varying and random environment branching processes. Tech. Rep. 68. Dep. of Statistics, U. of Florida, 1973.

19. Athreya, K. B. and Karlin, S., On branching processes with random environments, I: Extinction probabilities. *Ann. Math. Statist.* **42**, 1499–1520, 1971.

20. Athreya, K. B. and Karlin, S. Branching processes with random environments, II: Limit theorems. *Ann. Statist.* **42**, 1843–1858, 1971.

21. Church, J. D., On infinite composition products of probability generating functions. *Z. Wahrscheinlichkeitstheorie verw. Geb.* **19**, 243–256, 1971.

22. Fearn, D. H., Galton–Watson processes with generation dependence. *Proc. Sixth Berkeley Symp. Math. Statist. Prob. IV,* 159–172, 1972.

23. Labkovskiy, V. A., Predelnaya teorema dlya obobščonnyh vetvyaščihsya slučaynyh processov, *Teor. Veroyatnost. i Primenen* **17**, 71–83, 1972. Translated as A limit theorem for generalized random branching processes depending on the size of the population. *Theor. Prob. Appl.* **17**, 72–85, 1972.

24. Levina, L. V., Leontovič, A. M., and Pyatetskiy–Šapiro, I. I., Ob odnom reguliruyemom vetvyaščemsya processe. (A controllable branching process). *Problemy Peredači Informacii* **4**, 72–82, 1968.

25. Lindvall, T., Almost sure convergence of branching processes in varying and random environments, *Ann. Prob.* **2**, 344–346, 1974.

26. Jagers, P., Galton–Watson processes in varying environments. *J. Appl. Prob.* **11**, 174–178, 1974.

27. Chung, K. L., *A Course in Probability Theory*. Harcourt, Brace, and World, New York, 1968.
28. Kaplan, N., A theorem on compositions of random probability generating functions and applications to branching processes with random environments. *J. Appl. Prob.* **9**, 1–12, 1972.
29. Keiding, N. and Nielsen, J. E., The growth of supercritical branching processes with random environments. *Ann. Prob.* **1**, 1065–1067, 1973.
30. Loève, M. *Probability Theory,* Van Nostrand, New York, 1955.
31. Mountford, M. D. The significance of clutch-size. In: Bartlett, M. S. and Hiorns, R. W. (eds.) *The Mathematical Theory of the Dynamics of Biological Populations*. Pp. 315–324 Academic Press, London and New York, 1973.
32. Smith, W. L. and Wilkinson, W. E., On branching processes in random environments. *Ann. Math. Statist.* **40**, 814–827, 1969.
33. Turnbull, B. W., Inequalities for branching processes. *Ann. Prob.* **1**, 457–474, 1973.

Chapter 4

Results for Multi-type Processes

4.1 FUNDAMENTALS

An r-type Galton–Watson process is a branching process, where r kinds of individuals are distinguished between. These are to be called type $1, 2, \ldots, r$. Each individual has unit life–length and if it is of type k it splits into k_1 children of type one, k_2 of type 2, ..., up to k_r of type r with probability $p_k(k_1, k_2, \ldots, k_r)$.

The basis for a formal definition was provided by the few words about multitype processes in Chapter 1. An individual $(a, k_0; \ldots; j_n, k_n)$ is *of type k_n* and it is realized or not according to the definition in Chapter 1. The rest proceeds as in Section 2.1. Let r_x be one if x is realized and zero otherwise. Write for $1 \leq k, i \leq r$

$$z_n^{(i)}(k) = \sum_{x \in I_{(a,k)}(n, i)} r_x,$$

$$z_n(k) = \left[z_n^{(1)}(k), \ldots, z_n^{(r)}(k) \right].$$

The sequence $\{z_n(k); n \in Z_+\}$ is an *r-type Galton–Watson process with ancestor of type k.* As in the one-dimensional case $z_{n+1}(k)$ is of the form

$$\sum_{i=1}^{r} \sum_{j=1}^{z_n^i(k)} X_{nj}^{(i)} \tag{4.1.1}$$

with independent vectors $X_{nj}^{(i)}$,

$$P\left[X_{nj}^{(i)} = \left[k_1, \ldots, k_r \right] \right] = p_i(k_1, \ldots, k_r).$$

To ease the notation, a little at least, we shall use vector symbols:

$$\tilde{0} = [0, \ldots, 0],$$
$$\tilde{1} = [1, \ldots, 1],$$
$$\tilde{k} = [k_1, \ldots, k_r] \in Z_+^r,$$
$$s = [s_1, \ldots, s_r] \in [0, 1]^r,$$

87

$$s^{\bar{k}} = \prod_{j=1}^{r} s_j^{k_j},$$

$$f^{(k)}(s) = \sum_{\bar{k}} p_k(\bar{k}) s^{\bar{k}},$$

$$f = [f^{(1)}, \dots, f^{(r)}],$$

$$f_n^{(k)}(s) = E[s^{z_n(k)}],$$

$$f_n = [f_n^{(1)}, \dots, f_n^{(r)}],$$

$$m_{ij} = E[z_1^{(j)}(i)] = (\partial f^{(i)}/\partial s_j)(\tilde{1}),$$

$$m = \begin{bmatrix} m_{11} & \cdots & m_{1r} \\ m_{21} & \cdots & m_{2r} \\ \vdots & & \vdots \\ m_{r1} & \cdots & m_{rr} \end{bmatrix}.$$

With f_0 the identity map the recurrence scheme from one-type processes persists,

$$f_0(s) = s, \qquad\qquad (4.1.2)$$
$$f_{n+1}(s) = f_n \circ f(s),$$

and m^n is the matrix with elements

$$(m^n)_{ij} = E[z_n^{(j)}(i)], \qquad 1 \le i, j \le r. \qquad (4.1.3)$$

This means that

$$E[z_n(k)] = e_k m^n \qquad (4.1.4)$$

where e_k has the number one at place number k and zeros for the rest. Similarly

$$E[z_{n+1}(k)|z_n(k)] = z_n(k) m. \qquad (4.1.5)$$

The corresponding variance formulae are considerably more involved. Let $\Sigma_n(k)$ be the covariance matrix with entries

$$\Sigma_n^{ij}(k) = \operatorname{Cov}[z_n^{(i)}(k), z_0^{(j)}(k)].$$
$$\Sigma = \Sigma_1$$

By conditioning with respect to $z_{n-1}(k)$ we obtain

$$E[z_n^{(i)}(k) z_n^{(j)}(k)] = \sum_{v=1}^{r} E[z_{n-1}^{(v)}(k)] \Sigma^{ij}(v) + \sum_{\mu,v} m_{vi} E[z_{n-1}^{(v)}(k) z_{n-1}^{(\mu)}(k)] m_{\mu j}$$

or, equivalently, by Equation (4.1.4)

$$\text{Cov}\left[z_n^{(i)}(k), z_n^{(j)}(k)\right] = \sum_{v=1}^{r} E\left[z_{n-1}^{(v)}(k)\right] \Sigma^{ij}(v)$$
$$+ \sum_{\mu,v} m_{vi} \,\text{Cov}\left[z_{n-1}^{(v)}(k), z_{n-1}^{(\mu)}(k)\right] m_{\mu j},$$

i.e.

$$\Sigma_n(k) = \sum_{v=1}^{r} \Sigma(v)(m^{n-1}e_k)_v + m'\Sigma_{n-1}(k)\,^m$$

where $(m^{n-1}e_k)_v$ is the vth coordinate of the vector and the prime stands for transpose. Repetition yields

$$\Sigma_n(k) = \sum_{j=1}^{r} m'^{(n-j)} \left\{ \sum_{v=1}^{r} \Sigma(v)(m^{j-1}e_k)_v \right\} m^{n-j}. \qquad (4.1.6)$$

In this way much of the many-type theory reduces to one-type theory in matrix-vector notation. Unfortunately this is not all the truth. Processes with several types are more complex than the original Galton–Watson process because besides the branching pattern they contain another Markovian structure of movement between different states, the types. One extreme, pure branching is obtained if $r = 1$, the other extreme, a Markov chain with r states and no branching if

$$f^{(i)}(s) = \sum_{j=1}^{r} p_{ij}s_j, \qquad 1 \le i \le r,$$

for some numbers p_{ij}. Such processes are said to be *singular* and can be viewed as one individual, wandering between the r different states-types.

Thus all the well-known and irritating complications of Markov chains like different classes of states and periodicity reappear here. For a thorough treatment the reader is referred to Sevast'yanov's book [5]. In Western literature it has been common to avoid—or at least diminish—the difficulties by assuming the process positively regular: An r-type Galton–Watson process is *positively regular* if there is an $n \in N$ such that all entires in m^n are strictly positive. Then the Perron–Frobenius theory of positive matrices comes to help.

However, multi-type processes of interest for applications are often not positively regular. We shall give two examples and devote the next section to results for positively regular multi-type processes.

Example (4.1.1)

Cells divide in the manner shown in the diagram:

1. Chromosomes contract and take place along the equatorial plane of the nucleus.

2. Chromosomes divide and move towards the poles.

3. A membrane develops at the equator.

4. The cell divides.

Hence the usual is that a cell with $2n$ chromosomes ($2n$ for diploidy) divides into two similar ones. However sometimes the third and fourth steps do not occur (this phenomenon is known as *endomitosis*) and a diploid cell gives rise to only one but a tetraploid cell (i.e. one with $4n$ chromosomes). When the latter divides it behaves as a diploid cell—usually begetting two daughters of its own kind sometimes changing into a cell with $8n$ chromosomes.

Say that cells with $2^i n$ chromosomes are of type i and assume that the probability of a cell of type i undergoing endomitosis is p_i. Then we are led to a process with countably many types, each cell of type i resulting with probability $1 - p_i$ in two i-type individuals and with probability p_i in one of type $i + 1$.

In practice the incidence of higher ploidies than four is small and it may suffice to consider a two-type model: diploid cells are type one, the rest (polyploid cells) type two. If the probability of endomitosis is p (independent of the ploidy) the reproduction law is then

$$p_1(2, 0) = 1 - p$$
$$p_1(0, 1) = p$$

$$p_2(0, 1) = p$$
$$p_2(0, 2) = 1 - p$$

The mean is

$$m = \begin{bmatrix} 2(1 - p) & p \\ 0 & 2 - p \end{bmatrix}$$

and the zero remains in all powers of m.   □

Example (4.1.2)

A fashionable trick in some demographic circles has been to introduce an age-structure into Galton–Watson processes through a back door [3].

Assume that people can not attain age c and that women are fertile from (and including) age a to (but excluding) age b. Say that a woman, i years old, is an individual of type $i + 1$ and that a man aged i is of type $c + i + 1$, $i = 0, 1, ..., c - 1$. Let $X^{(i)}$ be random vectors like those in Equation (4.1.1), i.e. giving the reproduction of an i-type individual.

A man, as well as a non-reproductive woman, either dies during a year or survives giving place to a new individual of the immediately following type. Let $p_i, 0 \le i < c$ be the probability of a woman aged i attaining age $i + 1$ and $p_i, c \le i < 2c$ the corresponding survival probability for $i - c$ years old men. Then

$$P[X^{(i)} = e_{i+1}] = 1 - P[X^{(i)} = \tilde{0}] = p_{i-1}$$

for $i \le a$ or $i \le b$. Obviously $p_{c-1} = p_{2c-1} = 0$ whereas the other p_i can be estimated from life tables [1]. For women in reproductive age we write

$$P[X^{(i)} = e_{i+1} + je_1 + ke_{c+1}] = p_{i-1}(j, k),$$
$$P[X^{(i)} = je_1 + ke_{c+1}] = q_{i-1}(j, k)$$

for $a < i \le b$ and $0 \le j + k \le$ say 5 and assume

$$\sum_{0 \le j+k \le 5} p_{i-1}(j, k) + q_{i-1}(j, k) = 1,$$

$$\sum_{0 \le j+k \le 5} p_{i-1}(j, k) = p_{i-1}.$$

The enormous matrix m looks the following way (with

$$m_G^i = \sum_{0 \le j+k \le 5} jp_i(j, k)$$

and

$$m_B^{(i)} = \sum_{0 \le j+k \le 5} kp_i(j, k)):$$

row\column	1	2	3	4	... $a+2$	... $b+1$	... $c+1$	$c+2$	$c+3$	... $2c-1$
1	0	p_0	0	0	0	0	0	0	0	0
2	0	0	p_1	0	0	0	0	0	0	0
3	0	0	0	p_2	0	0	0	0	0	0
⋮										
$a+1$	m_G^a	0	0	0	p_a	0	m_B^a	0	0	0
⋮										
b	m_G^{b-1}	0	0	0	0	p_{b-1}	m_B^{b-1}	0	0	0
⋮										
c	0	0	0	0	0	0	0	0	0	0
$c+1$	0	0	0	0	0	0	0	p_c	0	0
$c+2$	0	0	0	0	0	0	0	0	p_{c+1}	0
⋮										
$2c-1$	0	0	0	0	0	0	0	0	0	0

□

Background

For the demographic theory touched upon, turn to References [1–3].

4.2 ANALOGUES OF CLASSICAL RESULTS

As observed in the preceding section the Perron–Frobenius theorems render it possible to obtain results for a large class of multi-type Galton–Watson processes which are very close in appearance to those of one-type theory. Here follows the basic theorem:

Theorem (4.2.1)

Let A be an $r \times r$ matrix with non-negative entries such that for some integer $n > 0$ all entries of A^n are strictly positive. Then there exists a positive eigenvalue ρ of A, which is greater than the absolute value of any other eigenvalue. The eigenvalue ρ (often called the spectral radius) is also a left eigenvalue. Indeed there are row vectors u and v with positive coordinates u_i, v_i such that

$$Au' = \rho u'$$

$$vA = \rho v$$

Normalize u and v so that $u\tilde{1}' = vu' = 1$, and form the matrix B with entries $u_i v_j$. Then, as $n \to \infty$,

$$A^n/\rho^n \to B.$$

Proofs can be found in Karlin's "A First Course in Stochastic Processes" or matrix textbooks like those of Bellman, Gantmacher or Varga.

In the positively regular case the matrix m meets the requirements of this theorem and ρ (with the eigenvector normalized as above) plays the rôle of the reproduction mean of one-type processes. In particular, processes may be termed supercritical, critical, or subcritical according as the spectral radius ρ of m is > 1, $= 1$, or < 1. However degeneracies corresponding to the case $p_1 = 1$ in one-type processes can sneak in on ways not covered by the singularity concept in Section (4.1). This already complicates the analysis of the extinction probabilities, $q_k = P[z_n(k) \to \tilde{0}]$. For, assume that some subset S of $T = \{1, 2, \dots, r\}$ has the property that any individual whose type is in S begets exactly one child, her type again being in S. Obviously then $q_k = 0$ at least for $k \in S$.

Theorem (4.2.2)

The probability of extinction $q = [q_1, q_2, \dots, q_r]$ of an r-type positively regular and non-singular Galton–Watson process is the solution of the equation

$$f(s) = s$$

that lies closest to the origin in the unit cube. If $\rho \leq 1$ then all $q_k = 1$ and if $\rho > 1$ all $q_k < 1$.

This has been generalized to not positively regular processes by Sevast'-yanov. He gives a decomposition of m into submatrices satisfying the requirements of the Perron–Frobenius theorem, defines ρ as the largest of all corresponding sub-spectral radii, and arrives at the result that all $q_k = 1$ if and only if there are no sets S as described before the theorem and $\rho \leq 1$. In the positively regular and non-singular case the characteristic explosion or extinction behaviour of branching processes can also be established: if z_n does not become zero, then all coordinates of z_n tend to infinity.

Throughout in the sequel $vm = \rho$, $mu' = \rho u'$, $u\tilde{1}' = vu' = 1$, and the processes are positively regular and non-singular. Then if $\rho \leq 1$ it can be shown (page 192 of Reference [1] that

$$\lim_{n \to \infty} \{\tilde{1} - f_n(s)\}/v\{\tilde{1} - f_n(s)\}' = u. \tag{4.2.1}$$

Furthermore, in the critical case, with

$$0 < a = \tfrac{1}{2} \sum_{i,j,k} \frac{\partial^2 f^{(i)}}{\partial s_j \, \partial s_k} (\tilde{1}) \, v_i u_j u_k < \infty,$$

it follows, much as for the one-type process, that

$$\lim_{n \to \infty} \frac{1}{n} \left\{ \frac{1}{v\{\tilde{1} - f_n(s)\}'} - \frac{1}{v\{\tilde{1} - s\}'} \right\} = 1/a$$

uniformly for s in the unit cube, $s \neq \tilde{1}$.

Hence for $1 \leq k \leq r$, still uniformly,

$$\tilde{1} - f_n^{(k)}(s) = \frac{u_k v(\tilde{1} - s)' \{1 + o(1)\}}{1 + anv(\tilde{1} - s)'}$$

as $n \to \infty$. From this there is but a short step to

Theorem (4.2.3)

If $\rho = 1$ and $0 < a < \infty$ in a positively regular process, then

(a) $\lim\limits_{n \to \infty} nP[z_n(k) \neq \tilde{0}] = u_k/a$,

(b) $\lim\limits_{n \to \infty} E[z_n(k)/n|z_n \neq \tilde{0}] = av$

(the expectation of a vector being the vector of expectations)
(c) The conditional law of $z_n(k)/n$ given that $z_n(k) \neq \tilde{0}$, converges as $n \to \infty$ to the distribution of $X \cdot v$, where X is scalar,

$$P[X > u] = e^{-au}.$$

The analysis of subcritical processes also parallells the classical Galton–Watson case. It is not difficult to show that there is a non-negative function φ such that
$$v\{\tilde{1} - f_n(s)\}/\rho^n \downarrow \varphi(s).$$

By Equation (4.2.1) $\{1 - f_n(s)\}/\rho^n \to \varphi(s) u$. The critical point is then, as before, to establish $c = \varphi(\tilde{0}) > 0 \Leftrightarrow E[z_1^{(i)}(k) \log z_1^{(i)}(k)] < \infty$ for all $1 \leq i$, $k \leq r$. The rest follows immediately:

Theorem (4.2.4)

In a subcritical positively regular process

$$\lim\limits_{n \to \infty} P[z_n(k) \neq \tilde{0}]/\rho^n$$

exists and is positive $(= c)$ if and only if all $E[z_n^{(i)}(k) \log z_n^{(i)}(k)] < \infty$.

Theorem (4.2.5)

For processes as in the preceding theorem

$$\lim\limits_{n \to \infty} P[z_n(k) = \tilde{k}|z_n(k) \neq \tilde{0}] = b_{\tilde{k}}$$

exists for $1 \leq k \leq r$ and $\tilde{k} = [k_1, k_2, \ldots, k_r] \in Z_+^r$,

$$\sum_{k \in Z_+^r} b_{\tilde{k}} = 1,$$

$$\sum_{[k_1, \ldots, k_r]} k_j b_{[\tilde{k}_1, \ldots, \tilde{k}_r]} = v_j/c.$$

The probabilities $b_{\tilde{k}}$ are independent of the type k of the ancestor and g,

$$g(s) = \sum_{\tilde{k}} b_{\tilde{k}} s^{\tilde{k}},$$

satisfies

$$g \circ f = \rho g + 1 - \rho.$$

This equation determines g provided all

$$E\left[z_1^{(i)}(k) \log z_1^{(i)}(k)\right] < \infty. \tag{4.2.2}$$

In the supercritical case first observe that

$$w_n(k) = z_n(k) u'/\rho^n$$

is a non-negative martingale with respect to the σ-algebrae $\mathscr{B}_n = \sigma(z_0(k), \ldots, z_n(k))$. The martingale has an a.s. limit $w(k)$ and since

$$E[w_n(k)] \to u_k,$$
$$E[w(k)] \le u_k.$$

Let φ_k be the moment generating function of w_k and write $\varphi = [\varphi_1, \ldots, \varphi_n]$. Dominated convergence shows that

$$\varphi(\rho u) = f \circ \varphi(u), \qquad u \ge 0. \tag{4.2.3}$$

A truncation method can be used to show that $E[w(k)] = u_k$ if and only if Equation (4.2.2) is satisfied. (I know of no way to apply the elegant arguments of Section 2.7). Also Inequality (4.2.2) can be shown to imply that a.s. $z_n(k)/\rho^n \to w(k) v$ as $n \to \infty$. For the sake of uniformity we restate this as

Theorem (4.2.6)

Let $z_n(k)\, 1 \le k \le r$ be a supercritical positively regular process. If Inequality (4.2.2) does not hold then $z_n(k)/\rho^n \to 0$ almost surely. Assume that Inequality (4.2.2) is valid. Then there is for each k a random variable $w(k)$, $E[w(k)] = u_k$, such that $z_n(k)/\rho^n \to w(k) v$ a.s. The moment generating functions of $w(k)$, $1 \le k \le r$, satisfy Equation (4.2.3) and $P[w(k) \ne \tilde{0}, z_n(k) \nrightarrow \tilde{0}] = 0$. Apart from a jump of size q_k at the origin $w(k)$ has a continuous density.

There is an immediate and appealing corollary:

Corollary (4.2.7)

Under the conditions of the preceding theorem it holds for $1 \le j, k \le r$ that

$$\lim_{n \to \infty} z_n^{(j)}(k)/\{z_n^{(1)}(k) + \ldots + z_n^{(r)}(k)\} = v_j/(v_1 + \ldots + v_r)$$

a.s. on the set where the process does not become extinct.

Background

Proofs of theorems in this section can be found in two exhaustive books [4, 5]. The first one is better for the supercritical case (Sevast'yanov only gives an L^2-treatment) but the latter reference preferable in the other two cases. Both have fairly much to say about the not positively regular processes which I have shunned here.

REFERENCES

1. Chiang, C. L., *Introduction to Stochastic Processes in Biostatistics.* John Wiley and Sons, New York, 1968.
2. Feichtinger, G., Stochastische Modelle demographischer Prozesse. *Lecture Notes in Operations Research and Mathematical Systems*, 44. Springer, Berlin, 1971.
3. Pollard, J. H., On the use of the direct matrix product in analysing certain stochastic population models. *Biometrika* **53**, 397–415, 1966.
4. Mode, Charles J., *Multitype branching processes.* Elsevier, New York, 1971.
5. Sevast'yanov, B. A., *Vetvyaščiesya processy* (Branching processes). Mir, Moscow, 1971.

Chapter 5

Interlude about Martingales, Renewal Theory and Point Processes

5.1 MARTINGALES

One of the most fundamental facts of analysis is that of the convergence of monotone sequences. The martingale convergence theorem can be viewed as a probabilistic counterpart to this and thus its frequent efficacy should not be surprising. Let $(\Omega, \mathscr{S}, P)$ be a probability space and $\{\mathscr{B}_n, n \in N\}$ an increasing sequence of subsigma algebrae of $\mathscr{S}$, $\mathscr{B}_n \subset \mathscr{B}_{n+1} \subset \mathscr{S}$. Consider a sequence $\{x_n, n \in N\}$ of random variables such that each x_n is measurable with respect to $\mathscr{B}_n$. If further the variables have expectations and $E[x_{n+1}|\mathscr{B}_n] \leq, =,$ or $\geq x_n$, the sequence is said to be a *supermartingale*, a *martingale*, or a *submartingale*, respectively, with respect to $\{\mathscr{B}_n\}$. In the important particular case $\mathscr{B}_n = \sigma\{x_1, \ldots, x_n\}$ we speak just of a (super/sub) martingale. The interpretation is that a supermartingale corresponds to a non-increasing sequence, in the sense that you should not expect it to increase, whatever information you have been given on the first part of the sequence.

A positive integer or infinite valued random variable v is called a *stopping time* relative to $\{\mathscr{B}_n\}$ if, for each n, $\{v = n\} \in \mathscr{B}_n$. The reason for this somewhat picturesque terminology may be found 'n the gambling background of the martingale concept. A stopping time is a time to leave the game, and its value must of course be based solely on the past of the game. The first important martingale theorem has the interpretation that in a game where the gain decreases no gambler can change this by a shrewd choice of stopping time. It is known as the optional sampling theorem:

Theorem (5.1.1)
Let $\{x_n\}$ be a non-negative supermartingale and $v \leq \mu$ stopping times relative to some $\{\mathscr{B}_n\}$. Define $x_\infty = 0$. Then $E[x_\mu] \leq E[x_v]$.

Proof. Since the x_n are non-negative $E[x_v; v \leq n] \uparrow E[x_v; v < \infty]$ by monotone convergence. It follows that

97

$$E[x_v] = E[x_v; v < \infty] = \lim_{n \to \infty} E[x_v; v \le n] = \lim_{n \to \infty} \sum_{k=1}^{n} E[x_k; v = k]$$

$$= \lim_{n \to \infty} \sum_{k=1}^{n} (E[x_k; v > k - 1] - E[x_k; v > k])$$

$$= E[x_1]$$

$$+ \lim_{n \to \infty} \left\{ \sum_{k=2}^{n} (E[x_k; v > k - 1] - E[x_{k-1}; v > k - 1]) - E[x_n; v > n] \right\}.$$

Since (prime for complement)

$$\{v > k - 1\} = \left(\bigcup_{j=1}^{k-1} \{v = j\} \right)' \in \mathscr{B}_{k-1},$$

we can proceed to

$$E[x_v] = E[x_1]$$

$$+ \lim_{n \to \infty} \left\{ \sum_{k=2}^{n} E[(E[x_k|\mathscr{B}_{k-1}] - x_{k-1}); v > k - 1] - E[x_n; v > n] \right\},$$

where the integrands $E[x_k|\mathscr{B}_{k-1}] - x_{k-1}$ are by assumption not positive. The same representation is valid for $E[x_\mu]$ and, as $v \le \mu$, the set $\{v > k - 1\} \subset \{\mu > k - 1\}$. Thus

$$E[(E[x_k|\mathscr{B}_{k-1}] - x_{k-1}); \mu > k - 1]$$

$$\le E[(E[x_k|\mathscr{B}_{k-1}] - x_{k-1}); v > k - 1]$$

and also

$$-E[x_n; \mu > n] \le -E[x_n; v > n].$$

The assertion follows. $\qquad\square$

The renowned martingale convergence theorem can be obtained from this.

Theorem (5.1.2)

Let $\{x_n\}$ be a non-negative supermartingale with respect to some increasing sequence of σ-algebrae. Then $\lim_{n \to \infty} x_n$ exists a.s. and is finite.

Proof. If $\lim_{n \to \infty} x_n$ were infinite on a set with probability $p > 0$, then for some n $P[x_n > 2E[x_1]/p]$ would be $> p/2$. By the non-negativity this would imply that $E[x_n] > E[x_1]$, contradicting the supermartingale property. Thus, if $\{x_n\}$ does not converge a.s., then there exist numbers $a < b$ such that the set

$$D = \left\{ \liminf_{n \to \infty} x_n < a < b < \limsup_{n \to \infty} x_n \right\}$$

has positive probability. We define two increasing sequences of stopping times inductively,

$$v_1 = \inf\{n; x_n < a\}$$

$$\mu_1 = \inf\{n; n > v_1, x_n > b\}$$

$$v_{k+1} = \inf\{n; n > \mu_k, x_n < a\}$$

$$\mu_{k+1} = \inf\{n; n > v_{k+1}, x_n > b\}.$$

This means that as long as $v_k < \infty$ $\{v_k\}$ is a sequence of x-visits below a, i.e. $x_{v_k} < a$ and analogously $x_{\mu_k} > b$. Since $v_k \leq \mu_k$, Theorem (5.1.1) implies that

$$bP[\mu_k < \infty] < E[x_{\mu_k}] \leq E[x_{v_k}] < aP[v_k < \infty]$$

(recall the convention $x_\infty = 0$). Further by monotonicity

$$P[v_k < \infty] \to P\left[\bigcap_{k=1}^{\infty} \{v_k < \infty\} \right],$$

$$P[\mu_k < \infty] \to P\left[\bigcap_{k=1}^{\infty} \{\mu_k < \infty\} \right],$$

and since

$$\bigcap_{k=1}^{\infty} \{v_k < \infty\} = D = \bigcap_{k=1}^{\infty} \{\mu_k < \infty\}$$

it follows that

$$bP[D] = \lim_{k \to \infty} bP[\mu_k < \infty] \leq \lim_{k \to \infty} aP[v_k < \infty] = aP[D],$$

which of course shows that there can not be any $a < b$ such that $P[D] > 0$. Thus the convergence follows. $\square$

Corollary (5.1.3)

If $\{x_n\}$ is a supermartingale with respect to $\{\mathscr{B}_n\}$ and $\sup_n E[|x_n|] < \infty$, then $\{x_n\}$ converges a.s. as $n \to \infty$.

Proof. We shall show that $x_n = y_n - z_n$ for two non-negative supermartingales. Then the theorem applies. First observe that

$$E[x_{n+1+k}|\mathscr{B}_k] = E[E[x_{n+1+k}|\mathscr{B}_{n+k}]|\mathscr{B}_k] \le E[x_{n+k}|\mathscr{B}_k]. \quad (5.1.1)$$

It follows for each $k \in N$ that as $n \to \infty$ $E[x_{n+k}|\mathscr{B}_k] \downarrow$ some ξ_k. The limit is measurable with respect to $\mathscr{B}_k$ and by Fatou's lemma and Jensen's inequality

$$E[|\xi_k|] = E\left[\liminf_{n \to \infty} |E[x_{n+k}|\mathscr{B}_k]|\right] \le \liminf_{n \to \infty} E[|E[x_{n+k}|\mathscr{B}_k]|]$$

$$\le \liminf_{n \to \infty} E[E[|x_{n+k}||\mathscr{B}_k]] = \liminf_{n \to \infty} E[|x_n|] < \infty.$$

Moreover by monotone convergence

$$E[\xi_{k+1}|\mathscr{B}_k] = \lim_{n \to \infty} E[E[x_{n+k+1}|\mathscr{B}_{k+1}]|\mathscr{B}_k]$$

$$= \lim_{n \to \infty} E[x_{n+k+1}|\mathscr{B}_k] = \xi_k,$$

showing that $\{\xi_k\}$ constitutes a martingale with $\sup E[|\xi_k|] < \infty$. From the construction it is also clear that

$$x_k \ge E[x_{1+k}|\mathscr{B}_k] \ge \xi_k.$$

As a second step we let $\{\xi_k\}$ undergo the same treatment, only with the difference of an inserted absolute value: By Jensen and the martingale property $E[|\xi_{n+k}||\mathscr{B}_k]$ increases in n to some limit $\eta_k \ge 0$ but also $\ge \xi_k$. The sequence $\{\eta_k\}$ is again a martingale (just argue as above) and

$$x_n = (x_n - \xi_n + \eta_n) - (\eta_n - \xi_n)$$

is the desired decomposition. $\quad\square$

As mentioned, the applications of this and the preceding theorem are many and diverse. An impressing series is related in Reference [2]. (See also Reference [1].) Here we only point at Section 2.7 and establish two results, needed for the renewal theory in the following sections.

Corollary (5.1.4)

Let x with $E[|x|] < \infty$ be a random variable, measurable with respect to $\mathscr{B}_\infty = \sigma(\bigcup_n \mathscr{B}_n)$ where $\{\mathscr{B}_n\}$ is an increasing sequence of σ-algebrae. Then, $\lim_{n \to \infty} E[x|\mathscr{B}_n] = x$ a.s. and in L^1.

Proof. The random variables $x_n = E[x|\mathscr{B}_n]$ constitute a martingale with

$E[|x_n|] \leq E[|x|] < \infty$. Hence $x_n \to$ some x_∞ a.s. Further,

$$E[|x_n|; |x_n| > c] = E[|E[x|\mathscr{B}_n]|; |x_n| > c]$$
$$\leq E[E[|x||\mathscr{B}_n]; |x_n| > c] = E[|x|; |x_n| > c].$$

But

$$P[|x_n| > c] \leq E[|x_n|]/c \leq E[|x|]/c.$$

Thanks to the finiteness of $E[|x|]$, $E[|x|; A]$ can be made arbitrarily little by choosing $P(A)$ little enough. Hence, to any $\varepsilon > 0$ there is a c such that

$$E[|x|; |x_n| > c] < \varepsilon$$

for all n, i.e. $\{x_n\}$ is uniformly integrable. Therefore, $x_n \to x_\infty$ also in L^1 and $E[x_\infty; A] = E[x; A]$ for all $A \in \bigcup_n \mathscr{B}_n$. Dynkin's theorem completes the proof, since if the equality holds for all A in $\mathscr{B}_\infty$, the random variables must be equal (consider e.g. $A = \{x < x_\infty\}$). $\square$

From this Kolmogorov's famed zero-one law is almost immediate:

Corollary (5.1.5)

If $\{x_n\}$ is a sequence of independent random variables, $\mathscr{B}_n = \sigma\{x_1, \ldots, x_n\}$ and y an integrable random variable measurable with respect to $\mathscr{B} = \sigma(\{x_n\})$ but independent of each $\mathscr{B}_n$, then y is a.s. constant.

Proof. Define

$$y_n = E[y|\mathscr{B}_n].$$

By the preceding

$$y_n \to y$$

a.s. but by the independence

$$y_n = E[y|\mathscr{B}_n] = E[y]. \qquad \square$$

We shall find employment for Hewitt's and Savage's zero-one law:

Corollary (5.1.6)

Let $f: R^\infty \to R$ be a bounded function, measurable with respect to the product Borel algebra on R^∞. Suppose that for any permutation π of finitely many coordinates $f \circ \pi = f$. Then, if $\{x_n\}$ is an i.i.d. sequence $f(x_1, x_2, \ldots)$ is a.s. constant. In particular, if it is the indicator function of an event, then the latter has probability zero or one.

Proof. By the measurability of f there are functions $f_n : R^n \to R$ such that $f_n \to f$ and $|f_n| \leq \sup |f| < \infty$. It follows that to any $\varepsilon > 0$ for n large enough it holds that

$$E[|f_n(x_1, \ldots, x_n) - f(x_1, x_2, \ldots)|] < \varepsilon.$$

But since the x_i are i.i.d, the left hand side equals

$$E[|f_n(x_{n+1}, \ldots, x_{2n}) - f(x_{n+1}, \ldots, x_{2n}, x_1, \ldots, x_n, x_{2n+1}, \ldots)|].$$

But thanks to the permutation invariance

$$f(x_{n+1}, \ldots, x_{2n}, x_1, \ldots, x_n, x_{2n+1}, \ldots) = f(x_1, x_2, \ldots)$$

and so

$$E[|f_n(x_{n+1}, \ldots, x_{2n}) - f(x_1, x_2, \ldots)|] < \varepsilon$$

for n large enough. Therefore $f_n(x_{n+1}, \ldots, x_{2n})$ converges in L^1 to $x = = f(x_1, x_2, \ldots)$. But clearly $f_n(x_{n+1}, \ldots, x_{2n})$ is independent of x_k for $k \leq n$ and so x is independent of each x_n and by Kolmogorov's law is a.s. constant. □

In Section 2.11 we made use of the Ballot Theorem. It has an easy martingale proof:

Lemma (5.1.7)

If $x_1, \ldots, x_n$ constitute a martingale with finitely many members with respect to $\mathscr{B}_1 \subset \ldots \subset \mathscr{B}_n$, i.e. $E[x_{k+1}|\mathscr{B}_k] = x_k$ for $1 \leq k < n$, and v is a stopping time with values between 1 and n, then

$$E[x_v|\mathscr{B}_1] = x_1.$$

Proof. Let $A \in \mathscr{B}_1$. Note that $x_k = E[x_n|\mathscr{B}_k]$ for $1 \leq k \leq n$ and $A \cap \cap \{v_n = k\} \in \mathscr{B}_k$.

$$E[x_v ; A] = \sum_{k=1}^{n} E[x_k ; A \cap \{v = k\}]$$

$$= \sum_{k=1}^{n} E[E[x_n|\mathscr{B}_k]; A \cap \{v = k\}]$$

$$= \sum_{k=1}^{n} E[x_n ; A \cap \{v = k\}] = E[x_n ; A]$$

$$= E[x_1 ; A]. \qquad \square$$

Theorem (5.1.8)

Let $y_1, \ldots, y_n$ be i.i.d. (or just exchangeable) non-negative and integer valued random variables with finite expectations. Then, with $s_k = \sum_{j=1}^{k} y_j$,

$$P[s_k < k \text{ for } 1 \leq k \leq n | s_n] = (1 - s_n/n)^+. \tag{5.1.2}$$

Proof. Define $\mathscr{B}_1 = \sigma(s_n)$, $\mathscr{B}_2 = \sigma(s_{n-1}, s_n), \ldots \mathscr{B}_k = \sigma(s_{n-k+1}, \ldots, s_n), \ldots,$ $\mathscr{B}_n = \sigma(s_1, \ldots, s_n)$ and $x_1 = s_n/n$, $x_2 = s_{n-1}/(n-1), \ldots, x_k = s_{n+1-k}/(n+1-k), \ldots, x_n = s_1$. Then $\mathscr{B}_k \subset \mathscr{B}_{k+1}$ and each x_k is measurable with respect to $\mathscr{B}_k$. Further, for $j \leq n - k + 1$

$$E[y_j | \mathscr{B}_k] = E[y_1 | \mathscr{B}_k]$$

by symmetry. Hence

$$(n - k + 1) E[y_1 | \mathscr{B}_k] = \sum_{j=1}^{n-k+1} E[y_j | \mathscr{B}_k]$$

$$= E[s_{n-k+1} | \mathscr{B}_k] = s_{n-k+1}$$

and

$$x_k = s_{n+1-k}/(n+1-k) = E[y_1 | \mathscr{B}_k].$$

It follows that $\{x_k\}_1^n$ constitutes a martingale.

Now define

$$v = \inf\{k; x_k \geq 1\} = n + 1 - \sup\{k; s_k \geq k\}$$

on the set where $\sup_k s_k/k \geq 1$ and $v = n$ otherwise. Note that $x_v = s_1 = 0$ in the latter case. If $s_n \geq n$ the assertion to be proved is evident. Assume that $s_n < n$ and that $\sup s_k/k \geq 1$. Then, with $\mu = n + 1 - v$, $s_\mu \geq \mu$ but $s_{\mu+1} < \mu + 1$. Hence $y_{\mu+1} = s_{\mu+1} - s_\mu < \mu + 1 - \mu = 1$, i.e. $y_{\mu+1} = 0$, $\mu \leq s_\mu = s_{\mu+1} < \mu + 1$, and $s_\mu = \mu$ implying that $x_v = 1$. By Lemma (5.1.7), for $s_n < n$,

$$P[s_k \geq k \text{ for some } 1 \leq k \leq n | s_n] = E[x_v | \mathscr{B}_n] = x_1 = s_n/n,$$

concluding the proof.   □

Corollary (5.1.9)

If $\{y_k; k \in N\}$ is a sequence with members as in Theorem (5.1.8) then

$$P[s_k < k \text{ for } \forall k \in N] = (1 - E[y_1])^+.$$

Proof. Taking expectations of Equation (5.1.2) for any fixed n, we see that

$$P[s_k < k \text{ for } \forall k < n] = E[(1 - s_n/n)^+]$$

Let $n \to \infty$. Then $s_n/n \to E[y_1]$ a.s. and in L^1.   □

Background

Martingale theory, developed essentially by J. L. Doob, is nowadays an established topic with a chapter of its own in most text books. Through the martingale $x_n = z_n/m^n$ its bearing on branching processes is immediately felt. The given treatment differs somewhat from the usual ones. I learnt it from K. Murali Rao [4]. The proof of the Ballot Theorem (due to L. Takács) using martingales is ascribed to G. Simons in Reference [1].

5.2 THE RENEWAL EQUATION

In general types of branching processes, as well as several other parts of probability theory, an important role is played by the so called renewal equation. For the background see Reference [6]. Some words on the interpretation can be found in Section 5.4. Here I shall present a somewhat harsh but straightforward treatment, built upon martingales and an approach due to Feller [7].

Let f and x be measurable functions and μ a measure on R_+. If μ is a probability measure, the relation

$$x(t) = f(t) + \int_0^t x(t - u)\,\mu(du), \qquad t \geq 0, \tag{5.2.1}$$

is called a *renewal equation*. In convolution notation we shall write $x = f + x * \mu$. If $\mu(\infty) < 1$, then Equation (5.2.1) is said to be a *defective renewal equation* and if $\mu(\infty) > 1$, it is *excessive*. These cases are rather simple (depending upon ambitions though) or can be brought back to $\mu(\infty) = 1$. They will be treated later. For Equation (5.2.1) with given f and μ, $\mu(\infty) = 1$, we shall establish the uniqueness of x, its explicit form, and asymptotics as $t \to \infty$. Throughout we assume that $\mu(0) < 1$. Otherwise the equation reduces to $x = f + x$ and if $f \neq 0$, x cannot be finite.

Lemma (5.2.1)

The sum

$$v = \sum_{n=0}^{\infty} \mu^{*n}$$

of convolution powers is a Radon (i.e. finite on bounded sets) measure.

Proof. Let $t > 0$ be given. Since $\mu(0) < 1$, there is (by the law of large numbers) a $k \in N$ such that $\mu^{*k}(t) < 1$. And, as $\mu^{*k} \leq \mu^{*j}$ for $j \leq k$, and $\mu^{*j} \leq \mu^j$,

$$v(t) = \sum_{n=0}^{\infty} \{\mu^{*nk}(t) + \mu^{*nk+1}(t) + \ldots + \mu^{*(n+1)k-1}(t)\} \leq$$

$$\leq \sum_{n=0}^{\infty} k\mu^{*nk}(t) \leq \sum_{n=0}^{\infty} k\{\mu^{*k}(t)\}^n \leq k/\{1 - \mu^{*k}(t)\} < \infty. \qquad \square$$

Theorem (5.2.2)

If f is bounded, then Equation (5.2.1) has a unique solution which is bounded on finite intervals, namely $x = f * v$.

Proof. By the lemma $f * v$ is well defined and

$$f + (f * v) * \mu = f + f * \sum_{n=1}^{\infty} \mu^{*n} = f * v.$$

If x and y are two solutions bounded on finite intervals, then

$$x - y = (x - y) * \mu = \ldots = (x - y) * \mu^{*n} \to 0. \qquad \square$$

Lemma (5.2.3)

There are constants A, B (depending upon μ) such that $v(t + u) - v(t) \leq \leq Au + B$ for all $u, t \geq 0$.

Proof. There is an $a > 0$ such that $\mu(a) < 1$. If $t \geq a$, then

$$1 = v(t) - \mu * v(t) = \int_0^t \{1 - \mu(t - u)\} \, v(du)$$

$$\geq \int_{t-a}^t \{1 - \mu(t - u)\} \, v(du) \geq \{1 - \mu(a)\} \{v(t) - v(t - a)\}.$$

Hence

$$v(t + u) - v(t) \leq \sum_{j=1}^{[u/a]+1} \{v(t + ja) - v(t + ja - a)\}$$

$$\leq \{(u/a) + 1\}/\{1 - \mu(a)\}.$$

If $t \leq a$, then $v(t) \leq v(a)$. $\qquad \square$

Next, let $\{r_n\}$ be any real sequence tending to infinity. By Lemma (5.2.3) and the Helly selection theorem (page 263 of Reference [7]), there is a subsequence such that

$$v_n = v(r_n + \cdot) \xrightarrow{v} v_\infty, \tag{5.2.2}$$

$v_n \xrightarrow{v} v_\infty$ (read vaguely) meaning that

$$\int g \, dv_n \to \int g \, dv_\infty \tag{5.2.3}$$

for all continuous g with compact support on R. We write $\{r_n\}$ for the subsequence as well. Turning back to the solution x of Equation (5.2.1) we see that

$$x(r_n + t) = \int_0^{r_n + t} f(r_n + t - u)\, v(du)$$

$$= \int_{-r_n}^{t} f(t - u)\, v_n(du) \to \int_{-\infty}^{t} f(t - u)\, v_\infty(du) = \varphi(t) \quad (5.2.4)$$

provided f is a decent enough, e.g. continuous with compact support on R_+ and either $f(0) = 0$ or $v_\infty\{t\} = 0$. We shall see that actually rather little is required.

To begin with assume f in Equation (5.2.1) continuous with support in $[0, a]$ and zero at the origin. Since $f(r_n + t) \to 0$ and

$$|x(r_n + t)| \le \sup_u |f(u)|\, \{v(r_n + t) - v(r_n + t - a)\} \le \sup_u |f(u)|\, (Aa + B)$$

dominated convergence applied to Equations (5.2.1) and (5.2.4) yields

$$\varphi(t) = \int_0^\infty \varphi(t - u)\, \mu(du). \quad (5.2.5)$$

We have already seen that φ must be bounded. But it is also continuous. Indeed if $\varepsilon > 0$, $t \ge 0$ are given choose $0 < \delta < a$ such that the (uniformly continuous) function f does not vary more than $\varepsilon/2(Aa + b)$ over intervals of length δ. Since $f(0) = 0$

$$|x(r_n + t + u) - x(r_n + t)| \le \int_0^{r_n + t} |f(r_n + t + u - y) -$$

$$- f(r_n + t - y)|\, v(dy) + \int_{r_n + t}^{r_n + t + u} |f(r_n + t + u - y)|\, v(dy) < \varepsilon,$$

if only $0 \le u < \delta$. Left continuity follows analogously. This means that Equation (5.2.5) is an equation of the type treated by the Choquet–Deny theorem, which follows (with a proof due to Doob, Snell and Williamson).

Theorem (5.2.4)

Assume that h is a bounded and continuous function on R and λ a probability measure. Then the following holds:

$$h(t) = \int_{-\infty}^{+\infty} h(t - u)\, \lambda(du) \quad (5.2.6)$$

for all t if and only if $h(t - u) = h(t)$ for all $t \in R$ and all u in the smallest closed group containing the support of λ.

The theorem can be given a more transparent formulation if we observe that the only closed subgroups of R are R itself and sets of the form $\{kd; k \in Z\}$ for some $d \geq 0$. We shall say that the probability measure λ is *lattice* with *span d* if

$$\lambda(\{kd; k \in Z\}) = 1$$

and d is the largest such number (which always exists). If λ is not lattice with any span d we say that it is *non-lattice*. The reformulation is given as an immediate

Corollary (5.2.5)

Let h and λ be as in Theorem (5.2.4). Then h is constant if λ is non-lattice and periodic with period d if λ is lattice with span $d > 0$.

Proof of the theorem. Assume that Equation (5.2.6) is true. Let $x_1, x_2, \ldots$ be independent random variables defined on some space Ω, all with distribution λ. Denote the partial sums by s_n, $s_0 = 0$, $s_{n+1} = s_n + x_{n+1}, n \in Z_+$. Let $\mathscr{B}_0$ be the trivial σ-algebra of the empty and full sets, $\varnothing$ and Ω, and $\mathscr{B}_n = \sigma(x_1, \ldots, x_n)$ for $n \in N$. For fixed t, $h(t - s_n)$ is a random variable measurable with respect to $\mathscr{B}_n$. Further, $|h(t - s_n)| \leq \sup_t |h(t)| < \infty$ and

$$E[h(t - s_{n+1})|\mathscr{B}_n] = \int_{-\infty}^{+\infty} h(t - s_n - u)\,\lambda(du) = h(t - s_n).$$

Hence the sequence $\{h(t - s_n)\}$ constitutes a bounded martingale and

$$y_t = \lim_{n \to \infty} h(t - s_n)$$

exists a.s. as well as in L^1. By Hewitt's and Savage's zero-one law [Corollary (5.1.6)] y_t is a.s. constant, $y_t = E[y_t]$ a.s. Therefore

$$h(t - x_1) = E[y_t|\mathscr{B}_1] = E[y_t] = E[h(t - s_0)] = h(t).$$

This means that $h(t - u) = h(t)$ for λ—almost all u (and for any t). Since h is a continuous function and the support is the smallest closed set with λ—measure one, it follows that $h(t - u) = h(t)$ for all u in the support. But if for all t $h(t - u) = h(t)$ and $h(t - v) = h(t)$, then $h(t - (u + v)) = h(t - u - v) = h(t - u) = h(t)$, proving the group property. Closedness follows from the continuity of h. $\qquad \square$

We return to Equation (5.2.5), considering first non-lattice μ. Then for all continuous f with compact support and $f(0) = 0$

$$\varphi(t) = \int_{-\infty}^{t} f(t - u)\,v_\infty(du)$$

is independent of t. This means that v_∞ is translation invariant and therefore a multiple of Lebesgue measure, say $v_\infty(du) = c\,du$. And therefore $v_n(t) \to$ $\to v_\infty(t)$ for all t, implying that the class of functions g for which Relation (5.2.3) holds can be widened by the continuous mapping theorem (page 30 of Reference [5]) at least so as to contain all bounded functions g with compact support which are a.e. continuous with respect to Lebesgue measure. (Recall that these are exactly the bounded functions with compact support that are Riemann integrable (page 70 of Reference [8]). In the latter formulation there is an easy elementary argument: approximate g from above and below by step functions.) The assumption of a compact support, can be relaxed to the requirement that g be Lebesgue integrable and integrable uniformly with respect to the v_n in the sense that

$$\int_{|t|>r} |g(t)|\, v_n(dt)$$

can be made arbitrarily small for all n by choosing r large. Then Relation (5.2.3) still holds.

Theorem (5.2.6)

Let μ be a non-lattice probability measure on R_+ and f a function on R_+ which is continuous a.e. with respect to Lebesgue measure and such that

$$\sum_{k=0}^{\infty} \sup_{0 \le t < 1} |f(k+t)| < \infty. \qquad (5.2.7)$$

Then the unique bounded solution x of $x = f + x * \mu$ satisfies

$$\lim_{t \to \infty} x(t) = \int_0^\infty f(t)\,dt \Big/ \int_0^\infty t\mu(dt),$$

where the quotient is interpreted as zero if μ has an infinite first moment.

Note. *If $|f|$ is dominated by an integrable decreasing function, then it complies with Condition (5.2.7). Functions doing so are said to be directly Riemann integrable if they are also Riemann integrable on bounded intervals.*

Proof. By Equation (5.2.4)

$$x(r_n + t) = \int_{-r_n}^t f(t-u)\, v_n(du).$$

Obviously $f(t - \cdot)$ is an a.e. continuous function on R and to establish the convergence

$$x(r_n + t) \to \int_{-\infty}^t f(t-u)\, v_\infty(du) = c \int_0^\infty f(u)\,du$$

we must only observe that [with A, B from Lemma (5.2.3)]

$$\int_{u < -r} |f(t - u)| \, v_n(du) = \int_0^{r_n - r} |f(r_n + t - u)| \, v(du)$$

$$\leq (A + B) \sum_{k=r}^{\infty} \sup_{0 \leq u < 1} |f(k + u)|$$

can be made arbitrarily small by a choice of r large by assumption. Since the r_n was a subsequence from an arbitrary sequence tending to infinity, we have shown that

$$\lim_{t \to \infty} x(t) = c \int_0^\infty f(u) \, du,$$

provided c is independent of the sequence. Then it remains to show this by calculating c. To that end integrate the renewal equation to obtain after a change of variable

$$\int_0^{r_n} x(t) \, dt = \int_0^{r_n} f(t) \, dt + \int_0^{r_n} \mu(du) \int_0^{r_n - u} x(t) \, dt.$$

Hence

$$\int_0^{r_n} f(t) \, dt = \int_0^{r_n} \mu(du) \int_{r_n - u}^{r_n} x(t) \, dt + \{1 - \mu(r_n)\} \int_0^{r_n} x(t) \, dt$$

$$= \int_0^{r_n} \mu(du) \int_0^u x(r_n - t) \, dt + \{1 - \mu(r_n)\} \int_0^{r_n} x(t) \, dt.$$

By dominated convergence and Theorem (5.2.2)

$$\int_0^{r_n} \mu(du) \int_0^u x(r_n - t) \, dt \to c \int_0^\infty u\mu(du) \int_0^\infty f(u) \, du,$$

if only the right hand integral is finite. Then also

$$\{1 - \mu(r_n)\} \int_0^{r_n} x(t) \, dt \to 0,$$

and hence

$$\int_0^\infty f(t) \, dt = c \int_0^\infty t\mu(dt) \int_0^\infty f(t) \, dt$$

as claimed.

Finally assume that

$$\int_0^\infty t\mu(dt) = \infty.$$

If $f \geq 0$, then so is x and let $r_n \geq r > 0$ to get

$$\infty > \int_0^\infty f(t)\,dt \geq \int_0^r \mu(du) \int_0^u x(r_n - t)\,dt \to c \int_0^r u\mu(du) \int_0^\infty f(u)\,du.$$

Since the last term can be made arbitrarily large provided $\int f(u)\,du \neq 0$, c must be zero. For general f, this proves that $|f| * v(t) \to 0$ and since $|x| = |f * v| \leq |f| * v$ the proof is complete. $\qquad\square$

We turn to the lattice case, supposing that μ has span $0 < d < \infty$. Then v is concentrated on the set $\{jd; j \in Z_+\}$. Write $a_j = v\{jd\} = v(jd) - v((j-1)d)$. Fix any $t \geq 0$. It satisfies $t = kd + s$ for unique $k \in Z_+$ and $0 \leq s < t$ and by Theorem (5.2.2)

$$x(nd + t) = x(nd + kd + s) = \sum_{j=0}^{n+k} f\{(n + k - j)\,d + s\}\,a_j$$

$$= \sum_{j=0}^{n+k} f(jd + s)\,a_{n+k-j}.$$

By Relation (5.2.2) $a_{n+k-j} \to$ some b_{k-j} as $n \to \infty$ via some subsequence. Lemma (5.2.3) gives the information that the numbers a_{n+k-j} remain bounded. Thus, if Inequality (5.2.7) holds

$$x(nd + t) \to \varphi(t) = \sum_{j=0}^\infty f(jd + s)\,b_{k-j},$$

as $n \to \infty$ in the subsequence. Since $t = kd + s$, Corollary (5.2.5) implies that $\varphi(t) = \varphi(s)$, which can be calculated by the approach from Theorem (5.2.6):

$$\sum_{j=0}^n f(jd + s) = \sum_{k=0}^n \mu\{kd\} \sum_{j=0}^{k-1} x(nd - jd + s) + \{1 - \mu(nd)\} \sum_{j=0}^n x(jd + s)$$

shows that

$$\varphi(s) = d \sum_{j=0}^\infty f(jd + s) \Big/ \int_0^\infty t\mu(dt)$$

when the denominator converges. Otherwise $\varphi(s) = 0$ as before. In conclusion, since the limit was the same for a subsequence of any subsequence of Z_+,

Theorem (5.2.7)

Suppose that the probability measure μ on R_+ is lattice with span d, $0 < < d < \infty$, and that

$$\sum_{j=0}^\infty f(jd + s)$$

converges absolutely for $0 \leq s < d$. *Then the unique bounded solution* x
of the renewal equation $x = f + x * \mu$ *satisfies*

$$\lim_{n \to \infty} x(nd + s) = d \sum_{j=0}^{\infty} f(jd + s) \Big/ \int_0^{\infty} t\mu(dt).$$

Finally some words about defective and excessive renewal equations.
If $\mu(0) < 1 < \mu(\infty) < \infty$ (or $\mu(t)$ does not increase too rapidly) there exists
an $\alpha > 0$ such that

$$\hat{\mu}(\alpha) = \int_0^{\infty} e^{-\alpha t} \mu(dt) = 1. \tag{5.2.8}$$

If $\mu(\infty) < 1$, this might still be the case, but then with $\alpha < 0$. Multiplication
of Equation (5.2.1) by $e^{-\alpha t}$ yields a proper renewal equation,

$$e^{-\alpha t} x(t) = e^{-\alpha t} f(t) + \int_0^t e^{-\alpha(t-u)} x(t-u) e^{-\alpha u} \mu(du).$$

In the excessive case the meeting of Condition (5.2.7) is facilitated by the
rapid decrease of $e^{-\alpha t} f(t)$. Indeed:

Theorem (5.2.8)

Let α *be defined by Equation (5.2.8) for a non-lattice measure* μ *on* R_+
with $\mu(0) < 1 < \mu(\infty) \leq \infty$. *If* f *is continuous and bounded a.e. with respect
to Lebesgue measure and* $x = f + x * \mu$, *then*

$$\lim_{t \to \infty} e^{-\alpha t} x(t) = \int_0^{\infty} e^{-\alpha t} f(t)\, dt \Big/ \int_0^{\infty} t\, e^{-\alpha t} \mu(dt). \tag{5.2.9}$$

The lattice case of this and the following theorem are left for the reader.

Theorem (5.2.9)

Let μ *be a measure on* R_+ *with total mass less than one, and* f *any bounded
measurable function such that* $\lim_{t \to \infty} f(t) = f(\infty)$ *exists. Then the solution*
x *of Equation (5.2.1) has the limit*

$$\lim_{t \to \infty} x(t) = f(\infty)/\{1 - \mu(\infty)\}.$$

If further the α *of Equation (5.2.8) exists,* μ *is not lattice,* f *is a.e. Lebesgue
continuous and* $e^{\alpha t} f(t)$ *satisfies Inequality (5.2.7), then Equation (5.2.9) holds.*

Proof. The latter assertion is immediate and the former follows from Theorem
(5.2.2) and Lemma (5.2.10) on the next page. □

Lemma (5.2.10)

Let f be a bounded, measurable function and λ a finite measure on R_+. Assume that $f(\infty) = \lim\limits_{t \to \infty} f(t)$ exists. Then, so does

$$\lim_{t \to \infty} f * \lambda(t) = f(\infty)\,\lambda(\infty).$$

Proof. Use dominated convergence. □

We shall also need some rates of divergence in renewal equations where $f(t) \not\to 0$.

Lemma (5.2.11)

If λ is a measure on R_+ such that $\lambda(t) \sim t$, as $t \to \infty$, then for all $n \in Z_+$

$$\int_0^t u^n \lambda(du) \sim t^{n+1}/(n+1)$$

as $t \to \infty$.

Proof. Integrate by parts to obtain

$$\int_0^t u^n \lambda(du) = t^n \lambda(t) - n \int_0^t \lambda(u)\,u^{n-1}\,du.$$

Let $\varepsilon > 0$ be given and choose v such that $(1 - \varepsilon)\,u \le \lambda(u) \le (1 + \varepsilon)\,u$ for $u \ge v$. Then, for $t \ge v$,

$$(1 - \varepsilon)\,(t^{n+1} - v^{n+1})/(n+1) \le \int_v^t \lambda(u)\,u^{n-1}\,du \le \int_0^t \lambda(u)\,u^{n-1}\,du$$

$$\le \lambda(v)\,v^n + (1 + \varepsilon)\,(t^{n+1} - v^{n+1})/(n+1).$$

Divide by t^{n+1} and let $t \to \infty$. □

Lemma (5.2.12)

Let λ be as in Lemma (5.2.11) and f a measurable function on R_+, bounded on finite intervals and such that $f(t) \sim t^n$, $n \in Z_+$ when $t \to \infty$. Then $f * \lambda(t) \sim$ $\sim t^{n+1}/(n+1)$. If $f(t) \to 0$, then $f * \lambda(t) = o(t)$.

Proof. Choose $\varepsilon > 0$ and v such that $(1 - \varepsilon)\,t^n \le f(t) \le (1 + \varepsilon)\,t^n$ for $t \ge v$. For t large and some constant C

$$\int_0^t f(t - u)\,\lambda(du) \le \sup_{0 \le u \le v} |f(u)|\,\{\lambda(t) - \lambda(t - v)\}$$

$$+ (1 + \varepsilon) \int_0^{t-v} (t - u)^n\,\lambda(du) \le Ct + (1 + \varepsilon) \int_0^{t-v} (t - u)^n\,\lambda(du)$$

with a corresponding lower estimate. Since, as $t \to \infty$

$$\int_0^t (t - u)^n \lambda(du) = \sum_{k=0}^n \binom{n}{k} t^{n-k} \int_0^t (-u)^k \lambda(du) \sim t^{n+1} \sum_{k=0}^n \binom{n}{k}$$

$$\cdot (-1)^k/(k + 1) = t^{n+1}/(n + 1),$$

the proof of the first case is complete. The second assertion is simple:

$$\left| \int_0^t f(t - u) \lambda(du) \right| \Big/ t \leq \sup_{u \geq v} |f(u)| \, \lambda(t)/t$$

$$+ \sup_{u \geq 0} |f(u)| \, \{\lambda(t) - \lambda(t - v)\}/t \to \sup_{u \geq v} |f(u)| \to 0$$

as first $t \to \infty$ then $v \to \infty$. $\qquad\square$

Theorem (5.2.13)

Let μ be a probability measure on R_+ with $0 < \beta = \int_0^\infty t\mu(dt)$. Then

$$v = \sum_{n=0}^\infty \mu^{*n}$$

satisfies

$$v(t) - v(t - v) \to v/\beta, \qquad (5.2.10)$$

as $t \to \infty$ (for any $v \geq$ in the non-lattice case, only when v is a multiple of the span if μ is lattice), and

$$v(t) \sim t/\beta, \qquad (5.2.11)$$

as $t \to \infty$.

Proof. The first assertion is just the renewal Theorem (5.2.6) or (5.2.7) with the choice $f = 1_{[0,v]}$. For the second, let d be the span if μ is lattice. Otherwise choose it arbitrarily > 0. Let $\varepsilon > 0$ and take u such that $|v(t) - v(t - d) - d/\beta| < \varepsilon$ for $t \geq u$. For such t, with $t - (r + 1)d \leq u < t - rd$,

$$v(t) - v(u) \leq \sum_{j=0}^r \{v(t - jd) - v(t - (j + 1)d)\} \leq (r + 1)(\varepsilon + d/\beta).$$

Since $r < (t - u)/d$ and $\varepsilon > 0$ was arbitrary,

$$\limsup_{t \to \infty} v(t)/t \leq 1/\beta.$$

Summation just to $r - 1$ yields a lower estimate. $\qquad\square$

In companionship with Lemma (5.2.12) this theorem yields

Corollary (5.2.14)

Consider $x = f + x * \mu$ *with* μ *a probability measure on* R_+, $0 < \beta = \int t\mu(dt) < \infty$, *and* f *measurable, bounded on finite intervals and such that* $f(t) \sim ct^n$ *for some* c *and* $n \in Z_+$. *Then, as* $t \to \infty$,

$$x(t) \sim ct^{n+1}/\beta(n + 1).$$

Background

The renewal theorems have an impressive and somewhat confused history during the last fifty years, see Reference [9]. The final results are due to Feller [7].

5.3 REFINED RENEWAL THEOREMS

With

$$0 < \beta = \int_0^\infty t\mu(dt) < \infty$$

we saw that $v(t)/t \to 1/\beta$, as $t \to \infty$. This can be refined when

$$\omega = \int_0^\infty t^2\mu(dt) < \infty.$$

To avoid repetitive arguments of little interest we assume throughout this section that μ is not lattice.

Lemma (5.3.1)

If μ *is not lattice but has a finite second moment* ω, *then*

$$0 \leq v(t) - t/\beta \to \omega/2\beta^2.$$

Proof (from Reference [7]). Consider

$$x(t) = v(t) - t/\beta.$$

Since $\int_0^t (t - u)\,\mu(du) = \int_0^t \mu(u)\,du$ after an integration by parts,

$$x(t) - x * \mu(t) = v(t) - t/\beta - v(t) + 1 + \int_0^t \mu(u)\,du/\beta$$

$$= 1 - \int_0^t \{1 - \mu(u)\}\,du/\beta = \int_t^\infty \{1 - \mu(u)\}\,du/\beta,$$

and x satisfies Equation (5.2.1) with $f(t)$ the last integral above. This $f \geq 0$.

Hence so is x. A reference to Theorem (5.2.6) and integration by parts,

$$\int_0^\infty dt \int_t^\infty \{1 - \mu(u)\}\, du = \omega/2,$$

yield the result (as f does not increase).   □

Next we shall make use of this and the following two lemmata to relax the Tauberian condition (5.2.7). Recall that a function is of locally bounded variation if and only if it can be written as the difference of two non-decreasing functions.

Lemma (5.3.2)

Let λ be a function of locally bounded variation on R_+. Assume that $\lambda\{\infty\} = \lim_{t\to\infty} \lambda(t)$ exists finitely. Then so does

$$\lim_{s\downarrow 0} \hat{\lambda}(s) = \lim_{s\downarrow 0} \int_0^\infty e^{-st} \lambda(dt) = \lambda(\infty).$$

Proof. To $\varepsilon > 0$ given choose t such that $|\lambda(u) - \lambda(t)| < \varepsilon$ for $u \geq t$. Then, for all $s > 0$, integration by parts yields

$$\left| \int_t^\infty e^{-su}\lambda(du) \right| = \left| \int_t^\infty \lambda(u)\, se^{-su}\, du - e^{-st}\lambda(t) \right|$$

$$\leq \int_t^\infty se^{-su}|\lambda(u) - \lambda(t)|\, du < \varepsilon.$$

From this it is evident that $\hat{\lambda}(s)$ converges for $s > 0$, but also that

$$\limsup_{s\downarrow 0} \left| \lambda(\infty) - \int_0^\infty e^{-su}\lambda(du) \right| \leq \limsup_{s\downarrow 0} \left| \lambda(\infty) - \int_0^t e^{-su}\lambda(du) \right| +$$

$$+ \limsup_{s\downarrow 0} \left| \int_t^\infty e^{-su}\lambda(du) \right| < |\lambda(\infty) - \lambda(t)| + \varepsilon < 2\varepsilon.$$

□

Lemma (5.3.3)

Let ϑ of locally bounded variation on R_+ satisfy $\lim_{s\to 0} \hat{\vartheta}(s) = 0$ and

$$\int_0^t u\vartheta(du)/t \to 0,$$

as $t \to \infty$. Then $\lim_{t\to\infty} \vartheta(t) = 0$, as well.

Proof. Write

$$\kappa(t) = \int_0^t u\vartheta(\mathrm{d}u).$$

Integration by parts (the only trick in this section!) yields

$$\int_1^\infty \mathrm{e}^{-su}\vartheta(\mathrm{d}u) = \int_1^\infty \{\mathrm{e}^{-su}/u\}\,\kappa(\mathrm{d}u)$$

$$= -\kappa(1)\,\mathrm{e}^{-s} + s\int_1^\infty \{\kappa(u)/u\}\,\mathrm{e}^{-su}\,\mathrm{d}u + \int_1^\infty \{\kappa(u)/u^2\}\,\mathrm{e}^{-su}\,\mathrm{d}u.$$

Since $\kappa(u)/u \to 0$,

$$s\int_1^\infty \{\kappa(u)/u\}\,\mathrm{e}^{-su}\,\mathrm{d}u \to 0,$$

as $s \to 0$ according to Lemma (5.3.2). But by assumption

$$\lim_{s\to 0}\int_1^\infty \mathrm{e}^{-su}\vartheta(\mathrm{d}u) = \lim_{s\to 0}\left\{\hat\vartheta(s) - \int_0^1 \mathrm{e}^{-su}\vartheta(\mathrm{d}u)\right\} = -\vartheta(1).$$

Hence, as $s \to 0$,

$$\int_1^\infty \{\kappa(u)/u^2\}\,\mathrm{e}^{-su}\,\mathrm{d}u \to \kappa(1) - \vartheta(1).$$

Now

$$\left|\int_1^t \{\kappa(u)/u^2\}\,\mathrm{d}u - \int_1^\infty \{\kappa(u)/u^2\}\,\mathrm{e}^{-u/t}\,\mathrm{d}u\right|$$

$$\le \int_1^t \{(1 - \mathrm{e}^{-u/t})\,|\kappa(u)|/u^2\}\,\mathrm{d}u + \int_t^\infty \left(|\kappa(u)/u|\,\mathrm{e}^{-u/t}/u\right)\mathrm{d}u$$

$$\le \int_1^t |\kappa(u)/u|\,\mathrm{d}u/t + \int_t^\infty \left(|\kappa(u)/u|\,\mathrm{e}^{-u/t}\,\mathrm{d}u\right)/t$$

by elementary inequalities. Since $\kappa(u)/u \to 0$ the same is true for both these integrals and thus

$$\int_1^\infty \{\kappa(u)/u^2\}\,\mathrm{d}u = \lim_{t\to\infty}\int_1^\infty \{\kappa(u)/u^2\}\,\mathrm{e}^{-u/t}\,\mathrm{d}u = \kappa(1) - \vartheta(1).$$

However, integrating again by parts,

$$\vartheta(t) = \vartheta(1) + \int_1^t u^{-1}\,\kappa(\mathrm{d}u) = \vartheta(1) + \kappa(t)/t - \kappa(1)$$

$$+ \int_1^t \{\kappa(u)/u^2\}\,\mathrm{d}u \to 0$$

as $t \to \infty$. $\qquad\qquad\qquad\qquad\qquad\qquad\qquad\qquad\qquad\qquad\square$

Theorem (5.3.4)

In the renewal equation (5.2.1) suppose that f is of locally bounded variation and such that $\lim_{t\to\infty} f(t) = 0$. Further let μ be non-lattice and with a finite second moment ω. Then

$$\lim_{t\to\infty} x(t) = \int_0^\infty f(t)\,\mathrm{d}t/\beta,$$

where β is the first moment of μ. Here the right hand integral is itself the limit

$$\lim_{t\to\infty} \int_0^t f(u)\,\mathrm{d}u$$

and the theorem is true as soon as this one has a meaning. (i.e. even if f is not in L_1 or $\int_0^\infty f(u)\,\mathrm{d}u = +\infty$).

Proof. Write $\lambda(t) = v(t) - t/\beta$. By Lemma (5.3.1) λ is bounded and $\lambda(t) \to \omega/2\beta^2$. Hence Lemma (5.3.2) yields $\hat\lambda(s) \to \omega/2\beta^2$, as $s \to 0$. In the same way $\hat{f}(s) \to 0$. Further,

$$\int_0^t uf * \lambda(\mathrm{d}u) = f(t) \int_0^t u\lambda(\mathrm{d}u) + \lambda(t) \int_0^t uf(\mathrm{d}u)$$

$$= f(t)\left\{ t\lambda(t) - \int_0^t \lambda(u)\,\mathrm{d}u \right\} + \lambda(t)\left\{ tf(t) - \int_0^t f(u)\,\mathrm{d}u \right\}.$$

By assumption this is $o(t)$, as $t \to \infty$. Lemma (5.3.3) applies to $f * \lambda$ and

$$\lim_{t\to\infty} f * \lambda(t) = \lim_{s\to 0} \widehat{f * \lambda}(s) = \lim_{s\to 0} \hat{f}(s)\,\hat\lambda(s) = 0.$$

But

$$x(t) = \int_0^t f(t-u)\,v(\mathrm{d}u) = \int_0^t f(t-u)\,\lambda(\mathrm{d}u) + \int_0^t f(u)\,\mathrm{d}u/\beta$$

after a change of variable in the last integral. $\qquad\qquad\qquad\qquad\square$

In a simple manner Theorem (5.3.4) leads to results on the rate of convergence in the renewal theorem. But first a third lemma.

Lemma (5.3.5)

Let f and μ be as in Theorem (5.3.4). Then

$$\lim_{t \to \infty} \int_0^t \{x(u) - x(\infty)\} \, du = \int_0^\infty \{x(u) - x(\infty)\} \, du$$

$$= \left\{ \int_0^\infty f(u) \, du\omega/2\beta - \int_0^\infty uf(u) \, du \right\} \Big/ \beta$$

exists, provided only

$$\int_0^\infty t|f(t)| \, dt < \infty. \tag{5.3.2}$$

Proof. By Inequality (5.3.2)

$$x(\infty) = \int_0^\infty f(u) \, du/\beta$$

is finite and

$$x(u) - x(\infty) = f(u) - \int_0^\infty f(v) \, dv \{1 - \mu(u)\} \Big/ \beta$$

$$+ \int_0^u \{x(u - v) - x(\infty)\} \mu(dv).$$

Integrate this from 0 to t to obtain

$$\int_0^t \{x(u) - x(\infty)\} \, du = \int_0^t f(u) \, du - \int_0^\infty f(v) \, dv \int_0^t \{1 - \mu(u)\} \, du/\beta$$

$$+ \int_0^t \int_0^{t-v} \{x(u) - x(\infty)\} \, du\mu(dv)$$

after a change in the order of integration in the double integral. This is a renewal equation and in order to apply Theorem (5.3.4) we must only check that

$$g(t) = \int_0^t f(u) \, du - \int_0^\infty f(v) \, dv \int_0^t \{1 - \mu(u)\} \, du/\beta \tag{5.3.3}$$

satisfies the requirements on f there and also that

$$\int_0^\infty g(t) \, dt$$

converges. But g must have a bounded local variation, and obviously it tends to zero, as $t \to \infty$. Further

$$g(t) = \int_0^\infty f(u)\,du \int_t^\infty \{1 - \mu(u)\}\,du/\beta - \int_t^\infty f(u)\,du.$$

Changing the order of integration we see that

$$\int_0^\infty dt \int_t^\infty |f(u)|\,du = \int_0^\infty u|f(u)|\,du < \infty$$

and

$$\int_0^\infty dt \int_t^\infty \{1 - \mu(u)\}\,du = \int_0^\infty t^2\mu(dt)/2 = \omega/2 < \infty.$$

Hence Fubini's theorem justifies a change of order of integration in $\int_0^\infty g$ as well,

$$\int_0^\infty g(t)\,dt = \int_0^\infty f(t)\,dt\omega/2\beta - \int_0^\infty tf(t). \qquad \square$$

Note. *As an exercise, try to prove this for an f not necessarily of locally bounded variation but satisfying Inequality (5.3.2) plus the requirements of Theorem (5.2.6).*

Theorem (5.3.6)

Let f and μ be as in Theorem (5.3.4). Assume that Inequality (5.3.2) holds and that $tf(t) \to 0$. Then

$$\lim_{t \to \infty} t\{x(t) - x(\infty)\} = 0.$$

Proof. Clearly

$$t\{x(t) - x(\infty)\} = tf(t) - \int_0^\infty f(u)\,dut\{1 - \mu(t)\}\,/\,\beta$$

$$+ \int_0^t \{x(t - u) - x(\infty)\}\, u\mu(du)$$

$$+ \int_0^t (t - u)\{x(t - u) - x(\infty)\}\, \mu(du).$$

Theorem (5.3.4) is applicable to this. It yields

$$\lim_{t \to \infty} t\{x(t) - x(\infty)\} =$$

$$\left\{ \int_0^\infty tf(t)\,dt - \int_0^\infty f(t)\,dt\omega/2\beta + \int_0^\infty \{x(t) - x(\infty)\}\,dt \right\} \Big/ \beta = 0,$$

since

$$\lim_{t \to \infty} \int_0^t \{x(t - u) - x(\infty)\}\, u\mu(du) = 0$$

and

$$\int_0^r dt \int_0^t \{x(t - u) - x(\infty)\}\, u\mu(du)$$

$$= \int_0^r u\mu(du) \int_0^{r-u} \{x(t) - x(\infty)\}\,dt \to \beta \int_0^\infty \{x(t) - x(\infty)\}\,dt$$

$$= \int_0^\infty f(t)\,dt\omega/2\beta = \int_0^\infty tf(t)\,dt$$

as $r \to \infty$, by Lemmata (5.2.10) and (5.3.5). $\qquad\square$

We conclude the section by an analogue of Lemma (5.3.5) for defective μ. Its proof is direct: integrate the equation and invoke Theorem (5.2.9).

Lemma (5.3.7)
If $\mu(\infty) < 1$, $x = f + x * \mu$ and

$$\int_0^\infty f(t)\,dt$$

converges, then so does

$$\int_0^\infty x(t)\,dt = \int_0^\infty f(t)\,dt \Big/ \{1 - \mu(\infty)\}.$$

Background
Lemma (5.3.1) is due to Smith [11]. Except the classical Lemmata (5.3.2) and (5.3.3) the rest is from Reference [10].

5.4 POINT PROCESSES

Consider a system of some kind, where a component has a random duration, say with distribution $F, F(0) < 1$. Assume that the system is started with a new component at time zero and that the component is renewed immediately at failure. Its development might then be described by a sequence of i.i.d. random variables $\{x_n\}_1^\infty$, x_n being the duration of the nth component. An integer valued stochastic process in $t \geq 0$, the *renewal process* ξ_t, is defined to be the number of renewals up to and including time t, $\xi_t = n \Leftrightarrow s_n \leq t < s_{n+1}, s_n = x_1 + \dots + x_n$. The expectation of this satisfies

$$E[\xi_t] = E[\xi_t; x_1 > t] + E[\xi_t; x_1 \leq t]$$

$$= 1 - F(t) + \int_0^t \{1 + E[\xi_{t-u}]\} F(\mathrm{d}u) = 1 + \int_0^t E[\xi_{t-u}] F(\mathrm{d}u),$$

which accounts for the name of the renewal equation. From Theorem (5.2.13) we see that in the long run some $t/\int_0^\infty u F(\mathrm{d}u)$ components are needed during t time units (at least in some suitable weak sense).

The renewal process is non-decreasing in t and can be defined as a process on arbitrary Borel sets, $\xi(A)$ being the number of renewals in the set A. Then $\xi([0, t]) = \xi_t$. Such processes are termed point processes. To be precise, a *point process* ξ on R_+ is a map from some probability space into the set of integer or infinite valued measures on R_+ (with the customary Borel algebra), such that the mass $\xi(A)$ given to a bounded Borel set is a finite random variable. (We shall write $\mathcal{N}(R_+)$ for the set of integer or infinite valued positive measures on R_+, that are finite on bounded sets. The set is then endowed with the cylinder σ-algebra, $\mathcal{B}(\mathcal{N})$, i.e. the one generated by all maps $\mu \to \mu(A)$, A a Borel set, and a point process is a measurable function into $\mathcal{N}(R_+)$.)

We defined the renewal process starting from a sequence of random variables. Conversely, from a point process we can define an associated sequence of intervals between atoms of the point process: The function $\xi([0, t])$ in t is a pure step function and if $s_0 = 0, s_n = \inf\{t \geq 0; \xi[0, t] \geq n\}$, then $x_n = s_n - s_{n-1}$, constitute the intervals sequence.

Point processes can be of various kinds, we have already considered renewal processes. Point processes ξ, whose distribution is

$$P[\xi(A) = k] = \mathrm{e}^{-v(A)} \{v(A)\}^k / k\,!$$

for $k \in Z_+$, some non-atomic Radon measure v (the *parameter measure*), and all Borel sets A, play a central rôle in the theory. They are known as *Poisson processes*. E.g. these are the only processes which give an independent number of points to disjoint sets and have no multiple points, i.e. their intervals satisfy $P[x_n = 0] = 0$ for all n. If for any measurable $A_1, \dots, A_n$ and $t > 0$ the two vectors $[\xi(A_1), \dots, \xi(A_n)]$ and $[\xi(A_1 + t), \dots, \xi(A_n + t)]$ have the same distribution, then the process is said to be *stationary*. The only stationary Poisson processes are the time homogeneous ones, where v is a multiple of Lebesgue measure.

We shall need almost nothing of the theory of point processes, merely the concept to formulate the general branching process. Here, as the reader might recall, individuals are supposed to give birth to random numbers of children at random time epochs during life, i.e. at the points of a point process ξ. Of particular interest is the degenerate case where ξ can give mass only to one random point λ, yielding splitting processes. Another one is that when ξ is obtained from a Poisson process η as

$$\xi(A) = \eta(A \cap [0, \lambda])$$

i.e. by disregarding children born after their mother's death. As a third example, the Poisson process could be allowed to operate only on some fertility subinterval of $[0, \lambda]$.

Background

A commendable survey of point processes is Reference [12].

REFERENCES

1. Chow, Y. S., Robbins, H., and Siegmund, D., *Great Expectations: The Theory of Optimal Stopping.* Houghton Mifflin, Boston, 1971.
2. Meyer, P. A., *Probabilités et potentiels.* Hermann, Paris, 1966. Translated as: Probability and Potentials. Blaisdell, Waltham, Mass., 1966.
3. Neveu, J., *Martingales à temps discret.* Masson, Paris, 1972.
4. Rao, K. M., *Lecture notes for Statistik 3.* Matematisk Institut, Aarhus Universitet, 1972.
5. Billingsley, P., *Convergence of Probability Measures.* John Wiley and Sons, New York, 1968.
6. Cox, D., *Renewal Theory.* Methuen, London, 1962.
7. Feller, W., *An Introduction to Probability Theory and its Applications II.* John Wiley and Sons, New York, 1966.
8. Royden, H. L., *Real Analysis,* Mac Millan, New York, 1963.
9. Smith, W. L., Renewal theory and its ramifications. *J. Roy. Statist. Soc. B,* **20**, 243–302, 1958.
10. Jagers, P., Renewal theory and the almost sure convergence of branching processes. *Ark. Mat.* 7, 495–504, 1968.
11. Smith, W. L., Asymptotic renewal theorems. *Proc. Royal Soc. Edinburgh A* **64**, 9–48, 1954.
12. Daley, D. J. and Vere-Jones, D., A summary of the theory of point processes. In: Lewis P. A. W. (ed.), *Stochastic Point Processes: Statistical Analysis, Theory, and Applications,* 299–383. J. Wiley and Sons, New York, 1972.

Chapter 6

The General Process

6.1 INTRODUCTION

We recall part of Section 1.2: Consider a population with one ancestor. With each individual $x \in I$, associate one non-negative random variable λ_x, the *life-length* of x, and one point process ξ_x, the *reproduction* of x. Assume that the pairs (λ_x, ξ_x) are i.i.d., say with probability distribution Q. (This law is a measure on the product space $R_+ \times \mathcal{N}(R_+)$ with the corresponding product σ-algebra.) Its margin on R_+, L,

$$L(u) = P[\lambda_x \leq u]$$

is the *life-length distribution* and its margin on $\mathcal{N}(R_+)$ the *reproduction law*. Q itself might be called the *individual law*. When the contrary is not explicitly stated we shall assume that

$$P[\xi_x(\lambda_x, \infty) = 0] = 1. \tag{6.1.1}$$

This means that no child can be born after its mother's death.

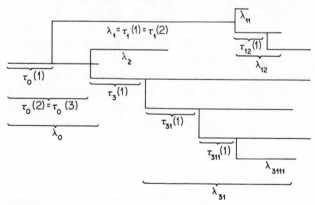

Figure 6.1.1 A possible realization of part of a family history. Here $\xi_0(\infty) = 3$, $\xi_1(\infty) = 2$, $\xi_{11}(\infty) = 0$, $\xi_2(\infty) = 0$, $\xi_{31}(\infty) = 1$, and $\xi_{3111}(\infty) = 0$.

123

Realization is defined as previously, the ancestor 0 is realized, (x, k) is realized if x is and $\xi_x(\infty) \geq k$—we write $\xi_x(t) = \xi_x[0, t], \xi_x(\infty) = \xi_x[0, \infty)$. A realized individual (x, k) is born when her mother x is aged

$$\tau_x(k) = \inf\{t; \xi_x(t) \geq k\}$$

and this occurs at time $\sigma_{(x,k)}$, the *birth time* of (x, k). If $x = (j_1, \ldots, j_n) \in I$, then

$$\sigma_x = \tau_0(j_1) + \ldots + \tau_{(j_1,\ldots,j_{n-1})}(j_n).$$

Define $\sigma_0 = 0$. The individual x is *alive* at time $t \geq 0$ if it has been born but has not died, i.e.

$$\sigma_x \leq t < \sigma_x + \lambda_x.$$

Its *age* is $t - \sigma_x$ and so, $t \geq 0, a \geq 0$

$$z_t^a(x) = \begin{cases} 1 & \text{if } t - a < \sigma_x \leq t < \sigma_x + \lambda_x \\ 0 & \text{otherwise} \end{cases}$$

is one exactly if x is alive and younger than a at time t. Define

$$z_t^a = \sum_{x \in I} z_t^a(x) \leq \infty.$$

The stochastic process $\{z_t^a; t, a \geq 0\}$ is the *general branching* process. Note that if $L(0) = 1$ with probability one no individual is ever alive. Thus we always assume that $L(0) < 1$. If $P[\xi_x\{\lambda_x\} = \xi_x(\infty)] = 1$, we shall talk of a *splitting* (or *Sevast'yanov*) process. If further ξ_x and λ_x are independent, of a *Bellman–Harris* process. As mentioned in Chapter 1, the customary name is that of an *age-dependent* process, mirroring the importance of age here as opposed to for the generation counting Galton–Watson process. Note that an individual x is realized if and only if $\sigma_x < \infty$. We write $z_t(= z_t^a$ for $a > t)$ for the size of the population at time t and $\{\zeta_n; n \in Z_+\}$ for the *imbedded Galton–Watson process*, $\zeta_n =$ the number of realized individuals belonging to the nth generation.

We shall now deduce a counterpart of the relation (2.1.1), representing the number of individuals at a given time as a sum of independent variables. The first step is a simple consequence of relation (1.2) in Chapter 1. Write $I_n = \{(n, x); x \in I\}$ for the set of descendants from the ancestor's nth child and define

$$z_t^{(n)a} = \sum_{x \in I_n} z_t^a(x).$$

Then

$$z_t^a = z_t^a(0) + \sum_{n=1}^{\infty} z_t^{(n)a} = z_t^a(0) + \sum_{n=1}^{\xi_0(t)} z_t^{(n)a}, \tag{6.1.2}$$

since $\xi_0(t) < n \Rightarrow z_t^{(n)} = 0$. Intuitively it is clear that $\{z_t^{(n)a}; t, a \geq 0\}$ is a new branching process, initiated at the birth of individual n. We shall give some basis for this.

The random variables $z_t^a(x)$, $x \in I$ and therefore also z_t^a and $z_t^{(n)a}$ are defined as functions of $\{(\lambda_y, \xi_y); y \in I\}$. To catch the relation between the latter two define for $x \in I$ an operator S_x on the range space Ω of all (λ_y, ξ_y) by

$$S_x(\{(\lambda_y, \xi_y); y \in I\}) = \{(\lambda_{(x,y)}, \xi_{(x,y)}), y \in I\}.$$

Then, since the pairs (λ_x, ξ_x), $x \in I$, are i.i.d., $S_x(\{(\lambda_y, \xi_y); y \in I\})$ has the same distribution as its argument $\{(\lambda_y, \xi_y); y \in I\}$. This implies that any composed function $f \circ S_x$ as a random variable on Ω follows the same probability law as f itself.

Lemma (6.1.1)

For any a, t,

$$z_t^{(n)a} = z_{t-\tau_0(n)}^a \circ S_n.$$

Note. *We define $z_t^a(x) = 0$ for negative t.*

Proof. If $\tau_0(n) > t$, then for $x \in I_n$ $\sigma_x \geq \tau_0(n) > t$ implying that $z_t^a(x) = 0$, for $x \in I_n$, and so $z_t^{(n)a} = 0$. Assume that $\tau_0(n) \leq t$. Then

$$z_{t-\tau_0(n)}^a(x) \circ S_n = 1 \Leftrightarrow t - \tau_0(n) - a < \sigma_x \circ S_n$$

$$= \sigma_{(n,x)} - \tau_0(n) \leq t - \tau_0(n) < \sigma_{(n,x)} - \tau_0(n) + \lambda_{(n,x)}$$

$$\Leftrightarrow t - a < \sigma_{(n,x)} \leq t < \sigma_{(n,x)} + \lambda_{(n,x)} \Leftrightarrow z_t^a(n, x) = 1.$$

Summation over all $x \in I$ completes the proof. ☐

We state the basic representation as a theorem:

Theorem (6.1.2)

For $t, a \geq 0$ z_t^a can be written as the sum of $z_t^a(0)$ and the number of individuals younger than a in $\xi_0(t)$ independent branching processes each with the same individual law as $\{z_t^a\}$ but started at times $\tau_0(1), \ldots, \tau_0 \circ \xi_0(t)$. Indeed,

$$z_t^a = z_t^a(0) + \sum_{n=1}^{\xi_0(t)} z_{t-\tau_0(n)}^a \circ S_n. \tag{6.1.3}$$

If the reproduction has no multiple points, i.e. $P[\forall u: \xi\{u\} \leq 1] = 1$, this can be given an enticing convolution form,

$$z_t^a = z_t^a(0) + \int_0^t z_{t-u}^a \circ S_{\xi_0(u)} \xi_0(\mathrm{d}u). \tag{6.1.4}$$

Note. *The process $\{z_t^a\}$ is generally not Markovian.*

Background

Variants of the model presented here were introduced independently in References [1–3]. Splitting branching processes have been known longer, the Bellman–Harris model since the forties, the Sevast'yanov model since 1964. For an explicit measure theoretic construction of the basic probability space, consult Reference [2].

6.2 THE FINITENESS OF THE PROCESS

Though we did not go into that subject explicitly, it was evident that Galton–Watson processes are always finite. In the general case this need not be so and what happens is (essentially) determined by the *reproduction function* (in the sequel we write ξ, λ, or $\tau(n)$ for ξ_x, λ_x, or $\tau_x(n)$ with unspecified $x \in I$)

$$\mu(t) = E[\xi(t)]$$

and practically, indeed, by its value at the origin, $\mu(0)$. Define y_t as the total number of individuals that have been born up to time t,

$$y_t(x) = \begin{cases} 1 & \text{when } \sigma_x \leq t, \\ 0 & \text{otherwise,} \end{cases}$$

$$y_t = \sum_x y_t(x).$$

We give criteria for the finiteness of $\{y_t\}$. Obviously $z_t \leq y_t$.

Theorem (6.2.1)

If $\mu(0) > 1$, then for all $t \geq 0$, $P[y_t = \infty] > 0$.

Proof. Let η_n be the number of individuals $x \in N^n$ with $\sigma_x = 0$. The sequence $\{\eta_n\}$ is a Galton–Watson process with reproduction mean $\mu(0) > 1$. Hence there is a positive chance that $\eta_n \to \infty$. But obviously $\eta_n \leq y_t$ for $t \geq 0$. $\square$

Note. *If $\xi(0)$ has the generating function g and a is the smallest non-negative root of $g(s) = s$, then obviously $P[y_t = \infty] \geq 1 - a$.*

In most natural circumstances $y_t = \infty$ implies that at least some $z_u = \infty$. However this hinges upon the life distribution. For example, if $L(0) = 1$ and $\mu(0) > 1$ (almost) no individuals are ever alive, though infinitely many might well be born already at $t = 0$ (to give a drastic illustration of the terminology used). On the other hand, if $\xi_x(0)$ and λ_x are independent random variables with $L(0) < 1$, then $\mu(0) > 1$ implies that y_0 may be infinite and if $y_0 = \infty$ also infinitely many individuals must survive some positive time. Thus $P[z_0 = \infty] = P[y_0 = \infty] > 0$.

Theorem (6.2.2)

If $\mu(0) < 1$ and $\mu(t)$ is finite for some $t > 0$, then

$$P[\forall t : y_t < \infty] = 1.$$

Hence so is $P[\forall t : z_t < \infty] = 1$.

Proof. Since $\mu(t) < \infty$ for some $t > 0$ it follows by monotone convergence that $\mu(t) \downarrow \mu(0) < 1$. Let $u > 0$ be such that $\mu(u) < 1$. Denote by η_n here the number of $x = (j_1, \ldots, j_n) \in N^n$ satisfying $\tau_0(j_1) \leq u, \ldots, \tau_{(j_1,\ldots,j_{n-1})}(j_n) \leq$ $\leq u$, i.e. we count only the individuals born before their mothers attain age u. The random variables $\{\eta_n\}$ again constitute a Galton–Watson process though a subcritical one; the reproduction mean is $\mu(u)$. Also for $t \leq u$

$$y_t \leq y_u \leq \sum_{n=0}^{\infty} \eta_n < \infty$$

with probability one.

Knowing that y_t is finite for $t \leq u$ we shall use the basic representation and induction to complete the proof. The arguments leading to Theorem (6.1.2) yield

$$y_t = 1 + \sum_{n=1}^{\xi_0(t)} y_{t-\tau_0(n)} \circ S_n. \tag{6.2.1}$$

Assume that y_t is a.s. finite for $t \leq s$. Then if $y_{s+u} = \infty$ at least one $y_{s+u-\tau_0(n)} \circ$ $\circ S_n$ must also be so. By hypothesis this must be one (neglecting sets of probability zero) where $\tau_0(n) \leq u$. Use the monotonity of y_t and the branching property to obtain

$$P[y_{s+u} = \infty] \leq E\left[P\left[\bigcup_{\tau_0(n) \leq u} \{y_{s+u-\tau_0(n)} \circ S_n = \infty\} | \lambda_0, \xi_0 \right] \right]$$

$$\leq E\left[\sum_{\tau_0(n) \leq u} P[y_{s+u-\tau_0(n)} \circ S_n = \infty | \lambda_0, \xi_0] \right]$$

$$\leq P[y_{s+u} = \infty] \mu(u).$$

Since $\mu(u) < 1$, $P[y_{s+u} = \infty] = 0$ and the proof is almost complete. We need only note that since y_t is nondecreasing in t

$$P[\forall t : y_t < \infty] = P[\forall n : y_n < \infty] = 1. \qquad \square$$

The remaining case, $\mu(0) = 1$, is by far the most intricate. To see one side of the coin, suppose that $\mu(u) = \mu(0) = 1$ for some $u > 0$. Then, if $P[\xi(u) = 1] < 1$ (or, what amounts to the same, $P[\xi(0) = 1] < 1$), the first half of the proof of Theorem (6.2.2) still applies to prove that $P[y_u < \infty] = 1$, since critical Galton–Watson processes also die out. To see that y_t remains finite for all t consider an individual x aged a. During the age interval $(a, a + u]$ it begets $\xi_x(a + u) - \xi_x(a)$ children, each of which initiates a new branching process, which remains finite for the first u time units of its course. Thus if at or before some time t only finitely many individuals have been born y_{t+u} remains finite. This inductive argument leads to another and a trifle stronger version of the finiteness criterion:

Theorem (6.2.3)

   If there is a $u > 0$ such that $\mu(u) \leq 1$ and $\xi(0)$ is not a.s. one, then

$$P[\forall t : z_t < \infty] = P[\forall t : y_t < \infty] = 1.$$

To see that processes with $\mu(0) = 1$ can explode within a finite time consider a μ increasing fairly rapidly. Assume that there is a sequence $t_k \downarrow 0$ such that the two series

$$\sum_{k=1}^{\infty} t_k = t, \qquad \sum_{n=1}^{\infty} 1 \bigg/ \left\{ \prod_{k=1}^{n} \mu(t_k)\, \mu(t_n) \right\}$$

converge. Suppose further that $\mathrm{Var}\,[\xi(t)] < \infty$. Then so is each $\mathrm{Var}\,[\xi(t_u)] \leq$ $\leq E[\xi^2(t_u)] \leq E[\xi^2(t)]$. Let η_0 be one, η_1 the number of the ancestor's children born before time t_1, η_2 the number of individuals in the second generation born by mothers not older than t_2, who themselves were born before time t_1 etc. .... The sequence $\{\eta_n\}$ is a Galton–Watson process in a varying environment,

$$y_t \geq \sum_{n=1}^{\infty} \eta_n$$

and by Theorem (3.6.1) the sum is infinite with positive probability.

Assumption. *From the following section on we consider only processes with $\mu(0) < 1$, $\mu(t) < \infty$ for some $t > 0$. In fact practically always we require that $m = \mu(\infty) < \infty$.*

Example (6.2.4)

   Here is a more explicit illustration of the construction above. Let all individuals have life span one and let the reproduction process be obtained from a Poisson process η as in Section 5.4, $\xi(A) = \eta(A \cap [0, 1])$, where $E[\eta(t)] = 1 + t^{1/4}$. Let $t_k = 1/(k + 1)^2$.

Then $\displaystyle\sum_1^{\infty} t_k < 1$, $1 \bigg/ \prod_2^{n} \mu(t_k) = \exp - \sum_2^{n+1} \log(1 + k^{-1/2}) \sim A\, e^{-B\sqrt{n}}, A, B >$ > 0, as $n \to \infty$ (approximate the sum by an integral).

Example (6.2.5)

In applications to cell proliferation the usual situation is that cells live through a time of random length, the so called *generation time* or *cell cycle*, after which they divide. However they may also die before their life cycle is completed. This phenomenon is called *cell death* or disintegration. To describe such populations assume that the probability of a cell dying without dividing is $1 - p$ and that the generation time has distribution G. The number $1 - p$ is called the frequency of cell death. Assume further that a cell dying before its generation time is completed does so after a random time with distribution F. Then the life length has distribution

$$L = (1 - p) F + pG$$

and the reproduction function is

$$\mu = 2pG.$$

Even though $F(0)$ might be strictly positive due to abortive divisions, it is difficult to imagine situations where $G(0) > 0$. This example will be referred to as *binary splitting* (with cell death). Note that the imbedded Galton–Watson process is indeed a binary splitting process with $p_2 = p, p_0 = 1 - p$.

Background

For splitting branching processes Sevast'yanov [5] has made an ambitious but still not exhaustive investigation of finiteness problems. Most results given are from References [1] or [2] (the second part of Theorem 1 in Reference [2] corresponding to Theorem (6.2.1) here contains a misprint, there should be a bar over all appearing z_t). Only Theorem (6.2.3) is new. The Galton–Watson processes in the proofs of Theorems (6.2.1) and (6.2.2) were suggested by T. Lindvall.

6.3 MOMENTS AND THE GENERATING FUNCTIONS

We introduce *the process generating functions φ_t^a*,

$$\varphi_t^a(s) = E[s^{z_t^a}], \qquad 0 \le s \le 1,$$

$$\varphi_t = \varphi_t^a \quad \text{for } a > t.$$

For age-dependent branching processes let f_u be the generating function of $\xi(u)$, given that $\lambda = u$, and generally let the *reproduction generating function f* be

$$f(s) = E[s^{\xi(\infty)}].$$

This f is the reproduction generating function of the imbedded Galton–Watson process.

From Theorem (6.1.2) it follows that

$$\varphi_t^a(s) = E\big[E\big[s^{z_t^a(0)+\sum_1^{\xi_0(t)} z_{t-\tau_0(n)}^a \circ S_n} \,|\, \lambda_0, \xi_0\big]\big] = E\left[s^{z_t^a(0)} \prod_{n=1}^{\xi_0(t)} \varphi_{t-\tau_0(n)}^a(s)\right]$$

$$= E\left[s^{z_t^a(0)} \exp \int \log \varphi_{t-u}^a(s)\, \xi_0(du)\right]. \tag{6.3.1}$$

For age-dependent processes this reduces to

$$\varphi_t^a(s) = E\big[s^{z_t^a(0)}; \lambda_0 > t\big] + E\big[f_{\lambda_0} \circ \varphi_{t-\lambda_0}^a(s); \lambda_0 \le t\big]$$

$$= s^{1_{[0,a)}(t)}\{1 - L(t)\} + \int_0^t f_u \circ \varphi_{t-u}^a(s)\, L(du), \tag{6.3.2}$$

For the Bellman–Harris model again we obtain

$$\varphi_t^a(s) = s^{1_{[0,a)}(t)}\{1 - L(t)\} + \int_0^t f \circ \varphi_{t-u}^a(s)\, L(du). \tag{6.3.3}$$

The relation (6.3.1) retains little of the simplicity of the recursive relations for generating function of Galton–Watson processes. However it does determine the distribution of z_t^a:

Theorem (6.3.1)

For each $0 \le s \le 1$ and $0 \le a \le \infty$, there is just one measurable function $R_+ \to [0, 1]$ satisfying

$$\varphi_t^a(s) = E\left[s^{z_t^a(0)} \exp \int \log \varphi_{t-u}^a(s)\, \xi_0(du)\right].$$

Proof. Let $u > 0$ be such that $\mu(u) < 1$, and suppose that the two functions g and h from R_+ into the unit interval both satisfy the equation for some s. Then for $0 \le t \le u$

$$|g(t) - h(t)| \le E\left[\left|\prod_{n=1}^{\xi_0(t)} g\{t - \tau_0(n)\} - \prod_{n=1}^{\xi_0(t)} h\{t - \tau_0(n)\}\right|\right]$$

$$\le E\left[\sum_{n=1}^{\xi_0(t)} |g\{t - \tau_0(n)\} - h\{t - \tau_0(n)\}|\right]$$

$$\le \sup_{0 \le t \le u} |g(t) - h(t)|\, \mu(u),$$

and

$$\sup_{0 \le t \le u} |g(t) - h(t)| \le \mu(u) \sup_{0 \le t \le u} |g(t) - h(t)|.$$

It follows that $h(t) = g(t)$ for $0 \leq t \leq u$. Next assume that the two functions have been shown to coincide on some interval $[0, t_0]$. We conclude the proof of showing that then they are the same on $[t_0, t_0 + u]$ as well. Indeed, then for $t_0 \leq t \leq t_0 + u$

$$|g(t) - h(t)| \leq E \left[\sum_{0 \leq \tau_0(n) \leq u} |g\{t - \tau_0(n)\} - h\{t - \tau_0(n)\}| \right]$$

$$\leq \sup_{0 \leq t \leq t_0 + u} |g(t) - h(t)| \, \mu(u)$$

and the argument used applies again.   □

Before deducing relations which determine moments, let us investigate when the expectation and variance of y_t are finite. Define for $x \in I$

$$\delta_x(t) = E[y_t(x)] = P[\sigma_x \leq t],$$

and observe that

$$\delta_{(x,n)}(t) = P[\sigma_x + \tau_x(n) \leq t] = \delta_x * \delta_n(t),$$

$$\sum_{n=1}^{\infty} \delta_n(t) = \sum_{n=1}^{\infty} P[\tau_0(n) \leq t] = \sum_{n=1}^{\infty} P[\xi(t) \geq n] = \mu(t).$$

By induction

$$E[y_t] = \sum_{x \in I} E[y_t(x)] = 1 + \sum_{n=1}^{\infty} \sum_{x \in N^n} \delta_x(t) = \sum_{n=0}^{\infty} \mu^{*n}(t) < \infty,$$

since Lemma (5.2.1) holds for any Radon measure such that $\mu(0) < 1$. This resolves the first problem:

Theorem (6.3.3)

If the reproduction function is finite, then so is $E[y_t]$ and therefore also $m_t = E[z_t]$ for all t. Further, $m_t^a = E[z_t^a]$ satisfies

$$m_t^a = 1_{[0,a)}(t) \{1 - L(t)\} + \int_0^t m_{t-u}^a \mu(du). \tag{6.3.4}$$

If $m = \mu(\infty) < 1$ (the subcritical *case), then as $t \to \infty$*

$$m_t \to 0.$$

If $m = 1$ (the critical *case) and μ is non-lattice, then for $0 \leq a < \infty$*

$$m_t^a \to \int_0^a \{1 - L(u)\} \, du \bigg/ \int_0^\infty u\mu(du).$$

If further

$$\int_0^\infty t L(\mathrm{d}t) < \infty,$$

then also

$$m_t \to \int_0^\infty u L(\mathrm{d}u) \Big/ \int_0^\infty u \mu(\mathrm{d}u).$$

When m > 1 (the supercritical *case), μ is not lattice, and α > 0 is the* Malthusian parameter *defined by* $\hat{\mu}(\alpha) = 1$, *then for* $0 \leq a \leq \infty$

$$m_t^a \sim \mathrm{e}^{\alpha t} \int_0^a \mathrm{e}^{-\alpha u} \{1 - L(u)\} \, \mathrm{d}u \Big/ \int_0^\infty u \, \mathrm{e}^{-\alpha u} \mu(\mathrm{d}u).$$

In the lattice cases corresponding assertions hold.

Proof. Being just the expectation of Equation (6.1.4), Equation (6.3.4) needs no derivation. The assertions on asymptotics are consequences of Theorems (5.2.9), (5.2.6), and (5.2.8).   □

Note. *In the age-dependent case*

$$\mu(t) = \int_0^t f_u'(1) \, L(\mathrm{d}u) = \int_0^t m(u) \, L(\mathrm{d}u)$$

with $m(u) = f_u'(1)$. *For Bellman–Harris processes this is simply* $mL(t)$.

Though more burdened with technicalities, the analysis of the second moments proceeds in the same vein. We introduce

$$\gamma_{ij}(u, v) = P[\xi(u) \geq i, \xi(v) \geq j],$$

$$\gamma_i(u, v) = \sum_{j=1}^\infty \gamma_{ij}(u, v) = E[\xi(v); \xi(u) \geq i],$$

$$\gamma(u, v) = \sum_{i,j} \gamma_{ij}(u, v) = E[\xi(u) \xi(v)],$$

and note that by Fubini's theorem these functions can also be viewed as measures. We shall use two-dimensional convolutions,

$$f * g(u, v) = \int_{\substack{0 \leq s \leq u \\ 0 \leq t \leq v}} f(u - s, v - t) \, g(\mathrm{d}s, \mathrm{d}t).$$

Lemma (6.3.4)

If $x, x' \in N^k$, then for $s, t \in R_+$

$$\sum_{x'' \in I} E[y_t(x) y_s(x', x'')] = \int_0^s E[y_t(x) y_{s-u}(x')] \sum_{n=0}^{\infty} \mu^{*n}(du).$$

Proof. Assume first that $x \in N^k$ and that $x' \in$ some $N^{k'}$ for $k' \geq k$. Then for $n \in N$. (Recall that (x, n) denotes the nth child of the individual x and that similarly (x, y) is the vector (individual) whose components are first those of x, then those of y.)

$$E[y_t(x) y_s(x', n)] = P[\sigma_x \leq t, \sigma_{x'} + \tau_{x'}(n) \leq s]$$

$$= \int_0^s P[\sigma_x \leq t, \sigma_{x'} \leq s - u] \delta_n(du)$$

$$= \int_0^s E[y_t(x) y_{s-u}(x')] \delta_n(du).$$

Hence for x, x' both in N^k

$$\sum_{x'' \in I} E[y_t(x) y_s(x', x'')] = E[y_t(x) y_s(x')] + \sum_{n=1}^{\infty} \sum_{x'' \in N^n} E[y_t(x) y_t(x', x'')]$$

$$= E[y_t(x) y_s(x')] + \sum_{n=1}^{\infty} \int_0^s E[y_t(x) y_{s-u}(x')] \mu^{*n}(du)$$

$$= \int_0^s E[y_t(x) y_{s-u}(x')] \sum_{n=0}^{\infty} \mu^{*n}(du). \qquad \square$$

Lemma (6.3.5)

Suppose that $x = (j_1, \ldots, j_n) \in N^n$ and that $1 \leq k \leq n$. Then

$$\sum_{x' \in N^k} E[y_t(x) y_s(x')] = \int_0^t E[y_{t-u}(j_{k+1}, \ldots, j_n)] \gamma_{j_1} * \ldots * \gamma_{j_k}(du, s).$$

Also

$$\sum_{x' \in N^n} E[y_t(x) y_s(x')] = \gamma_{j_1} * \ldots * \gamma_{j_n}(s, t).$$

Proof. If $j \in N$,

$$E[y_t(x) y_s(j)] = P[\sigma_x \leq t, \tau_0(j) \leq s]$$

$$= P[\tau_0(j_1) + \tau_{j_1}(j_2) + \ldots + \tau_{(j_1, \ldots, j_{n-1})}(j_n) \leq t, \tau_0(j) \leq s]$$

$$= \int_0^t E[y_{t-u}(j_2, \ldots, j_n)] \gamma_{j_1 j}(du, s).$$

Thus

$$\sum_{j=1}^{\infty} E[y_t(x)\, y_s(j)] = \int_0^t E[y_{t-u}(j_2, \ldots, j_n)]\, \gamma_{j_1}(du, s).$$

But if $x' \in N^{k-1}$, $1 < k \leq n$, then similarly

$$\sum_{j=1}^{\infty} E[y_t(x)\, y_s(j, x')] = \int_{\substack{0 \leq u \leq t \\ 0 \leq v \leq s}} E[y_t(j_2, \ldots, j_n)\, y_{s-v}(x')]\, \gamma_{j_1}(du, dv)$$

and the first assertion follows by induction. Since

$$E[y_t(i)\, y_s(j)] = \gamma_{ij}(t, s)$$

the second one follows, too. $\qquad\square$

After this drill we are ready for a theorem:

Theorem (6.3.6)
   If $E[\xi^2(0)] < 1$ *and* $E[\xi^2(t)]$ *is finite for all* $t \geq 0$, *then* $c_u(t) = E[z_t z_{t+u}] \leq$ $\leq E[y_t y_{t+u}] < \infty$ *for* $u, t \geq 0$.

Proof. By the Schwarz inequality it is enough to show the finiteness of $E[y_t^2]$. But the latter satisfies, with $n(x) = n \Leftrightarrow x \in N^n$,

$$E[y_t^2] = \sum_{x, x' \in I} E[y_t(x)\, y_t(x')] = E[y_t(0)\, y_t] + \sum_{x \neq 0} \Big\{ E[y_t(x)\, y_t(0)]$$

$$+ \sum_{k=1}^{n(x)-1} \sum_{x' \in N^k} E[y_t(x)\, y_t(x')] + \sum_{k \geq n(x)} \sum_{x' \in N^k} E[y_t(x)\, y_t(x')] \Big\}$$

The lemmata yield

$$E[y_t^2] = 2E[y_t] - 1$$

$$+ \sum_{n=1}^{\infty} \sum_{j_1, \ldots, j_n} \Big\{ \sum_{k=1}^{n-1} \int_0^t E[y_{t-u}(j_{k+1}, \ldots, j_n)]\, \gamma_{j_1} * \ldots * \gamma_{j_k}(du, t)$$

$$+ \int_0^t \gamma_{j_1} \ldots \gamma_{j_n}(t, t-u) \sum_{j=0}^{\infty} \mu^{*j}(du) \Big\}$$

$$= 2E[y_t] - 1 + \sum_{n=1}^{\infty} \sum_{k=1}^{n-1} \sum_{x \in N^{n-k}} \int_0^t E[y_{t-u}(x)]\, \gamma^{*k}(du, t)$$

$$+ \int_0^t \sum_{n=1}^{\infty} \gamma^{*n}(t, t-u) \sum_{j=0}^{\infty} \mu^{*j}(du) = 2E[y_t] - 1 + \int_0^t E[y_{t-u}] \sum_{k=1}^{\infty} \gamma^{*k}(du, t)$$

$$+ \int_0^t \sum_{n=1}^{\infty} \gamma^{*n}(t, t-u) \sum_{j=0}^{\infty} \mu^{*j}(du).$$

Since $\xi(0)$ is integervalued, $E[\xi^2(0)] < 1$ implies that so is $E[\xi(0)]$ (which besides is assumed throughout), $E[y_t] < \infty$ and the finiteness of

$$\sum \gamma^{*n}(t, s)$$

follows like in Lemma (5.2.1).   □

Example (6.3.7)
In the Bellman–Harris case

$$E[\xi^2(t)] = (\sigma^2 + m^2) L(t)$$

where σ^2 is the reproduction variance of the imbedded Galton–Watson process. For binary splitting with cell death frequency $1 - p$ and generation time distribution G

$$E[\xi^2(t)] = 4pG(t) < \infty.$$ □

Like the means m_t^a the second moments

$$c_u^{a,b}(t) = E[z_t^a z_{t+u}^b]$$

satisfy a convolution equation. It is an abominable one, though, with a still worse proof, and so deferred to a section of its own.

Background
The material in this section can be found in Crump's and Mode's [1] or my own paper [2] [except the new Theorem (6.3.1)].

6.4 MORE ABOUT THE SECOND MOMENTS

Consider first the case where never more than one individual can be born at any one instant. Then Equation (6.1.4) holds and

$$c_u^{a,b}(t) = E[z_t^a z_{t+u}^b] = E[z_t^a(0)\, z_{t+u}^b(0)] + E\left[z_t^a(0) \int_0^{t+u} z_{t+u-v}^b \circ S_{\xi_0(v)}\xi_0(dv)\right]$$

$$+ E\left[z_{t+u}^b(0) \int_0^t z_{t-v}^a \circ S_{\xi_0(v)}\xi_0(dv)\right]$$

$$+ E\left[\int_{\substack{0 \le v \le t \\ 0 \le w \le t+u \\ v \ne w}} z_{t-v}^a \circ S_{\xi_0(v)} z_{t+u-w}^b \circ S_{\xi_0(w)}\xi_0(dv)\, \xi_0(dw)\right]$$

$$+ E\left[\int_0^t z_{t-v}^a \circ S_{\xi_0(v)} z_{t+u-v}^b \circ S_{\xi_0(v)}\xi_0\{v\}\, \xi_0(dv)\right]. \qquad (6.4.1)$$

Write

$$\mu_t(v) = E\big[\xi(v); \lambda > t\big] \qquad (6.4.2)$$

and note that $\xi\{v\} = 1$ in the points where ξ places any mass. It follows that

$$c_u^{a,b}(t) = 1_{[0,a)}(t)\,1_{[0,b)}(t + u)\,\{1 - L(t + u)\} + 1_{[0,a)}(t)\int_0^{t+u} m_{t+u-v}^b\mu_t(dv)$$

$$+ 1_{[0,b)}(t + u)\int_0^t m_{t-v}^a\mu_{t+u}(dv) + \int_{\substack{0 \le v \le t \\ 0 \le w \le t+u}} m_{t-v}^a m_{t+u-w}^b\gamma(dv,\,dw)$$

$$- \int_0^t m_{t-v}^a m_{t+u-v}^b\mu(dv) + \int_0^t c_u^{a,b}(t - v)\,\mu(dv). \qquad (6.4.3)$$

Here we have used also that

$$\int_{\substack{0 \le v \le t \\ 0 \le w \le t+u \\ v = w}} m_{t-v}^a m_{t+u-w}^b\gamma(dv,\,dw) = \int_0^t m_{t-v}^a m_{t+u-v}^b E\big[\xi\{v\}\,\xi(dv)\big]$$

$$= \int_0^t m_{t-v}^a m_{t+u-v}^b\mu(dv).$$

Observe that Equation (6.4.1) is of the compact form

$$c_u^{a,b} = r_u^{a,b} + c_u^{a,b} * \mu$$

where $r_u^{a,b}$ does not involve higher moments of the process. The fact that general reproduction processes can be approximated by processes without multiple points makes it clear, that Equation (6.4.3) should hold generally:

Theorem (6.4.1)

If $E\big[\xi^2(0)\big] < 1$ and $E\big[\xi^2(t)\big] < \infty$ for $t \ge 0$, then $c_u^{a,b}$ satisfies Equation (6.4.3).

Proof. In the general case we have to rely upon Equation (6.1.2), obtaining

$$c_u^{a,b}(t) = E\big[z_t^a(0)\,z_{t+u}^b(0)\big] + \sum_{n=1}^\infty E\big[z_t^a(0)\,z_{t+u}^{(n)b}\big] + \sum_{n=1}^\infty E\big[z_{t+u}^b(0)\,z_t^{(n)a}\big]$$

$$+ \sum_{i \ne j} E\big[z_t^{(i)a}z_{t+u}^{(j)b}\big] + \sum_{n=1}^\infty E\big[z_t^{(n)a}z_{t+u}^{(n)b}\big].$$

The first of these terms was analyzed above. In the second (and similarly the third) one

$$E\left[z_t^a(0)\, z_{t+u}^{(n)b}\right] = 1_{[0,a)}(t) \int_0^{t+u} m_{t+u-v}^b P\left[\tau(n) \in dv, \lambda > t\right],$$

by Lemma (6.1.1). Since

$$\sum_{n=1}^{\infty} P\left[\tau(n) \le v, \lambda > t\right] = \sum_{n=1}^{\infty} P\left[\xi(v) \ge n, \lambda > t\right] = E\left[\xi(v); \lambda > t\right] = \mu_t(v),$$

these two terms equal

$$1_{[0,a)}(t) \int_0^{t+u} m_{t+u-v}^b \mu_t(dv)$$

and

$$1_{[0,b)}(t+u) \int_0^t m_{t-v}^a \mu_{t+u}(dv),$$

respectively. If $i \ne j$

$$E\left[z_t^{(i)a} z_{t+u}^{(j)b}\right] = \int_{\substack{0 \le v \le t \\ 0 \le w \le t+u}} m_{t-v}^a m_{t+u-w}^b P\left[\tau(i) \in dv, \tau(j) \in dw\right].$$

This time

$$\sum_{i,j} P\left[\tau(i) \le v, \tau(j) \le w\right] = \sum_{i,j} P\left[\xi(v) \ge i, \xi(w) \ge j\right] = \gamma(v, w),$$

whereas

$$\sum_{n=1}^{\infty} P\left[\tau(n) \le v, \tau(n) \le w\right] = \sum_{n=1}^{\infty} P\left[\xi(v \wedge w) \ge n\right] = \mu(v \wedge w).$$

If $\mu_0([0,v] \times [0,w]) = \mu(v \wedge w)$ is extended to a measure, this one will satisfy $\mu_0(A \times B) = \mu(A \cap B)$ for all intervals A, B. Hence

$$\int 1_{A \times B}(v, w)\, \mu_0(dv, dw) = \int 1_{A \cap B}(v)\, \mu(dv) = \int 1_{A \times B}(v, v)\, \mu(dv).$$

By Dynkin's theorem* any bounded measurable function of two real variables can replace $1_{A \times B}$ in this relation. We obtain

$$\sum_{i \ne j} E\left[z_t^{(i)a} z_{t+u}^{(j)b}\right] = \int_{\substack{0 \le v \le t \\ 0 \le w \le t+u}} m_{t-v}^a m_{t+u-w}^b \gamma(dv, dw) - \int_0^t m_{t-v}^a m_{t+u-v}^b \mu(dv).$$

Finally,

$$E\left[z_t^{(n)a} z_{t+u}^{(n)b}\right] = \int_0^t c_u^{a,b}(t-v)\, P\left[\tau_0(n) \in dv\right].$$

Summation over $n \in N$ concludes the proof. $\square$

* See Appendix.

Corollary (6.4.2)

For splitting processes Equation (6.4.3) *reduces to*

$$c_u^{a,b}(t) = 1_{[0,a)}(t)\, 1_{[0,b)}(t+u)\{1 - L(t+u)\} + 1_{[0,a)}(t) \int_t^{t+u} m(v)\, m_{t+u-v}^b L(dv)$$

$$+ \int_0^t f_u''(1)\, m_{t-v}^a m_{t+u-v}^b L(dv) + \int_0^t m(v)\, c_u^{a,b}(t-v)\, L(dv). \qquad (6.4.4)$$

Proof. We have already seen that $\mu(dv) = m(v)\, L(dv)$. Further

$$\gamma(v,w) = E[\xi(v)\,\xi(w)] = \int_0^{v \wedge w} \{f_u''(1) + m(u)\}\, L(du). \qquad \square$$

Finally the asymptotics:

Theorem (6.4.3)

Consider a non-lattice process with finite reproduction variance $\sigma^2 = $
$= \text{Var}[\xi(\infty)]$. *Then,*

$$\lim_{t \to \infty} c_u(t) = 0,$$

if $m < 1$, *and*

$$c_u^{a,b}(t) \sim t\sigma^2 \int_0^a \{1 - L(v)\}\, dv \int_0^b \{1 - L(v)\}\, dv \Big/ \int_0^\infty v\mu(dv),$$

if $m = 1$ *and* $0 \le a, b < \infty$. *If further* $\int tL(dt) < \infty$, *this is also true for a or*
$b = \infty$. *When* $m > 1$ *and* α *is the Malthusian parameter, then for* $0 \le a, b \le \infty$

$$c_u^{a,b}(t) \sim (\vartheta - \varphi)\, e^{2\alpha t + \alpha u} \int_0^a e^{-\alpha v} \{1 - L(v)\}\, dv \int_0^b e^{-\alpha v} \{1 - L(v)\}\, dv/(1 - \varphi),$$

as $t \to \infty$. *Here*

$$\vartheta = \int e^{-\alpha(u+v)}\, \gamma(du, dv),$$

$$\varphi = \hat\mu(2\alpha),$$

and $\vartheta \ge 1 > \varphi$ *with the first inequality also strict provided* $\xi(\alpha)$ *is not a.s.* $= 1$.

Proof. Consider Equation (6.4.3) and Theorem (6.3.3).
   (i) $m < 1$. Theorem (5.2.9) ends the case.
   (ii) $m = 1$. The appropriate reference is Corollary (5.2.14).
   (iii) $m > 1$. This uses Theorem (5.2.9) after multiplication of Equation

(6.4.3) by $e^{-2\alpha t - \alpha u}$, which yields a defective renewal equation,

$$e^{-2\alpha t - \alpha u} c_u^{a,b}(t) = e^{-2\alpha t - \alpha u} r_u^{a,b}(t) + \int_0^t e^{-2\alpha(t-v) - \alpha u} c_u^{a,b}(t-v) e^{-2\alpha v} \mu(dv).$$

The equation is defective since $\varphi = \hat{\mu}(2\alpha) < \hat{\mu}(\alpha) = 1$. The asymptotics of $r_u^{a,b}$ are left for the reader.

It remains to check the inequality between ϑ and 1:

$$\vartheta = E\left[\int e^{-\alpha(u+v)} \xi(du)\, \xi(dv)\right] = E\left[\left\{\int_0^\infty e^{-\alpha u} \xi(du)\right\}^2\right]$$

$$\geq \left\{E\left[\int_0^\infty e^{-\alpha u} \xi(du)\right]\right\}^2 = \{\hat{\mu}(\alpha)\}^2 = 1 > \varphi,$$

where the Schwarz inequality is strict exactly for non-constant $\hat{\xi}(\alpha)$. ☐

Note. *The reader can check that the requirement of a finite reproduction variance can be weakened somewhat in the supercritical case to $E[\hat{\xi}^2(\alpha)] < \infty$.*

Background

This is from Reference [2], the note being pointed out by Reference [6].

6.5 THE EXTINCTION PROBABILITY

To be able to talk about extinction, we should convince ourselves that the state zero is indeed absorbing, that is that

$$z_t = 0 \Rightarrow z_{t+u} = 0$$

for all $u \geq 0$. But certainly $z_{t+u}(0) = 1 \Rightarrow z_t(0) = 1$. If $z_{t+u}(j_1, \ldots, j_n) = 1$, this means that

$$\sigma_{(j_1, \ldots, j_n)} \leq t + u < \sigma_{(j_1, \ldots, j_n)} + \lambda_{(j_1, \ldots, j_n)}.$$

Obviously then either $\tau_0(j_1) > t$ (implying that $z_t(0) = 1$), $\sigma_{(j_1, \ldots, j_n)} \leq t$ (implying that $z_t(j_1, \ldots, j_n) = 1$), or for some $1 \leq k < n$

$$\sigma_{(j_1, \ldots, j_k)} \leq t < \sigma_{(j_1, \ldots, j_{k+1})}.$$

By the assumption (6.1.1) of no births after death this implies that a.s.

$$\sigma_{(j_1, \ldots, j_k)} \leq t < \sigma_{(j_1, \ldots, j_k)} + \lambda_{(j_1, \ldots, j_k)},$$

that is that $z_t(j_1, \ldots, j_k) = 1$. Thus, if $z_{t+u}(x) = 1$ for some x, $z_t(x') = 1$ for some x'. In other words, $z_{t+u} > 0 \Rightarrow z_t > 0$.

The *extinction* Q of the process can now be defined like in Section 2.3,

$$Q = \{\exists\, t \in R_+ \;;\; z_t = 0\} = \{\exists\, n \in Z_+ \;;\; z_n = 0\}$$

and

$$q = P(Q) = \lim_{t \to 0} P[z_t = 0] = \lim_{t \to \infty} \varphi_t(0).$$

$$(6.5.1)$$

The limits are monotone. And q is independent of the time structure of the process:

Theorem (6.5.1)

The extinction probability q is the smallest non-negative root of the equation $f(s) = s$. The size of q is determined by $m = f'(1) = \mu(\infty)$ as in Theorem (2.3.1).

Note. *Here and throughout we assume that all life-lengths are finite, $L(\infty) = 1$. Otherwise q would be determined by*

$$q = E[q^{\xi(\infty)}; \lambda < \infty].$$

Proof. Let $t \to \infty$ in Equation (6.3.1) for $a = \infty$ and $s = 0$. Since $z_t(0) \to 0$ a.s.,

$$q = E[q^{\xi(\infty)}] = f(q) \tag{6.5.2}$$

by dominated convergence. If now the imbedded Galton–Watson process $\zeta_n \not\to 0$, then infinitely many individuals are realized and therefore $y_t \to \infty$. But if $z_t \to 0$, say $z_t = 0$ for $t \geq \tau$, random $< \infty$, then $y_t \leq y_\tau < \infty$ by Theorem (6.2.2). Hence $z_t \to 0 \Rightarrow \zeta_n \to 0$ and $q = P[z_t \to 0] \leq P[\zeta_n \to 0]$ which is the smallest solution of the equation according to Theorem (2.3.1). $\square$

The theorem shows that m, which we shall call the *reproduction mean* plays the same crucial rôle here as for Galton–Watson processes. A first indication of this was given already in Theorem (6.3.3).

Theorem (6.5.2)

Except in the degenerate case $P[\xi(\infty) = 1] = 1$,

$$P[z_t \to 0 \text{ or } \infty] = 1.$$

Proof. When $q = 1$, there is nothing to prove. So assume $q < 1$. Call an individual *fecund* if it initiates an infinite line of descent. Let ζ_n^* be the number of *fecund* individuals in the nth generation. Given ζ_n, ζ_n^* is binomial with parameters $1 - q$ and ζ_n. Further, if $\varepsilon > 0$ is any number and X is binomial $1 - q, n$, we can find n_0 such that $P[X \leq (1 - q)\,n/2] < \varepsilon$ for $n \geq n_0$.

Hence, with ε given, and k large enough,

$$P[\zeta_n^* \leq k] = P[\zeta_n^* \leq k, \zeta_n \leq 2k/(1-q)] + P[\zeta_n^* \leq k, \zeta_n > 2k/(1-q)]$$
$$\leq P[\zeta_n \leq 2k/(1-q)] + \varepsilon P[\zeta_n > 2k/(1-q)],$$
$$\limsup_{n \to \infty} P[\zeta_n^* \leq k] \leq q + \varepsilon(1-q).$$

Obviously the latter relation holds also for small k and since ε was arbitrary it is true for any $k \in Z_+$ that

$$\limsup_{n \to \infty} P[\zeta_n^* \leq k] \leq q.$$

However ζ_n^* does not decrease. Therefore $P[\zeta_n^* \to \infty] = 1 - q$. (This yields a new proof of Theorem (2.3.2), since $\zeta_n \geq \zeta_n^*$.)

Thus, it is enough to prove that $\zeta_n^* \to \infty \Rightarrow z_t \to \infty$. But if k is any number choose $v = \inf\{n; \zeta_n^* \geq k\}$, finite on $\{\zeta_n^* \to \infty\}$. The random variable ζ_v^* is finite and so all individuals that contribute to it have been born after some finite random time τ. Obviously $z_t \geq k$ for $t \geq \tau$. ☐

We end this section by investigating the asymptotics of φ_t, as $t \to \infty$. Clearly, by Theorem (6.5.2) and dominated convergence,

$$\varphi_t(s) = E[s^{z_t}] \to q$$

This is full analogy with the Galton–Watson case. However for general processes it is not necessarily true that the convergence is monotone. We can prove the following, though:

Theorem (6.5.3)
For a Bellman–Harris process with $L(0) = 0$, $s \geq \varphi_t(s) \downarrow q$ or $s \leq \varphi_t(s) \uparrow q$, as $t \to \infty$, according as $s \geq q$ or $s \leq q$.

We start by proving a lemma.

Lemma (6.5.4)
If $L(0) = 0$ in a Bellman–Harris process, $s \leq \varphi_t(s) \leq q$ or $q \leq \varphi_t(s) \leq s$.

Proof. Suppose that $0 \leq s \leq q$ and take $\varepsilon > 0$. Set $t_0 = \inf\{t; \varphi_t(s) \leq s - \varepsilon\}$. We wish to prove that $\varphi_t(s) > s - \varepsilon$ for all t, or equivalently that $t_0 = \infty$. As z_t is right-continuous, so is the function

$$t \to \varphi_t(s) = E[s^{z_t}]$$

by dominated convergence. Hence $\varphi_{t_0}(s) \leq s - \varepsilon$ and $L(t_0)$ must be > 0.

But if $t_0 < \infty$ Equation (6.3.3) yields

$$s - \varepsilon \geq \varphi_{t_0}(s) = s\{1 - L(t_0)\} + \int_0^{t_0} f \circ \varphi_{t_0-u}(s)\, L(du) > s\{1 - L(t_0)\}$$

$$+ f(s - \varepsilon)\, L(t_0) > s\{1 - L(t_0)\} + (s - \varepsilon)\, L(t_0) \geq s - \varepsilon.$$

This contradiction shows that $t_0 = \infty$. On the other hand, if $t_1 = \inf\{t; \varphi_t(s) \geq q\}$ and $0 \leq s < q$, then $L(t_1) > 0$ and

$$q \leq \varphi_{t_1}(s) < s\{1 - L(t_1)\} + qL(t_1) \leq q.$$

Hence not only is $\varphi_t(s) \geq s$ but also $\varphi_t(s) < q$ for $0 \leq s < q$, the first inequality proved to hold also for $s = q$. However, then by continuity $\varphi_t(q) = q$. For $s \geq q$, finally, $\varphi_t(s) \geq \varphi_t(q) = q$ and an argument like the one given shows that $\varphi_t(s) \leq s$. □

Proof of the theorem. Fix $s \leq q$ and put

$$M(u) = \sup_{0 \leq t \leq u} \varphi_t(s).$$

Equation (6.3.3) can be written, $0 \leq t \leq u$,

$$\varphi_t(s) = s + \int_0^t \{f \circ \varphi_{t-v}(s) - s\}\, L(dv) \leq s + \int_0^t \{f \circ M(u - v) - s\}\, L(dv)$$

$$\leq s + \int_0^u \{f \circ M(u - v) - s\}\, L(dv),$$

since $f \circ M(u - v) \geq f \circ \varphi_{t-v}(s) \geq f(s) \geq s$ by Lemma (6.5.6). Taking the supremum at the left yields

$$M(u) \leq s\{1 - L(u)\} + \int_0^u f \circ M(u - v)\, L(dv).$$

Now define $M_0 = M$ and recursively $\{M_n\}_0^\infty$ by

$$M_{n+1}(u) = s\{1 - L(u)\} + \int_0^u f \circ M_n(u - v)\, L(dv)$$

By induction $\varphi_u(s) \leq M(u) \leq M_n(u) \leq M_{n+1}(u) \leq 1$ for all n. Therefore $M_\infty = \lim_{n+\infty} M_n$ must exist and solve Equation (6.3.3) (for $a = \infty$). By Theorem (6.3.1) $M_\infty(u) = \varphi_u(s)$ and since $\varphi_u(s) \leq M(u) \leq M_\infty(u)$ we have shown that $\varphi_u(s)$ increases in u for $s \leq q$.

In the case $s > q$, work analogously with $I(u) = \inf_{0 \leq t \leq u} \varphi_u(s)$. □

Background

In the preceding Theorem (6.5.1) is from References [1] or [2], the monotonicity theorem and the lemma from Reference [7]. Theorem (6.5.2) is new. 'Fecund' is Kingman's terminology.

6.6 CRITICAL PROCESSES

A generalized branching process was defined as critical if $m = \mu(\infty) = E[\xi(\infty)] = 1$. In this section we establish counterparts to the results for critical Galton–Watson processes. Unfortunately the results are not complete.* What complicates the investigation is, essentially, that mothers can live long after giving their last birth and that the reproduction may depend on the life-length. The variability of the latter is more easily taken care of. The idea is to approximate φ_t by a sequence $\{\varphi_t^{(k)}\}$ defined by

$$\varphi_t^{(0)}(s) = s \tag{6.6.1}$$

$$\varphi_t^{(k+1)}(s) = E\left[s^{z_t(0)} \prod_{n=1}^{\xi_0(t)} \varphi_{t-\tau_0(n)}^{(k)}(s) \right].$$

In its turn φ_t^k is not too different from the kth iterate f_k of the reproduction generating function. But the behaviour of f_k is known (Section 2.4). Probabilistically, $\varphi_t^{(k)}$ is the generating function of the number of individuals alive at t in the $k-1$ first generations plus the number of individuals in the kth generation who have been born. In other words, the $\varphi_t^{(k)}(s) = E[s^{z_t^*(k)}]$, where $\{z_t^*(k); t \geq 0\}$ is a process obtained from the original one by freezing at the kth generation. Strictly, with $n(x) = j \Leftrightarrow x \in N^j$

$$z_t^*(k) = \sum_{n(x) < k} z_t(x) + \sum_{x \in N^k} y_t(x). \tag{6.6.2}$$

From this, clearly

$$\lim_{t \to \infty} z_t^*(k) = \sum_{x \in N^k} 1_{[0,\infty)} \circ \sigma_x = \zeta_k \tag{6.6.3}$$

whereas

$$\lim_{k \to \infty} z_t^*(k) = \sum_{x \in I} z_t(x) = z_t. \tag{6.6.4}$$

Thus

$$\lim_{t \to \infty} \varphi_t^{(k)} = f_k \tag{6.6.5}$$

and

$$\lim_{k \to \infty} \varphi_t^{(k)} = \varphi_t. \tag{6.6.6}$$

* At the end of the section some recent, more exhaustive results will be outlined.

We shall go deeper into these convergences by analytic tools, seeing that if the argument $s = 0$ the situation is considerably simpler than generally. The first three lemmata compare φ_t with $\varphi_t^{(k)}$, the subsequent ones $\varphi_t^{(k)}$ with f_k.

Lemma (6.6.1)

It holds that

$$\varphi_t(s) - \varphi_t^{(k)}(s) \le (1 - s)\,\mu^{*k}(t)$$

and

$$\left|\varphi_t(s) - \varphi_t^{(k)}(s)\right| \le \mu^{*k}(t) \quad \text{for all } k \in Z_+.$$

Proof. We shall use, here and in the sequel, the relations

$$\prod_{n=1}^{k} a_n - \prod_{n=1}^{k} b_n \le c^{k-1} \sum_{n=1}^{k} c_n \tag{6.6.7}$$

valid for $k \in Z_+$, $0 \le a_n, b_n \le c$, $a_n - b_n \le c_n$, which is ≥ 0. In particular c_n can be chosen as $(a_n - b_n)^+$ or $|a_n - b_n|$. Note first that

$$\varphi_t(s) - \varphi_t^{(0)}(s) = \varphi_t(s) - s \le 1 - s$$

Use Equation (6.6.1) to see that

$$\varphi_t(s) - \varphi_t^{(k+1)}(s) = E\left[s^{z_t(0)} \left\{ \prod_{n=1}^{\xi_0(t)} \varphi_{t-\tau_0(n)}(s) - \prod_{n=1}^{\xi_0(t)} \varphi_{t-\tau_0(n)}^{(k)}(s) \right\} \right]$$

$$\le E\left[s^{z_t(0)} \sum_{n=1}^{\xi_0(t)} (1 - s)\,\mu^{*k}\big(t - \tau_0(n)\big) \right]$$

$$= E\left[s^{z_t(0)} \int_0^t (1 - s)\,\mu^{*k}(t - u)\,\xi(du) \right] \le (1 - s)\,\mu^{*(k+1)}(t),$$

if only

$$\varphi_t(s) - \varphi_t^{(k)}(s) \le (1 - s)\,\mu^{*k}(t).$$

Induction completes the proof of the former assertion. The latter follows analogously from

$$\left| \prod_{n=1}^{k} a_n - \prod_{n=1}^{k} b_n \right| \le \sum_{n=1}^{k} |a_n - b_n|, \tag{6.6.8}$$

valid provided $0 \le |a_n|, |b_n| \le 1$. □

Lemma (6.6.2)

   For all $k \in Z_+$

$$\varphi_t(0) \geq \varphi_t^{(k)}(0).$$

Proof.

$$\varphi_t^{(1)}(0) = P[z_t(0) + \xi_0(t) = 0] \geq 0 = \varphi_t^{(0)}(0).$$

By induction it follows that $\varphi_t^{(k)}(0)$ does not decrease in k. By Equation (6.6.6) [or Lemma (6.6.1)], $\varphi_t^{(k)}(0) \uparrow \varphi_t(0)$ and so $\varphi_t(0) \geq \varphi_t^{(k)}(0)$. A probability argument is still shorter: Evidently $z_t^*(k) = 0 \Rightarrow z_t = 0$ and so $P[z_t = 0] \geq \geq P[z_t^*(k) = 0]$. $\square$

Lemma (6.6.3)

   If $L \geq \mu$, then

$$\varphi_t \geq \varphi_t^{(k)}$$

for all t and k.

Proof. Since for all $n \in Z_+$ and $0 \leq s \leq 1$

$$1 - s^n \leq (1 - s)\, n,$$

$$\varphi_t^{(1)}(s) = E[s^{z_t(0) + \xi_0(t)}] \geq E[1 - (1 - s)\{z_t(0) + \xi_0(t)\}]$$

$$= s + (1 - s)\{L(t) - \mu(t)\} \geq s = \varphi_t^{(0)}(s).$$

Invoking induction and Equation (6.6.6) again, we conclude that $\varphi_t^{(k)} \uparrow \varphi_t$.

$\square$

Lemma (6.6.4)

   For any process and $k \in Z_+$

$$\varphi_t^{(k)}(0) \leq f_k(0).$$

Proof. The relation being true for $k = 0$, assume it to hold for a fixed k. Then

$$\varphi_t^{(k+1)}(0) = E\left[\prod_{n=1}^{\xi(t)} \varphi_{t-\tau(n)}^{(k)}(0); \lambda \leq t\right] \leq E[\{f_k(0)\}^{\xi(t)}; \lambda \leq t]$$

$$= E[\{f_k(0)\}^{\xi(\infty)}; \lambda \leq t] \leq f_{k+1}(0).   \square$$

Lemma (6.6.5)

   If $m = E[\xi(\infty)] = 1$ but for any $t \geq 0$ $E[\xi(t); \lambda \leq t] \geq L(t)$, then for $k \in Z_+$

$$\varphi_t^{(k)} \leq f_k.$$

Note. *These awkward conditions are satisfied by critical Bellman–Harris processes.*

Proof. Obviously

$$\varphi_t^{(0)} = f_0.$$

Assuming $\varphi_t^{(k)} \le f_k$ and using $s \le f_k(s)$ (since $m = 1$), we continue to see that

$$\varphi_t^{(k+1)}(s) \le E\left[s^{z_t(0)}\{f_k(s)\}^{\xi_0(t)}\right] \le E\left[\{f_k(s)\}^{z_t(0)+\xi_0(t)}\right]$$

$$= f_{k+1}(s) + E\left[\{f_k(s)\}^{z_t(0)+\xi_0(t)} - \{f_k(s)\}^{\xi_0(\infty)}\right]$$

$$= f_{k+1}(s) + E\left[\{f_k(s)\}^{z_t(0)+\xi_0(t)} - \{f_k(s)\}^{\xi_0(\infty)}; \lambda_0 > t\right]$$

$$\le f_{k+1}(s) + E\left[f_k(s) - \{f_k(s)\}^{\xi(\infty)}; \lambda > t\right]$$

$$\le f_{k+1}(s) + \{1 - f_k(s)\}\, E\left[\xi(\infty) - 1; \lambda > t\right] \le f_{k+1}(s).$$

Here we used that $s - s^n \le (1 - s)(n - 1)$ for $0 \le s \le 1$, $n \in Z_+$ and that

$$E[\xi(\infty); \lambda > t] = E[\xi(\infty)] - E[\xi(\infty); \lambda \le t]$$

$$= 1 - E[\xi(t); \lambda \le t] \le 1 - L(t)$$

by assumption.   □

Lemma (6.6.6)

For all k, t and s

$$f_k(s) - \varphi_t^{(k)}(s) \le (1 - s)(1 - L) * \sum_{n=0}^{k-1} \mu^{*n}(t).$$

Proof. Since $\varphi_t^{(0)} = f_0$ the relation holds for $k = 0$ ($\sum_0^{-1}$ as usual interpreted as zero). Assuming it for k, we obtain

$$f_{k+1}(s) - \varphi_t^{(k+1)}(s) = E\left[s^{z_t(0)}\left\{(f_k(s))^{\xi_0(t)} - \prod_{n=1}^{\xi_0(t)} \varphi_{t-\tau_0(n)}^{(k)}(s)\right\}\right]$$

$$+ E\left[\{f_k(s)\}^{\xi_0(\infty)} - s^{z_t(0)}\{f_k(s)\}^{\xi_0(t)}\right]$$

$$\le (1 - s)(1 - L) * \sum_{n=1}^{k} \mu^{*n}(t)$$

$$+ E\left[\{f_k(s)\}^{\xi_0(\infty)}\{1 - s^{z_t(0)}\}\right]$$

$$\le (1 - s)(1 - L) * \sum_{n=1}^{k} \mu^{*n}(t) + (1 - s)\{1 - L(t)\}.   □$$

Let us sum the preceding lemmata up:

Theorem (6.6.7)

If either $L \geq \mu$, $m = 1$, and $E[\xi(t); \lambda \leq t] \geq L(t)$ for all $t \geq 0$, or just $s = 0$, then it holds for all k and t that

$$- (1 - s) \mu^{*k}(t) \leq f_k(s) - \varphi_t(s) \leq (1 - s)(1 - L) * \sum_{n=0}^{k-1} \mu^{*n}(t).$$

Note that in case the first two conditions are satisfied, then

$$L(t) \leq E[\xi(t); \lambda \leq t] \leq E[\xi(t)] = \mu(t) \leq L(t).$$

Hence $L = \mu$, $m = \mu(\infty) = 1$, and the right hand side of the asserted inequality reduces to $(1 - s)\{1 - \mu^{*k}(t)\}$. But further, since $E[\xi(t); \lambda \leq t] = E[\xi(t)]$ for all t, $\xi(\infty) = \xi(\{\lambda\})$ a.s., i.e. the process is of the splitting type and by $L = \mu$ indeed a critical Bellman–Harris process.

The subsequent analysis relies heavily upon a strong form of the weak law of large numbers [8].

Theorem (6.6.8)

Assume that $\{x_k\}_1^\infty$ is a sequence of i.i.d. random variables with expectation zero. Let $r \geq 0$ and denote the partial sums by $s_n = \sum_{k=1}^n x_k$. Then

$$n^{1+r} P[|x_1| > n] \to 0 \Leftrightarrow \forall \varepsilon > 0 : n^r P[|s_n| > \varepsilon n] \to 0.$$

Proof. We shall only deduce the implication needed in our context, from the left to the right. The converse can be found in Reference [8]. By the well known symmetrization inequalities (page 147 of Reference [10]) it is enough to give the proof for symmetric x_1. Thus, assume x_1 to be symmetric and such that $n^{r+1} P[|x_1| > n] \to 0$. Write t_n for the truncation function, $t_n(x) = -n$ for $x < -n$, $t_n(x) = x$, $|x| \leq n$, and $t_n(x) = n$ for $x > n$. Since

$$n^r P[|s_n| > \varepsilon n] \leq n^r P\left[\bigcup_{k=1}^n \{|x_k| > \varepsilon n\}\right] + n^r P\left[\left|\sum_{k=1}^n t_n \circ x_k\right| > \varepsilon n\right]$$

$$\leq n^{r+1} P[|x_1| > \varepsilon n] + n^r P\left[\left|\sum_{k=1}^n t_n \circ x_k\right| > \varepsilon n\right]$$

where the first term tends to zero, it is enough to show that so does the latter. Take j to be an even integer $> 2r + 1$. By Markov's inequality

$$n^r P\left[\left|\sum_{k=1}^n t_n \circ x_k\right| > \varepsilon n\right] \leq n^{r-j} \varepsilon^{-j} E\left[\left(\sum_{k=1}^n t_n \circ x_k\right)^j\right]$$

$$\leq c \sum n^{r+m-j} \prod_{i=1}^m E[(t_n \circ x_1)^{k_i}]$$

for some constant c. Here the summation is over all vectors $(k_1, \ldots, k_m)$ of positive even numbers with sum j, $1 \le m \le j/2$. By symmetry all odd moments are zero. The factor n^m bounds the number of ways m labels can be distributed over n places. However for any $k \in N$

$$E[|t_n \circ x_1|^k] \le n^k P[|x_1| \ge n] + k \int_{-n}^{n} u^{k-1} P[|x_1| > u]\, du$$

after an integration by parts. Hence $E[|t_n \circ x_1|^k]$ remains bounded if $k < r + 1$, is $o(\log n)$ if $k = r + 1$, and is $o(n^{k-r-1})$ if $k > r + 1$. Consider some term

$$n^{r+m-j} \prod_{i=1}^{m} E[(t_n \circ x_1)^{k_i}] = n^{r+m-j} o\left(n^{\sum_{k_i > r+1} (k_i - r - 1)} \right) o(\log^v n),$$

where v is the number of k_i equalling $r + 1$ and the first ordo term should be interpreted as $O(1)$ if the sum in the exponent is empty. If further

$$u = \sum_{0 < k_i < r+1} (k_i - 1)$$

and

$$w = \text{no. of } k_i > r + 1,$$

we note that

$$\sum_{k_i > r+1} (k_i - r - 1) = j - u - m - (v + w)r$$

and so

$$n^{r+m-j} \prod_{i=1}^{m} E[(t_n \circ x_1)^{k_i}] = O(n^{-u})\, o(n^{r(1-v-w)} \log^v n)$$

as $n \to \infty$. Here $u \ge 0$ and thus if $w \ge 1$ the expression tends to zero. If $w = 0$, the sum $\sum_{k_i > r+1} (k_i - r - 1)$ is empty and thus directly

$$n^{r+m-j} \prod_{i=1}^{m} E[(t_n \circ x_1)^{k_i}] = n^{r+m-j} o(\log^v n) = n^{r+m-j} o(\log^m n)$$

since $v \le m$. But as $m \le j/2$ also this tends to zero, as $n \to \infty$. $\qquad\square$

Corollary (6.6.9)

Let G be a distribution function on R_+. Assume that

$$\int_0^{\infty} t G(dt) = \beta > 0 \tag{6.6.9}$$

$$\lim_{t \to \infty} t^{1+r} \{1 - G(t)\} = 0 \tag{6.6.10}$$

for some $r \geq 0$. *Let* $0 < \varepsilon < 1$ *and*

$$n_t = [t(1 + \varepsilon)/\beta]$$
$$k_t = [t(1 - \varepsilon)/\beta].$$

Then,

$$\lim_{t \to \infty} t^r G^{*n_t}(t) = 0. \tag{6.6.11}$$

$$\lim_{t \to \infty} t^r \{1 - G^{*k_t}(t)\} = 0. \tag{6.6.12}$$

Proof. If $x_k + \beta$ are independent with distribution G, $k \in N$, it holds for any $\delta > 0$ that

$$n^r G^{*n}(n\beta - n\delta) = n^r P[s_n \leq -n\delta] \to 0$$

by the theorem. Similarly

$$n^r \{1 - G^{*n}(n\beta + n\delta)\} \to 0.$$

To prove Equation (6.6.11) it is therefore sufficient to show that for some $\delta > 0$, $t \leq n_t(\beta - \delta)$, for t large at least. But since $n_t(\beta - \delta)/t \to \to (1 + \varepsilon)(\beta - \delta)/\beta > \beta$ for $0 < \delta < \beta/(1 + \varepsilon)$, this is evident. To deduce Equation (6.6.12), observe that $t \geq k_t(\beta + \delta)$ for large t if only $0 < \delta < \beta/(1 - \varepsilon)$. Hence,

$$t^r \{1 - G^{*k_t}(t)\} \to 0. \qquad \square$$

Lemma (6.6.10)

With G, ε, and k_t as in the preceding lemma, let f be measurable and satisfy Condition (5.2.7) and $0 \leq f(t) = o(t^{-r-1})$. Then

$$\lim_{t \to \infty} t^r f * \sum_{k=0}^{k_t - 1} G^{*k}(t) = 0.$$

Proof. Let $c = (1 - \varepsilon)/(1 - \varepsilon/2)$. Then $0 < c < 1$ and $k_t = [t(1 - \varepsilon)/\beta] = [(1 - \varepsilon/2) ct/\beta]$ and

$$f * \sum_{k=0}^{k_t - 1} G^{*k}(t) = \sum_{k < k_t} \int_0^{ct} f(t - u) G^{*k}(du) + \sum_{k < k_t} \int_{ct}^t f(t - u) G^{*k}(du).$$

The first sum does not exceed

$$k_t \sup_{u \geq (1-c)t} f(u) = \left\{ t^{r+1} \sup_{u \geq (1-c)t} f(u) \right\} (k_t/t^{r+1}) = o(1) O(t^{-r}) = o(t^{-r}).$$

For the second sum note that with $0 \le t_1 \le t_2$ and any n

$$1 - G^{*n}(t_2) = \sum_{k=0}^{n-1} \{G^{*k}(t_2) - G^{*(k+1)}(t_2)\} = \sum_{k=0}^{n-1} \int_0^{t_2} \{1 - G(t_2 - u)\} G^{*k}(du)$$

$$\ge \sum_{k=0}^{n-1} \int_{t_2-t_1}^{t_2} \{1 - G(t_2 - u)\} G^{*k}(du)$$

$$\ge \{1 - G(t_1)\} \sum_{k=0}^{n-1} \{G^{*k}(t_2) - G^{*k}(t_2 - t_1)\},$$

i.e. that

$$\sum_{k=0}^{n-1} \{G^{*k}(t_2) - G^{*k}(t_2 - t_1)\} \le \{1 - G^{*n}(t_2)\}/\{1 - G(t_1)\}.$$

Further define $b = \beta/2$. This b necessarily satisfies $0 < b$ and $G(b) < 1$. Finally, let $d = [(1 - c) t/b]$ and $f_j = \sup_{jb \le u \le (j+1)b} f(u)$, $0 \le j \le d$. Then, since $t - db \ge ct$, the second sum

$$\sum_{k<k_t} \int_{ct}^{t} f(t - u) G^{*k}(du) \le \sum_{k<k_t} \sum_{j=0}^{d} \int_{t-(j+1)b}^{t-jb} f(t - u) G^{*k}(du)$$

$$\le \sum_{k<k_t} \sum_{j=0}^{d} f_j \{G^{*k}(t - jb) - G^{*k}(t - (j + 1) b)\}$$

$$\le \sum_{j=0}^{d} f_j \sum_{k=0}^{k_t-1} \{G^{*k}(t - jb) - G^{*k}(t - jb - b)\}$$

$$\le \sum_{j=0}^{d} f_j \{1 - G^{*k_t}(t - jb)\}/\{1 - G(b)\}$$

$$\le \sum_{j=0}^{d} f_j \{1 - G^{*k_t}(ct)\}/\{1 - G(b)\}.$$

But

$$t^r \{1 - G^{*k_t}(ct)\} \to 0$$

according to Corollary (6.6.9) whereas

$$\sum_{j=0}^{\infty} f_j < \infty$$

by assumption. $\square$

Theorem (6.6.11)

   Consider a critical process with finite positive reproduction variance σ^2 and such that

$$\lim_{t \to \infty} t^2 \{1 - \mu(t)\} = 0,$$

and

$$\lim_{t \to \infty} t^2 \{1 - L(t)\} = 0.$$

Write

$$\beta = \int_0^\infty t\mu(dt).$$

Then:

   (a) $\lim_{t \to \infty} tP[z_t > 0] = 2\beta/\sigma^2.$

If μ is non-lattice
   (b) $\lim_{t \to \infty} E[z_t/t | z_t > 0] = l\sigma^2/2\beta^2,$

where

$$l = \int_0^\infty tL(dt).$$

If the process is of the Bellman–Harris type
   (c) $\lim_{t \to \infty} P[z_t/t \le u | z_t > 0] = 1 - \exp(-2\beta u/\sigma^2).$

Proof. (a) Let $\varepsilon > 0$. From Theorem (2.4.2) (a) it follows that

$$t\{1 - f_{n_t}(0)\} \to 2\beta/\sigma^2(1 + \varepsilon),$$

as $t \to \infty$, and from the left hand side of Theorem (6.6.7) that

$$\liminf_{t \to \infty} t\{1 - \varphi_t(0)\} - t\{1 - f_{n_t}(0)\} \ge \liminf_{t \to \infty} - t\mu^{*n_t}(t) = 0.$$

Since ε was arbitrary,

$$\liminf_{t \to \infty} t\{1 - \varphi_t(0)\} \ge 2\beta/\sigma^2.$$

Using for fixed $\varepsilon > 0$, the remaining inequality in Theorem (6.6.7), we obtain by Lemma (6.6.10)

$$\limsup_{t \to \infty} t\{1 - \varphi_t(0)\} - t\{1 - f_{k_t}(0)\} \le \limsup_{t \to \infty} t(1 - L) * \sum_{n=0}^{k_t - 1} \mu^{*n}(t) = 0.$$

$$(6.6.14)$$

Since

$$t\{1 - f_{k_t}(0)\} \to 2\beta/\sigma^2(1 - \varepsilon)$$

(a) is proved.

(b) $$E[z_t/t|z_t > 0] = m_t/(tP[z_t > 0]).$$

According to Equation (6.3.4)

$$m_t = 1 - L(t) + \int_0^t m_{t-u}\mu(du)$$

and by the renewal theorem $m_t \to l/\beta$.

(c) First recall that critical Bellman–Harris processes (and only such) satisfy the conditions in Theorem (6.6.7) for arbitrary s. Next, for $u, \varepsilon > 0$

$$t\{f_{n_t}(e^{-u/t}) - f_{n_t}(0)\} = t\{1 - f_{n_t}(0)\}\, E[\exp(-u(n_t/t)(\zeta_{n_t}/n_t))|\zeta_{n_t} > 0]$$

$$\to 2\beta/\{\sigma^2(1 + \varepsilon)(1 + u(1 + \varepsilon)\sigma^2/2\beta)\}$$

and

$$t\{f_{k_t}(e^{-u/t}) - f_{k_t}(0)\} \to 2\beta/\{\sigma^2(1 - \varepsilon)(1 + u(1 - \varepsilon)\sigma^2/2\beta)\},$$

As $\varepsilon > 0$ was again arbitrary

$$\lim_{t\to\infty} E[e^{-uz_t/t}|z_t > 0] = \lim_{t\to\infty} t\{\varphi_t(e^{-u/t}) - \varphi_t(0)\}/t\{1 - \varphi_t(0)\}$$

$$= 1/\{1 + u\sigma^2/2\beta\},$$

involving the same type of arguments as those used to prove (a).   □

The methods used thus work well to catch the extinction probability, but fail without mercy when it comes to establishing exponentiality in the general case. In a recent thesis Holte [13] managed to solve that problem almost completely. I shall sketch his approach:

Lemma (6.6.12)

Consider a critical process with $\beta = \int_0^\infty t\mu(dt) < \infty$ *and* $l = \int_0^\infty tL(dt) < \infty$. *Then, for* $(1 - 1/\sup m_t)^+ \le s \le 1$ *and* $n \in Z_+, t \in R_+$

$$f_n(1 - v(1 - s)) - \varphi_t(s) \le (1 - s)(1 - L) * \sum_{k=0}^{n-1} \mu^{*k}(t)$$

$$+ (1 - s)\int_0^t (m_{t-u} - v)^+ \mu^{*n}(du),$$

where v *is any number between* $\underline{m} = \inf_t m_t$ *and* $\bar{m} = \sup_t m_t$. [Note that the latter is finite by Theorem (6.3.3).] In the Bellman–Harris case $m_t = 1$

for all t and the lemma reduces to the right hand side of Theorem (6.6.7).

The proof starts by defining a sequence $\psi_t^{(n)}$ by the initial condition $\psi_t^{(0)}(s) = 1 - v(1 - s)$ and the recurrence relation in Equation (6.6.1). Induction plus Inequality (6.6.7) first yield that

$$\psi_t^{(n)}(s) - \varphi_t(s) \le (1 - s) \int_0^t (m_{t-u} - v)^+ \, \mu^{*n}(du),$$

then that

$$f_n\big(1 - v(1 - s)\big) - \psi_t^{(n)}(s) \le (1 - s)(1 - L) * \sum_{k=0}^{n-1} \mu^{*k}.$$

Lemma (6.6.13)

Under the circumstances of the preceding lemma, define

$$r_k(t, s) = E\left[s^{z_t(0)} \prod_{j=1}^{\xi_0(t)} f_k\big(1 - m_{t-\tau_0(j)}(1 - s)\big) \right] - f_{k+1}\big(1 - m_t(1 - s)\big)$$

and

$$\xi_n(s, t) = \sum_{k=0}^{n-1} r_k(\cdot, s)^+ * \mu^{*(n-k-1)}(t).$$

In this notation it holds that

$$\varphi_t(s) - f_n\big(1 - m_t(1 - s)\big) \le (1 - s) \int_0^t m_{t-u} \mu^{*n}(du) + \xi_n(s, t).$$

[Again, for Bellman–Harris processes this is nothing but the left hand side of Theorem (6.6.7).]

The proof follows the established pattern this time as well. The next step shows that Taylor's formula applies to yield an estimate of r_k for s close to one:

Lemma (6.6.14)

If things are as in the preceding lemmata and further the reproduction variance $\sigma^2 < \infty$, then for $(1 - v/\overline{m})^+ \le s < 1$

$$r_k(s, t) = (1 - s)^2 \, \partial^2 r_k(\bar{s}, t)/2\partial s^2,$$

where $s < \bar{s} < 1$.

The gist of Holte's method consists in a careful and very cumbersome estimation of $\partial^2 r/\partial s^2$, where Inequalities (6.6.7) and (6.6.8) are used over and over again.

Lemma (6.6.15)

Under the assumptions above

$$\frac{\partial^2 r_k(s,t)}{\partial s^2} \leq (2k\sigma^2 \overline{m}^2/\underline{m}) \int_0^t |m_t - m_{t-u}|\, \mu(\mathrm{d}u)$$

$$+ (6\overline{m}^2/\underline{m})\, E\left[\{\xi(t) - 1\} \int_0^t |m_t - m_{t-u}|\, \xi(\mathrm{d}u)\right]$$

$$+ (\overline{m}^3/\underline{m})\, E[\xi(t)\{\xi(t) + 1\}; \lambda > t].$$

Once this is obtained Lemma (6.6.13) can be rewritten into stating that

$$\varphi_t(s) - f_n(1 - m_t(1 - s)) \leq (1 - s) \int_0^t m_{t-u}\mu^{*n}(\mathrm{d}u)$$

$$+ (1 - s)^2 \left\{ R_1 * \sum_{k=0}^{n-1} k\mu^{*(n-k-1)}(t) + R_2 * \sum_{k=0}^{n-1} \mu^{*k}(t) \right\}, \qquad (6.6.15)$$

where

$$R_1(t) = (\sigma^2 \overline{m}^2/\underline{m}) \int_0^t |m_t - m_{t-u}|\, \mu(\mathrm{d}u)$$

and

$$R_2(t) = (3\overline{m}^2/\underline{m})\, E\left[\{\xi(t) - 1\} \int_0^t |m_t - m_{t-u}|\, \xi(\mathrm{d}u)\right]$$

$$+ (\overline{m}^3/2\underline{m})\, E[\xi(t)\{\xi(\infty) + 1\}; \lambda > t].$$

Here it is assumed that $\sigma^2 < \infty$. Clearly not only $\overline{m} < \infty$ but also $\underline{m} > 0$ at least for critical processes. (Check that!) Again the reader may make sure that in case of a Bellman–Harris branching process the relation $\varphi_t(s) - f_n(s) \leq (1 - s)\, \mu^{*n}(t)$ is recovered (under the new condition $\sigma^2 < \infty$, though). The next step is to obtain some information on R_1 and R_2. The difficult part is

Lemma (6.6.16)

Let μ be a probability measure on R_+ with finite mean $\beta > 0$ and $f(t) = = o(t^{-1})$ a bounded Borel function. Let $\varepsilon > 0$, $n_t = [t(1 + \varepsilon)/\beta]$ and assume that f and μ meet the requirements of the renewal theorem, i.e. either f is directly Riemann integrable or the conditions of Theorem (5.3.4) are fulfilled. Then, with

$$\varphi(t) = \int_0^t |f(t) - f(t - u)|\, \mu(\mathrm{d}u).$$

$$\varlimsup_{t \to \infty} \left\{ \varphi * \sum_{k=1}^{n_t - 1} k\mu^{*(n_t - k - 1)}(t) \right\} \Big/ t \leq 4\varepsilon \int_0^\infty |f(t)|\, \mathrm{d}t/\beta^2.$$

The idea is then, in the non-lattice case, to apply this with $f(t) = m_t - l/\beta$. By Theorem (5.3.6) we know that $m_t - l/\beta = o(t^{-1})$ under natural conditions and by Lemma (5.3.5) that

$$\int_0^\infty (m_t - l/\beta)\,dt$$

converges. However, to make use of the preceding we would need

$$\int_0^\infty |m_t - l/\beta|\,dt < \infty$$

and that $|m_t - l/\beta|$ satisfy Inequality (5.2.7). This is known to be the case only under the not very transparent condition that

$$\lim_{u \to \infty} \left| 1 - \int_0^\infty e^{iut}\mu(dt) \right| > 0 \qquad (6.6.16)$$

[13] and provided μ and L have a finite variance. The remaining estimates are considerably simpler and the result can be summarized in

Theorem (6.6.17)

Consider a critical non-lattice branching process with finite reproduction variance, such that

$$0 < \beta = \int_0^\infty t\mu(dt),$$

$$0 < l = \int_0^\infty tL(dt),$$

$$t^2\{1 - \mu(t)\} \to 0,$$

$$t^2\{1 - L(t)\} \to 0,$$

$$t\,|m_t - l/\beta| \to 0,$$

(by Theorem (5.3.6) this is the case if only L and μ have finite second moments), and $|m_t - l/\beta|$ is directly Riemann integrable (5.2.7) (which is thus true if Equation (6.6.16) holds besides L and μ being of finite variance. Of course it is also true for Bellman–Harris processes where $m_t = l/\beta = 1$). Then,

$$\lim_{t \to \infty} P[z_t/t \le u|z_t > 0] = 1 - e^{2\beta^2 u/\sigma^2 l}.$$

Holte also gives a corresponding result for the lattice case where, interestingly enough, different limits may arise for different sequences of $t \to \infty$. However, if the span of μ divides the span of L the limit remains unique.

Background

The two-step approximation method used here is due to Goldstein [11], who deviced it to treat Bellman–Harris processes. Lemma (6.6.10) is from Holte [12], who also found Theorem (6.6.17), as mentioned in the text. There are other ways of working: Sevast'yanov [14] has a criticality theorem for his splitting model but under harder moment restrictions. Durham [9] has proved the exponential limit law for general process, but using Weiner's moment method, which requires that all moments of the reproduction law be finite. For a good survey of critical branching processes see Reference [15].

6.7 THE SUBCRITICAL CASE

By Theorem (6.3.3) $m_t^a \to 0, 0 \le a \le \infty$, as $t \to \infty$, provided $m < 1$. Throughout the present section we assume this to be the case, but also that the processes considered are Malthusian with parameter α (necessarily < 0). We shall argue only for non-lattice μ. The asymptotic results remain true in the lattice case provided limits are taken only over lattice points.

The latter part of Theorem (5.2.9) yields that m_t tends exponentially to zero. Indeed,

$$\lim_{t \to \infty} e^{-\alpha t} m_t^a = \int_0^a e^{-\alpha t} \{1 - L(t)\} \, dt \Big/ \int_0^\infty t e^{-\alpha t} \mu(dt)$$

provided only the integrand in the numerator satisfies the requirements of Theorem (5.2.6). The a.e. continuity is obvious as well as the complying with the Tauberian Condition (5.2.7) when a is finite. When this is not the case,

$$\sum_{k=0}^\infty \sup_{0 \le t \le 1} e^{-\alpha(k+t)} \{1 - L(k + t)\} \le \sum_{k=0}^\infty e^{-\alpha(k+1)} \{1 - L(k)\}$$

$$\le \sum_{k=0}^\infty \int_{k-1}^k e^{-\alpha(t+2)} \{1 - L(t)\} \, dt = e^{-2\alpha} \int_{-1}^\infty e^{-\alpha t} \{1 - L(t)\} \, dt < \infty$$

$$\Leftrightarrow \int_0^\infty t e^{-\alpha t} L(dt) < \infty,$$

as seen after an integration by parts (of course $L(t) = 0$ for $t < 0$). In conclusion:

Theorem (6.7.1)

In a non-lattice, subcritical process admitting the Malthusian parameter α,

$$m_t^a \sim e^{\alpha t} \int_0^a e^{-\alpha t} \{1 - L(t)\} \, dt \Big/ \int_0^\infty t e^{-\alpha t} \mu(dt) \qquad (6.7.1)$$

for $0 \le a < \infty$, *as* $t \to \infty$. *For* $a = \infty$ *the relation still holds, provided*

$$\int_0^\infty t e^{-\alpha t} L(\mathrm{d}t) < \infty. \tag{6.7.2}$$

If further

$$\beta = \int_0^\infty t e^{-\alpha t} \mu(\mathrm{d}t) < \infty, \tag{6.7.3}$$

then,

$$\lim_{t \to \infty} m_t^a / m_t = \int_0^a e^{-\alpha t} \{1 - L(t)\} \, \mathrm{d}t \Big/ \int_0^\infty e^{-\alpha t} \{1 - L(t)\} \, \mathrm{d}t. \tag{6.7.4}$$

For lattice μ corresponding results hold.

If we write Equation (6.7.1)

$$e^{-\alpha t} m_t^a \to k(a),$$

it is worth bearing in mind that the limit $k(a)$ is > 0, if only $a > 0$ and Relation (6.7.3) holds.

Writing $z_t^{(0)} = z_t(0)$ and recalling that $z_t^{(n)}$ is the number of descendants from n, we turn to the asymptotics of

$$P[z_t > 0] = P[\exists n \in Z_+ ; z_t^{(n)} > 0]$$

$$= 1 - L(t) - \left\{ \sum_{n=0}^\infty P[z_t^{(n)} > 0] - P[\exists n \in Z_+ ; z_t^{(n)} > 0] \right\} + \sum_{n=1}^\infty P[z_t^{(n)} > 0]$$

$$= 1 - L(t) - f(t) + E\left[\sum_{n=1}^{\xi_0(t)} P[z_t^{(n)} > 0 | \lambda_0, \xi_0] \right]$$

$$= 1 - L(t) - f(t) + \int_0^t P[z_{t-u} > 0] \, \mu(\mathrm{d}u). \tag{6.7.5}$$

Here

$$0 \le f(t) = \sum_{n=0}^\infty P[z_t^{(n)} > 0] - P[\exists n \in Z_+ ; z_t^{(n)} > 0]. \tag{6.7.6}$$

To apply the renewal theorem some knowledge about this f is needed. First, note that

$$e^{-\alpha t} P[z_t > 0] \le e^{-\alpha t} m_t \tag{6.7.7}$$

which is bounded under the conditions of Theorem (6.7.1). Second,

$$\int_0^\infty e^{-st} f(t) \, dt = \int_0^\infty e^{-st} P[z_t > 0] \, dt \{\hat{\mu}(s) - 1\}$$

$$+ \int_0^\infty e^{-st} \{1 - L(t)\} \, dt \le \int_0^\infty e^{-st} \{1 - L(t)\} \, dt, \quad (6.7.8)$$

for $\alpha < s < 0$; we can take $s > \alpha$ by Conditions (6.7.2) and (6.7.3). Letting $s \downarrow \alpha$ we see that

$$0 \le \int_0^\infty e^{-\alpha t} f(t) \, dt < \infty. \quad (6.7.9)$$

To establish the existence of

$$\lim_{t \to \infty} e^{-\alpha t} P[z_t > 0] = \int_0^\infty e^{-\alpha t} \{1 - L(t) - f(t)\} \, dt / \beta$$

by applying Theorem (5.2.9) to Equation (6.7.5) we need only check that

$$\sum_{k=0}^\infty \sup_{0 \le t \le 1} e^{-\alpha(k+t)} f(k + t) < \infty.$$

However, if we write

$$v_t = z_t^{(0)} + \sum_{n=1}^\infty 1_{\{z_t^{(n)} > 0\}},$$

then the last sum is the number of first generation individuals whose progeny exist at t,

$$\sum_{n=0}^\infty P[z_t^{(n)} > 0] = E[v_t],$$

and

$$P[\exists n \in Z_+; z_t^{(n)} > 0] = P[v_t > 0].$$

It follows that

$$f(t) = E[(v_t - 1)^+].$$

But for any t, u, $v_{u+t} \le v_u + \xi_0(u + t) - \xi_0(u)$, and hence for $k \in Z_+$, $0 \le \le t \le 1$

$$f(k + t) \le f(k) + \mu(k + 1) - \mu(k)$$

and

$$\sum_{k=0}^\infty \sup_{0 \le t \le 1} e^{-\alpha(k+t)} f(k + t) \le \sum_{k=0}^\infty e^{-\alpha(k+1)} \{f(k) + \mu(k + 1) - \mu(k)\} < \infty,$$

by an argument analogous to that leading to Equation (6.7.9). Thus we have proved the existence part of the following:

Theorem (6.7.2)

Consider a subcritical non-lattice process with Malthusian parameter α. Assume that Conditions (6.7.2) and (6.7.3) are satisfied. Then

$$\lim_{t \to \infty} e^{-\alpha t} P[z_t > 0]$$

exists. The limit is strictly positive provided

$$E[\hat{\xi}(\alpha) \log \xi(\infty)] < \infty. \tag{6.7.10}$$

Note. *For Bellman–Harris processes this last condition reduces to*

$$\sum_{k=1}^{\infty} p_k k \log k < \infty \tag{6.7.11}$$

known from the Galton–Watson case [Theorem (2.6.1)]. As there, Inequality (6.7.11) is also necessary for the positivity (page 166 of Reference [16]). For the interpretation recall that $0(-\infty) = 0$.

Proof (of the positivity). Write $q_n = P[\zeta_n = 0]$ for the probability of the nth generation being empty. Let $r_n(t)$ be the probability of the nth generation consisting of exactly one individual who is alive at t whereas $\zeta_{n+1} = 0$,

$$r_n(t) = P[\zeta_n = z_{t,n} = 1, \zeta_{n+1} = 0]$$

where

$$z_{t,n} = \sum_{x \in N^n} z_t(x).$$

Clearly

$$P[z_t > 0] \geq \sum_{n=0}^{\infty} r_n(t).$$

We shall see that

$$e^{-\alpha t} \sum_{n=0}^{\infty} r_n(t)$$

remains bounded away from zero as $t \to \infty$. Set

$$z_{t,n}^{(k)} = \sum_{x \in N^{n-1}} z_t(k, x).$$

For $n \geq 1$

$$r_n(t) = \sum_{k=1}^{\infty} P[z_{t,n}^{(k)} = \zeta_n = 1, \zeta_{n+1} = 0]$$

$$= E\left[\sum_{\tau_0(k) \le t} P\left[z_{t-\tau_0(k), n-1} \circ S_k = \zeta_{n-1} \circ S_k = 1, \zeta_n \circ S_k \right.\right.$$

$$\left.\left. = 0 \big| \xi_0, \lambda_0\right] q_{n-1}^{\xi_0(\infty) - 1}; \xi_0(\infty) \ge 1 \right]$$

$$= \int_0^t r_{n-1}(t-u)\,\mu(du) - \int_0^t r_{n-1}(t-u)\,E[(1 - q_{n-1}^{\xi(\infty)-1})\,\xi(du); \xi(\infty) \ge 1].$$

After cancellation of a positive term repetition of this yields

$$r_0 \ge r_{n-2} * \mu^{*2} - r_{n-2} * \mu * E[(1 - q_{n-1}^{\xi(\infty)-1} + 1 - q_{n-2}^{\xi(\infty)-1})\,\xi; \xi(\infty) \ge 1]$$

for $n > 2$. Generally if $n > k$

$$r_n \ge r_k * \mu^{*(n-k)} - r_k * \mu^{*(n-k-1)} * E\left[\left\{\sum_{j=1}^{n-k} (1 - q_{n-j}^{\xi(\infty)-1})\right\}\xi; \xi(\infty) \ge 1\right]$$

$$\ge r_k * \mu^{*(n-k)} - r_k * \mu^{*(n-k-1)} * E\left[\left\{\sum_{j=k}^{\infty} (1 - q_j^{\xi(\infty)-1})\right\}\xi; \xi(\infty) \ge 1\right].$$

Therefore, in the notation

$$r_{k\alpha}(t) = e^{-\alpha t} r_k(t),$$

$$v = \sum_{n=0}^{\infty} \mu^{*n},$$

$$v_\alpha(t) = \int_0^t e^{-\alpha u} v(du),$$

$$\lambda_j(t) = E\left[(1 - q_j^{\xi(\infty)-1}) \int_0^t e^{-\alpha u} \xi(du); \xi(\infty) \ge 1\right],$$

it holds that

$$e^{-\alpha t} \sum_{n>1} r_n(t) \ge r_k * v_\alpha(t) - r_{k\alpha}(t) - r_{k\alpha} * v_\alpha * \sum_{j \ge k} \lambda_j(t).$$

Since

$$0 \le r_k(t) \le E[z_{t,k}] = \int_0^t \{1 - L(t-u)\}\,\mu^{*k}(du),$$

clearly

$$\lim_{t \to \infty} r_k(t) = 0.$$

Also $r_{k\alpha}$ satisfies the requirements for the renewal theorem. Hence

$$\lim_{t \to \infty} r_{k\alpha} * v_\alpha(t) = \int_0^\infty e^{-\alpha t} r_k(t)\,dt \bigg/ \int_0^\infty t e^{-\alpha t} \mu(dt)$$

to be called b exists and

$$\liminf_{t \to \infty} e^{-\alpha t} \sum_{n > k} r_n(t) \geq b \left\{ 1 - \sum_{j \geq k} \lambda_j(\infty) \right\}.$$

Thus we are done if we can prove the positivity of b or by Inequality (6.7.3) that

$$\int_0^\infty e^{-\alpha t} r_k(t) > 0$$

for any k and that

$$\sum_j \lambda_j(\infty) < \infty.$$

The latter is easy under Condition (6.7.10):

$$\sum_{j=1}^\infty \lambda_j(\infty) = \sum_{j=1}^\infty E\left[\hat{\xi}(\alpha)(1 - q_j^{\xi(\infty)-1}); \xi(\infty) \geq 1\right]$$

$$= E\left[\hat{\xi}(\alpha) \sum_{j=1}^{\log \xi(\infty)/-\log m} (1 - q_j^{\xi(\infty)-1}); \xi(\infty) \geq 1 \right]$$

$$+ E\left[\hat{\xi}(\alpha) \sum_{j > \log \xi(\infty)/-\log m} (1 - q_j^{\xi(\infty)-1}); \xi(\infty) \geq 1 \right].$$

Neither term is negative and the former not greater than a constant times $E[\hat{\xi}(\alpha) \log \xi(\infty)]$, which is finite by assumption. By Theorem (2.6.1) $1 - q_j \sim cm^j$ as $j \to \infty$. For that reason there is a C such that

$$1 - q_j^{\xi(\infty)-1} \leq \{\xi(\infty) - 1\}(1 - q_j) \leq C\xi(\infty) m^j$$

on the set where $\xi(\infty) \geq 1$. It follows that on this same set

$$\sum_{j > \log \xi(\infty)/-\log m} (1 - q_j^{\xi(\infty)-1}) \leq \sum_{j \geq \log \xi(\infty)/-\log m} C\xi(\infty) m^j$$

$$= C\xi(\infty) m^{\{\log \xi(\infty)/-\log m\}}/(1 - m) = C/(1 - m)$$

and the latter term above is majorized by

$$CE[\hat{\xi}(\alpha)]/(1 - m) < \infty.$$

Finally, we turn to the integral

$$\int_0^\infty e^{-\alpha t} r_k(t) \, dt.$$

If $p_1 = P[\xi(\infty) = 1]$ and $P[\lambda > 0 | \xi(\infty) = 0]$ both are positive, so must $r_k(t)$ be for t in some interval and the positivity of the integral follows. If $p_1 > 0$ but $P[\lambda > 0 | \xi(\infty) = 0] = 0$, construct a new process in the following manner:

When $\xi_x(\infty) > 0$, let $(\lambda'_x, \xi'_x) = (\lambda_x, \xi_x)$.

When $\xi_x(\infty) = 0$, let $(\lambda'_x, \xi'_x) = (1, \xi_x)$.

On $\{(\lambda'_x, \xi'_x); x \in I\}$ define a branching process z'_t. The reproduction function $\mu', \mu'(t) = E[\xi'_x(t)]$ is the same as the old one, whereas

$$
L'(t) = P[\lambda' \le t] = \begin{cases} P[\xi(\infty) \ge 1, \lambda \le t], & 0 \le t < 1, \\[2mm] L(t) + P[\xi(\infty) = 0, \lambda > t], & t \ge 1. \end{cases}
$$

Since $P[\lambda' > 0 | \xi(\infty) = 0] = 1$, $\xi' = \xi$, and L' satisfies Inequality (6.7.2),

$$
e^{-\alpha t} P[z'_t > 0] \to \text{some constant} > 0.
$$

However $z_t = 0 \Rightarrow z'_{t+1} = 0$ and so

$$
e^{-\alpha t} P[z_t > 0] \ge e^{\alpha} e^{-\alpha(t+1)} P[z'_{t+1} > 0].
$$

Thus the limit is positive as soon as $p_1 > 0$. In the remaining case $p_1 = 0$, we define a new process such that the ancestor in the old process dies at once with probability $1/2$ and with probability $1/2$ it does whatever it would have done in the original process. Descendants also die at once with probability $1/2$ and display the behaviour of individuals in the original process with probability $1/2$. The reader may check the formalities to convince her/himself that $\tilde{z}_t =$ the number of individuals existing at time t is that of a branching process with reproduction function

$$
\tilde{\mu} = 1/2 + \mu/2.
$$

Therefore the new process has the old Malthusian parameter α and

$$
E[\tilde{\xi}(\alpha) \log \tilde{\xi}(\infty)] = \tfrac{1}{2} 0 + \tfrac{1}{2} E[\hat{\xi}(\alpha) \log \xi(\infty)] < \infty.
$$

Finally the life-length distribution $\tilde{L}$ satisfies $\tilde{L} = 1/2 + L/2$ proving that Inequality (6.7.2) still holds. But since the two processes are the same, except for the numbering of individuals,

$$
\lim_{t \to \infty} e^{-\alpha t} P[z_t > 0] = \lim_{t \to \infty} e^{-\alpha t} P[\tilde{z}_t > 0] > 0. \qquad \square
$$

We turn directly to the last theorem about subcritical processes, which is easily proved after the work already done.

Theorem (6.7.3)

Under the conditions of Theorem (6.7.2)

$$
\lim_{t \to \infty} P[z_t^a = k | z_t > 0] = b_k^a
$$

exists for $k \in Z_+$ and $0 \leq a \leq \infty$. For all a

$$\sum_{k=0}^{\infty} b_k^a = 1.$$

Proof. Take $a = \infty$ for simplicity and write $z_t^{(0)} = z_t(0)$.

$$P[z_t = k] = P[\exists n : z_t^{(n)} = z_t = k] + P[z_t = k, \text{two or more } z_t^{(n)} > 0]$$

$$= \sum_{n=0}^{\infty} P[z_t^{(n)} = k, z_t^{(j)} = 0, j \neq n] +$$

$$+ P[z_t = k, \text{two or more } z_t^{(n)} > 0] = P[z_t(0) = k]$$

$$+ \sum_{n=1}^{\infty} P[z_t^{(n)} = k] - \sum_{n=0}^{\infty} P[z_t^{(n)} = k, z_t^{(j)} > 0 \text{ for some } j \neq n]$$

$$+ P[z_t = k, \text{two or more } z_t^{(n)} > 0] = f_k(t) + \int_0^t P[z_{t-u} = k] \mu(du),$$

where

$$f_k(t) = P[z_t(0) = k] - \sum_{n=0}^{\infty} P[z_t^{(n)} = k, z_t^{(j)} > 0 \text{ for some } j \neq n]$$

$$+ P[z_t = k, \text{two or more } z_t^{(n)} > 0].$$

This f_k satisfies Assumption (5.2.7) for the renewal theorem by arguments analogous to the prelude of Theorem (6.7.2), and

$$\lim_{t \to \infty} e^{-at} P[z_t = k]$$

exists. Since

$$P[z_t = k | z_t > 0] = P[z_t = k] / P[z_t > 0]$$

the first assertion is proved. The second follows since

$$E[z_t | z_t > 0] = e^{-at} m_t / e^{-at} P[z_t > 0],$$

which has a finite limit as $t \to \infty$. (The reader may wish to check the following: Let $\lim_{n \to \infty} p_{nk} = p_k$, $0 \leq p_{nk}$, $\sum_k p_{nk} = 1$, $\limsup_{n \to \infty} \sum_k k p_{nk} < \infty$. Then $\sum_k p_k = 1$.)

We conclude with a simple result, which makes it possible to replace Condition (6.7.10) by a requirement only on the total reproduction. □

Theorem (6.7.4)

If

$$\int_0^\infty te^{-\alpha t}\mu(dt) < \infty$$

and

$$E\big[\xi(\infty)^{1+\varepsilon}\big] < \infty$$

for some $\varepsilon > 0$, then

$$E\big[\hat{\xi}(\alpha)\log\xi(\infty)\big] < \infty.$$

Proof. Let $c = -\varepsilon/2\alpha$ and recall that $\alpha < 0$;

$$E\big[\hat{\xi}(\alpha)\log\xi(\dot\infty)\big] = E\left[\int_0^{c\{\log\xi(\infty)\}^+} e^{-\alpha t}\xi(dt)\log\xi(\infty)\right]$$

$$+ E\left[\int_{c\{\log\xi(\infty)\}^+}^\infty e^{-\alpha t}\xi(dt)\log\xi(\infty)\right]$$

$$\leq E\big[e^{-\alpha c\log\xi(\infty)}\xi(\infty)\log\xi(\infty)\big] + E\left[\int_{c\{\log\xi(\infty)\}^+}^\infty e^{-\alpha t}t\xi(dt)/c\right]$$

$$\leq E\big[\xi(\infty)^{1+\varepsilon/2}C\xi(\infty)^{\varepsilon/2}\big] + E\left[\int_0^\infty e^{-\alpha t}t\xi(dt)\right]\Big/ c$$

for some constant C.   □

Background

These results, and their proofs, are from Reference [17].

6.8 SUPERCRITICAL PROCESSES

As pointed out in Section 6.3, in the present supercritical case the Malthusian parameter α is always well defined and positive by $\hat\mu(\alpha) = 1$. Recall that Theorem (6.3.3) displayed an exponential increase of m_t^a, as $t \to \infty$. Turning to second moments and Equation (6.4.3) we shall see that this exponential growth persists in a considerably more essential sense. We write

$$k(a) = \int_0^a e^{-\alpha u}\{1 - L(u)\}\,du/\beta. \qquad (6.8.1)$$

$$K(a) = \int_0^a e^{-\alpha u}\{1 - L(u)\}\,du\Big/\int_0^\infty e^{-\alpha u}\{1 - L(u)\}\,du.$$

and, as in Theorem (6.4.3),

$$\vartheta = \int e^{-\alpha(u+v)}\gamma(du, dv) = E[\hat{\xi}^2(\alpha)]$$

$$\varphi = \hat{\mu}(2\alpha) = E[\hat{\xi}(2\alpha)].$$

Theorem (6.8.1)

Assume that $m > 1$, $\sigma^2 < \infty$ (or just $\vartheta < \infty$) and that μ is non-lattice. Then, there is a random variable $z \geq 0$ such that for all a, $0 \leq a \leq \infty$, $w_t^a = e^{-\alpha t}z_t^a \to k(a)z$ in mean square, as $t \to \infty$. Further $P[z = 0, z_t \not\to 0] = 0$, $E[z] = 1$, $\text{Var}[z] = (\vartheta - 1)/(1 - \varphi)$, and z is constant $(=1)$ if and only if $\hat{\xi}(\alpha) = 1$ with probability one.

Proof. By Theorem (6.4.3)

$$\lim_{t \to \infty} e^{-2\alpha t - \alpha u}c_u^{a,b}(t) = k(a)k(b)(\vartheta - \varphi)/(1 - \varphi).$$

Using the approach from Lemma (2.9.1) we see that

$$E[(w_{t+u}^a - w_t^a)^2] = e^{-2\alpha(t+u)}c_0^{a,a}(t + u)$$
$$- 2e^{-2\alpha t - \alpha u}c_u^{a,a}(t) + e^{-2\alpha t}c_0^{a,a}(t) \to 0, \qquad (6.8.2)$$

as $t \to \infty$. Hence by the completeness of L^2 there is a random variable $w(a)$ such that $w_t^a \to w(a)$ in mean square. Furthermore, $E[w(a)] = \lim_{t \to \infty} E[w_t^a] = k(a)$, $\text{Var}[w(a)] = \lim_{t \to \infty} \text{Var}[w_t^a] = k^2(a)(\vartheta - 1)/(1 - \varphi)$, and, with $w_t = e^{-\alpha t}z_t$,

$$E[\{w_t^a - k(a)w_t\}^2] = E[(w_t^a)^2] - 2k(a)E[w_t^a w_t] + k^2(a)E[w_t^2]$$
$$\to k^2(a)(\vartheta - \varphi)/(1 - \varphi) - 2k^2(a)(\vartheta - \varphi)/(1 - \varphi) + k^2(a)(\vartheta - \varphi)/(1 - \varphi) = 0.$$

Hence, $w(a) = k(a)w(\infty)$.

Note that $k(\infty) > 0$ and indeed $k(a) > 0$ for all $a > 0$, since $\beta < \infty$, and set $z = w(\infty)/k(\infty)$. Then $E[z] = 1$ and $\text{Var}[z] = (\vartheta - 1)/(1 - \varphi)$. But $\vartheta = 1 \Leftrightarrow \hat{\xi}(\alpha) = 1$ a.s. by Theorem (6.4.3). Finally,

$$P[e^{-\alpha t}z_t \to 0] = E\left[\prod_{n=1}^{\xi_0(\infty)} P[e^{-\alpha t}z_{t-\tau_0(n)} \circ S_n \to 0 | \lambda_0, \xi_0]\right]$$

$$= E[(P[z_t \to 0])^{\xi_0(\infty)}] = f(P[e^{-\alpha t}z_t \to 0]).$$

Since $w(\infty)$ is not identically zero, $P[w_t \to 0] = q$ and the theorem is completely proved. $\qquad \square$

Corollary (6.8.2)

Under the conditions of the theorem it holds that $z_t^a/z_t \to K(a)$ in probability on the set where $z_t \not\to 0$, as $t \to \infty$.

Proof. Mean square convergence implies convergence in probability and if $x_n \to x$ and $y_n \to y > 0$ in probability, it follows that $x_n/y_n \to x/y$ in probability.

Theorem (6.8.3)

Under the assumptions of Theorem (6.8.1) the moment generating function φ of z satisfies

$$\varphi(u) = E\left[\prod_{n=1}^{\xi(\infty)} \varphi(ue^{-\tau(n)}) \right], \qquad u \geq 0, \tag{6.8.3}$$

alone among bounded functions whose right derivative at the origin is -1.

Proof. Since convergence in mean square enhances convergence in distribution,

$$\varphi(u) = \lim_{t \to \infty} E\left[e^{-uw_t/k(\infty)} \right]$$

$$= \lim_{t \to \infty} E\left[e^{-ue^{-\alpha t}z_t(0)/k(\infty)} \prod_{n=1}^{\xi_0(t)} E\left[e^{-ue^{-\alpha\tau(n)}w_{t-\tau(n)}\circ S_n/k(\infty)} \Big| \lambda_0, \xi_0 \right] \right]$$

$$= E\left[\prod_{n=1}^{\xi(\infty)} \varphi(ue^{-\alpha\tau(n)}) \right],$$

by dominated convergence. Suppose that φ and ψ both satisfy Equation (6.8.3) and write $\delta(u) = |\varphi(u) - \psi(u)|/u$, $u > 0$. Then, by the usual [Equation (6.6.8)] argument,

$$\delta(u) \leq \int_0^\infty \delta(ue^{-\alpha t}) e^{-\alpha t}\mu(dt).$$

Thus, if $x_1, x_2, \ldots$ are i.i.d. random variables with distribution

$$P[x_1 \leq t] = \int_0^t e^{-\alpha u}\mu(du)$$

and $s_n = x_1 + \ldots + x_n$, repetition yields that

$$\delta(u) \leq E[\delta(ue^{-s_n})].$$

But if $\varphi'(0) = \psi'(0)$ exists δ remains bounded and $\delta(0+) = 0$. As $P[x_1 = 0] = \mu(0) < 1$, $s_n \to \infty$ by the law of large numbers and $\delta(u) = 0$ for all $u > 0$. $\qquad \square$

Background

The methods of this section go back to Harris's treatment of Bellman–Harris processes but apply well in the general setup. We have only discussed the case with a finite reproduction variance or at least $\vartheta < \infty$. However, it has been shown (by Athreya for Bellman–Harris processes and by Doney generally) that:

For non-lattice, supercritical processes it holds, as $t \to \infty$, that $w_t \overset{\text{d}}{\to} w(\infty)$, which is identically zero if and only if $E[\hat{\xi}(\alpha)|\log \hat{\xi}(\alpha)|] = \infty$. The limit random variable has a continuous density, except for a jump in the distribution at the origin.

For further comments see the next section and the footnote on p. 12.

6.9 POPULATIONS COUNTED WITH RANDOM CHARACTERISTICS

In several circumstances a generalization of the setup in Theorem (6.8.1) is needed. Instead of the number of individuals in some fixed age interval, we might be interested in the part of the population in some random phase of their life. This phase might be unrelated to life length or reproduction or heavily dependent upon them.

Such considerations lead to associating with each individual $x \in I$, instead of a pair, a random triple $(\lambda_x, \xi_x, \chi_x)$, where $\chi_x : R \to \{0, 1\}$, $\chi_x(t) = 0$ for $t < 0$ and $\chi_x(t) = 1$ if and only if x is in the phase in question at age t. The random process

$$z_t^\chi = \sum_x \chi_x(t - \sigma_x)$$

would then be the number of individuals at time t having the property (being in the phase) χ. If we assume that $\chi_x(t) = 0$ for $t \geq \lambda_x$, no dead individuals are counted. On the other hand, this will turn out to be an irrelevant restriction.

And so is, indeed, the limitation to stochastic processes χ_x taking only the value zero or one. Thus we are led to the following formulation (where usually but not necessarily χ_x will be non-negative): To each $x \in I$ let there be given, besides λ_x and ξ_x, a stochastic process on R, χ_x, such that $\chi_x(t) = 0$ for $-\infty \leq t \leq 0$. Let the $(\lambda_x, \xi_x, \chi_x)$ for different x be i.i.d. Define

$$z_t^\chi = \sum_x \chi_x(t - \sigma_x)$$

The random process z_t^χ; $t \geq 0$ will be referred to as a branching process *counted with characteristic* χ. We adopt the convention of deleting the index x of the *random characteristic* χ, as well. Since χ is finite z_t^χ is a.s. finite by Theorem (6.2.2).

Example (6.9.1)

Let $\chi = 1_{[0, \tau(1) \wedge \lambda)}$. Then z_t^χ is just the number of individuals who are alive at time t but have no children. More generally, let a, b be two random variables, $0 \leq a \leq b \leq \lambda$. Then, with $\chi = 1_{[a,b)}$, z_t^χ counts the individuals in the random phase $[a, b)$ of their lives.

If $\chi(t) = 1$ for $t \geq 0$, obviously $z_t = y_t$ and if $\chi = 1_{[\lambda, \infty]}$, then $z_t = d_t$, the number of realized individuals who have died by t. Similarly $\chi(t) = 1_{[0, a \wedge \lambda)}$ yields $z_t^{\chi} = z_t^{a}$.

But less trite processes can also be obtained. If $\chi(t)$ is simply t for $0 \leq t \leq \lambda$ and λ for $t \geq \lambda$, then

$$z_t^{\chi} = \int_0^t z_u \, du$$

This process has some applied interest; more about that later. A more general version is letting χ_x be a random non-decreasing function say measuring some cumulative activity of the individual x (like the toxin production of bacteria).   □

Now let χ be any random characteristic. Assume that for all t $E[\chi^2(t)] < \infty$ and that the functions $g(t) = E[\chi(t)]$ and $h_u(t) = E[\chi(t)\chi(t+u)]$ are measurable in t. Write $m_t = E[z_t^{\chi}]$, and $c_u^{\chi}(t) = E[z_t z_{t+u}^{\chi}]$ (why do they exist?). Since

$$z_t^{\chi} = \chi_0(t) + \sum_{n=1}^{\xi_0(t)} z_{t-\tau_0(n)}^{\chi} \circ S_n,$$

the known line of arguing leads to

$$m_t^{\chi} = g(t) + \int_0^t m_{t-u}^{\chi} \mu(du) \tag{6.9.1}$$

$$c_u^{\chi}(t) = h_u(t) + \int_0^{t+u} m_{t+u-v}^{\chi} E[\chi(t)\,\xi(dv)] + \int_0^t m_{t-v} E[\chi(t+u)\,\xi(dv)]$$

$$+ \int_{\substack{0 \leq v \leq t \\ 0 \leq w \leq t+u}} m_{t-v}^{\chi} m_{t+u-v}^{\chi} \gamma(dv, dw)$$

$$- \int_0^t m_{t-v}^{\chi} m_{t+u-v}^{\chi} \mu(dv) + \int_0^t c_u^{\chi}(t-v)\,\mu(dv) \tag{6.9.2}$$

The latter equation could also have been deduced for two different characteristics. In particular $c(t) = E[z_t^{\chi} z_t]$ satisfies, with $\mu_t(u) = E[\xi(u); \lambda > t]$,

$$c(t) = E[\chi(t); \lambda > t] + \int_0^t m_{t-u}^{\chi} \mu_t(du)$$

$$+ \int_0^t m_{t-u}^{\chi} E[\chi(t)\,\xi(du)] + \int_{0 \leq u, v \leq t} m_{t-u}^{\chi} m_{t-v}\gamma(du, dv)$$

$$- \int_0^t m_{t-u}^{\chi} m_{t-u}\mu(du) + \int_0^t c(t-u)\,\mu(du). \tag{6.9.3}$$

The arguments of Section 6.8 work out into

Theorem (6.9.2)

Consider a supercritical, non-lattice branching process with Malthusian parameter α. Let χ be a random characteristic such that $g(t) = E[\chi(t)]$ is locally of bounded variation, $e^{-\alpha t}g(t) \to 0$ and $\int e^{-\alpha t}g(t)\,dt$ converges. Then, as $t \to \infty$,

$$m_t^\chi \sim e^{\alpha t} \int_0^\infty e^{-\alpha u}g(u)\,du \Big/ \int_0^\infty e^{-\alpha u}u\mu(du) = e^{\alpha t}c_g.$$

If further $h_u(t) = E[\chi(t)\chi(t + u)]$ and $E[\chi(t)\xi(u)]$ are measurable in t for any $u > 0$, $E[\chi^2(t)] = o(e^{2\alpha t})$ as $t \to \infty$ and is bounded on finite intervals, and the reproduction variance is finite (or just $\vartheta < \infty$), then

$$c_u^\chi(t) \sim e^{2\alpha t + u}c_g^2(\vartheta - \varphi)/(1 - \varphi)$$

and, in mean square,

$$\lim_{t \to \infty} e^{-\alpha t}z_t = c_g z,$$

where z is the limit random variable of Theorem (6.8.1).

Note. The conditions for the assertion about m_t where given so as to fit with Theorem (5.3.3). If g is instead required to be a.e. (Lebesgue) continous and bounded, Theorem (5.2.8) could be appealed to. Direct recourse to Theorem (5.2.6) would lead to a third set of conditions.

Background

The number of individuals having some random age dependent property (i.e. χ zero or one) was first considered, for Bellman–Harris processes by Weiner [21], who studied means and several examples. The general concept of a random characteristic is from Reference [20].

6.10 ALMOST SURE CONVERGENCE IN THE SUPERCRITICAL CASE

We start right off:

Theorem (6.10.1)

Let all conditions of Theorem (6.9.2) be satisfied. Further assume that for some constant c

$$P[\forall t, u \geq 0 : \chi(t + u) \geq e^{-cu}\chi(t)] = 1, \tag{6.10.1}$$

$$\int_0^\infty e^{-\alpha t}tg(t)\,dt < \infty, \tag{6.10.2}$$

$$\int_0^\infty e^{-2\alpha t} h_0(t)\, dt < \infty,\qquad\qquad (6.10.3)$$

$$t\, e^{-\alpha t}\, g(t) \to 0.\qquad\qquad (6.10.4)$$

Then, a.s.,

$$\lim_{t\to\infty} e^{-\alpha t} z_t^\chi = c_g z.$$

Since for any real numbers a, b

$$z_t^{a\chi + b\chi'} = a z_t^\chi + b z_t^{\chi'},$$

we obtain immediately the proper main result, requiring Equations (6.10.1–4) not from χ but from its positive and negative parts:

Corollary (6.10.2)

If χ is a linear combination of characteristics such that the assumptions of Theorem (6.10.1) hold, then the conclusion is true as well.

Before proving the theorem we shall see its strength from a chain of corollaries. In all of them the process is supercritical and non-lattice with Malthusian parameter α and finite reproduction variance (or $\vartheta < \infty$). In none of them we need to take $c \neq 0$.

Corollary (6.10.3)

With probability one and in mean square

$$\lim_{t\to\infty} e^{-\alpha t} y_t = z/\alpha\beta.$$

Proof. As pointed out in Example (6.9.2) $y_t = z_t^\chi$ for $\chi = 1$.   □

Corollary (6.10.4)

$$P\left[\lim_{t\to\infty} e^{-\alpha t} z_t^a = k(a)\, z \text{ uniformly in } 0 \le a \le \infty \right] = 1$$

and with $K(a) = k(a)/k(\infty)$ (the stable age distribution)

$$P\left[\lim_{t\to\infty} z_t^a/z_t \to K(a) \text{ uniformly in } 0 \le a \le \infty \,|z_t \neq 0 \right] = 1.$$

Proof. The pointwise convergence to a continuous limit of non-decreasing bounded functions on $[0, \infty]$ is always uniform (check that!) and since there is convergence for all $0 \le a \le \infty$ if there is for rational ones (including ∞) we must only prove that $P[e^{-\alpha t} z_t^a \to k(a)\, z] = 1$ for fixed a. But the characteristic $1_{[a,\infty)}$ meets the requirements of the theorem. So do $1_{[\lambda,\infty)}$

and $1_{[\lambda \vee a, \infty)}$. But $z_t^a = z_t^\chi$ with $\chi = 1_{[0, a \wedge \lambda)} = 1 - 1_{(a, \infty)} - 1_{[\lambda, \infty)} + 1_{[\lambda \vee a, \infty)}$.
Hence Corollary (6.10.2) applies.   □

Corollary (6.10.5)

If $0 \le a_x \le b_x \le \lambda_x$ are random variables with $(a_x, b_x) x \in I$ i.i.d., define
$p(u) = P[a \le u \le b | \lambda > u]$ and $\chi_x = 1_{[a_x, b_x)}$. Then z_t^χ, the number of individuals in phase $[a, b)$, satisfies a.s. and in mean square

$$\lim_{t \to \infty} e^{-\alpha t} z_t^\chi = z \int_0^\infty p(u) \, k(du)/\beta$$

and on the set where the process does not become extinct

$$\lim_{t \to \infty} z_t^\chi / z_t = \int_0^\infty p(u) \, K(du),$$

a.s.

Proof. The characteristic is $1_{[a,b)} = 1_{[a, \infty)} - 1_{[b, \infty)}$. Use Corollary (6.10.2).   □

Corollary (6.10.5) could of course as well have been formulated in terms
of the unconditioned probability $P[a \le u \le b]$ and integration with
respect to $e^{-\alpha u} du$. Here follow two more corollaries, the first without proof:

Corollary (6.10.6)

Let χ be a random measure such that $E[\chi^2(\infty)] < \infty$. Write $g(u) = E[\chi(u)]$.
Then a.s. and in mean square

$$e^{-\alpha t} z_t^\chi \to z \int_0^\infty e^{-\alpha u} g(u) \, du/\beta.$$

Corollary (6.10.7)

With probability one and in mean square

$$\lim_{t \to \infty} e^{-\alpha t} \int_0^t z_t \, du = k(\infty) z/\alpha.$$

Proof. By Example (6.9.1) the appropriate characteristic is $\chi(t) = t \wedge \lambda$,
complying with the theorem,

$$g(t) = \int_0^t u L(du) + t\{1 - L(t)\},$$

and

$$\int_0^\infty e^{-\alpha t} g(t) \, dt = \int_0^\infty e^{-\alpha t} \{1 - L(t)\} \, dt/\alpha.$$ □

We turn to proving the theorem (i.e. consider $\chi \geq 0$).

Lemma (6.10.8)

If Equation (6.10.1) holds and

$$\int_0^\infty E\left[\{e^{-\alpha t}z_t^\chi - c_g z\}^2\right]\,dt < \infty, \tag{6.10.3}$$

then

$$e^{-\alpha t}z_t^\chi \to c_g z \quad a.s.$$

Proof. Suppose that $\limsup\limits_{t \to \infty} e^{-\alpha t} z_t^\chi > c_g z$, which latter variable we shall write just w in these proofs. Then there is a (random) $\varepsilon > 0$ and a sequence of (random) numbers $\tau_1 < \tau_2 < \dots$ such that $\tau_{n+1} - \tau_n > \varepsilon/(\alpha + c)(1 + \varepsilon)$ and $e^{-\alpha \tau_n}z_{\tau_n}^\chi > w(1 + \varepsilon)$. For $u \geq 0$

$$e^{-\alpha(\tau_n + u)}z_{\tau_n + u}^\chi = e^{-\alpha(\tau_n + u)}\sum_x \chi_x(\tau_n + u - \sigma_x)$$

$$\geq e^{-u(\alpha + c)}e^{-\alpha \tau_n}\sum_x \chi_x(\tau_n - \sigma_x) > \{1 - u(\alpha + c)\}w(1 + \varepsilon).$$

Here we used that $e^{-t} \geq 1 - t$ for $t \geq 0$ and Equation (6.10.1). It follows that

$$\int_{\tau_n}^{\tau_n + 1}\{e^{-\alpha t}z_t^\chi - w\}^2\,dt \geq \int_{\tau_n}^{\tau_n + \varepsilon/(\alpha + c)(1 + \varepsilon)}\{e^{-\alpha t}z_t^\chi - w\}^2\,dt$$

$$\geq w^2 \int_0^{\varepsilon/(\alpha + c)(1 + \varepsilon)}\{\varepsilon - (1 + \varepsilon)(\alpha + c)u\}^2\,du = w^2\varepsilon^3/3(1 + \varepsilon)(\alpha + c).$$

Thus, on the set where $\limsup\limits_{t \to \infty} e^{-\alpha t}z_t^\chi > w > 0$, the integral

$$\int_0^\infty \{e^{-\alpha t}z_t^\chi - w\}^2\,dt$$

diverges. But by Inequality (6.10.3) this divergence has no probability.

On the set where $w = 0$ but $\limsup e^{-\alpha t}z_t^\chi = \varepsilon > 0$ choose τ_n such that $\tau_{n+1} - \tau_n > 1/(\alpha + c)$ and $e^{-\alpha \tau_n}z_{\tau_n}^\chi > \varepsilon/2$. Argue as before to obtain

$$\int_{\tau_n}^{\tau_{n+1}}\{e^{-\alpha t}z_t^\chi\}^2\,dt \geq \varepsilon^2 \int_0^{1/(\alpha + c)}\{1 - u(\alpha + c)\}^2/4\,du = \varepsilon^2/12(\alpha + c),$$

and from that divergence of the integral

Now suppose that $\liminf\limits_{t \to \infty} e^{-\alpha t} z_t^\chi < w$. There are then $0 < \varepsilon < 1$ and

$\tau_1 < \tau_2 < \dots$ (still random) such that $\tau_{n+1} - \tau_n > 1/(\alpha + c)$ and $e^{-\alpha \tau_n} z^{\chi}_{\tau_n} <$ $< w(1 - \varepsilon)$. For $0 \le u < 1/(\alpha + c)$

$$e^{-\alpha(\tau_n - u)} z^{\chi}_{\tau_n - u} \le e^{-\alpha \tau_n} z^{\chi}_{\tau_n} e^{u(\alpha + c)} \le w(1 - \varepsilon)/\{1 - u(\alpha + c)\}.$$

The rest goes as before.   □

The next, and final, step is to prove that Inequality (6.10.3) always holds. This will be done through a sequence of lemmata and results from Section 5.3.

Lemma (6.10.9)

Consider a supercritical and non-lattice process with a non-negative characteristic such that $tg(t) = tE[\chi(t)]$ is of locally bounded variation and $o(e^{\alpha t})$, as $t \to \infty$. Suppose that Condition (6.10.2) holds. Then

$$e^{-\alpha t} m^{\chi}_t - c_g = o(t^{-1})$$

as $t \to \infty$, and the integral

$$\int_0^\infty \{e^{-\alpha t} m^{\chi}_t - c_g\}\, dt$$

converges (in the elementary meaning that $\lim\limits_{t \to \infty} \int_0^t$ exists finitely).

Proof. Multiply Equation (6.9.1) by $e^{-\alpha t}$ and subtract c_g to obtain an equation to which Lemma (5.3.5) and Theorem (5.3.6) are applicable; indeed $e^{-\alpha u}\mu(du)$ has finite moments of all orders.   □

Lemma (6.10.10)

In a process complying with all requirements of Theorem (6.10.1) (except possibly Equation (6.10.1)) it holds that

$$\int_0^\infty \left\{ e^{-2\alpha t - \alpha u}\, c^{\chi}_u(t) - c_g^2(\vartheta - \varphi)/(1 - \varphi) \right\} dt$$

converges.

Proof. We write Equation (6.9.2)

$$e^{-2\alpha t - \alpha u} c^{\chi}_u(t) - c_g^2(\vartheta - \varphi)/(1 - \varphi) = e^{-2\alpha t - \alpha u} h_u(t)$$

$$+ e^{-\alpha t} \int_0^{t+u} e^{-\alpha(t+u-v)} m^{\chi}_{t+u-v} e^{-\alpha v} E[\chi(t)\, \xi(dv)]$$

$$+ e^{-\alpha(t+u)} \int_0^t e^{-\alpha(t-v)} m^{\chi}_{t-v} e^{-\alpha v} E[\chi(t+u)\, \xi(dv)]$$

$$+ \int_{\substack{0 \le v \le t \\ 0 \le w \le t+u}} \{e^{-\alpha(t-v)}m_{t-v}e^{-\alpha(t+u-w)}m_{t+u-w} - c_g^2\}$$

$$e^{-\alpha(v+w)}\gamma(dv, dw) - c_g^2 \int_{\substack{v > t \text{ or} \\ w > t+u}} e^{-\alpha(v+w)}\gamma(dv, dw)$$

$$- \int_0^t \{e^{-\alpha(t-v)}m_{t-v}^\chi e^{-\alpha(t+u-v)}m_{t+u-v}^\chi - c_g^2\} e^{-2\alpha v}\mu(dv)$$

$$- c_g^2(\vartheta - 1) \int_t^\infty e^{-2\alpha v}\mu(dv)/(1 - \varphi)$$

$$+ \int_0^t \{e^{-2\alpha(t-v)-\alpha u}c_u^\chi(t - v) - c_g^2(\vartheta - \varphi)/(1 - \varphi)\} e^{-2\alpha v}\mu(dv). \qquad (6.10.5)$$

To this we apply Lemma (5.3.7).

The integrals of all terms at the right, except the fourth and the sixth (and the last) clearly converge. The convergence of the remaining two follows from Lemma (6.10.9) and the identity $ab - c^2 = (a - b)(b - c) + c(a - c) + b(b - c)$. For details see Reference [2].   □

Lemma (6.10.11)
Let the assumptions of Lemma (6.10.9) be true and define

$$c(t) = E[e^{-\alpha t}z_t^\chi c_g z] = \lim_{u \to \infty} e^{-2\alpha t - \alpha u}c_u^\chi(t)$$

Then,

$$\lim_{t \to \infty} c(t) = c_g^2(\vartheta - \varphi)/(1 - \varphi)$$

and the integral

$$\int_0^\infty \{c(t) - c_g^2(\vartheta - \varphi)/(1 - \varphi)\} \, dt$$

converges.

Proof. Let $u \to \infty$ in Equation (6.10.5). Argue as in Lemma (6.10.10).   □

Lemma (6.10.12)
Under the conditions of Lemma (6.10.9)

$$\int_0^\infty E[\{e^{-\alpha t}z_t^\chi - c_g z_\infty\}^2] \, dt < \infty.$$

Proof.

$$E[\{e^{-\alpha t}z_t^{\chi} - c_g z_{\infty}\}^2]$$

$$= \{e^{-2\alpha t}c_0^{\chi}(t) - c_g^2(\vartheta - \varphi)/(1 - \varphi)\} - 2\{c(t) - c_g(\vartheta - \varphi)/(1 - \varphi)\}. \quad \square$$

The two preceding lemmata complete the proof of the theorem.

Example (6.10.13)

Consider binary splitting with cell death. Assume that $p > 1/2$ and that G is not lattice. The last period of a cell's life, when it prepares for division is called mitosis. It is clearly discernible and the fraction of cells in mitosis, called the mitotic index, is a parameter of cell populations frequently studied. Assume that only not disintegrating cells can enter mitosis which starts at a random age v_x with distribution H, independently for different cells. Then $\lambda_x - v_x$ is the length of mitosis, with

$$\chi_x = \begin{cases} 0, & \text{if the cell disintegrates,} \\ 1_{[v_x, \lambda_x]}, & \text{if it completes its life cycle,} \end{cases}$$

z_t^{χ} is the number of cells in mitosis at t, and

$$p(u) = E[\chi(u)|\lambda > u] = P[\chi(u) = 1|\lambda > u] = p\{H(u) - G(u)\}/\{1 - L(u)\}$$

is the probability of a cell aged u being in mitosis. By Corollary (6.10.5) the mitotic index of a stabilized (to use a vague word) population should be,

$$\int_0^{\infty} p(u) K(du) = \{2p\hat{H}(\alpha) - 1\}/\{1 - 2(1 - p)\hat{F}(\alpha)\}.$$

If there is no cell loss, $p = 1$ and the mitotic index reduces to $2\hat{H}(\alpha) - 1$. $\quad \square$

Example (6.10.14)

Let $\chi_x(u) = 1$ if $u \leq \lambda_x \leq a + u$. Then $E[z_t^{\chi}/z_t|z_t > 0]$ is the probability that a randomly sampled individual dies within a time units. Here

$$g(u) = P[\chi(u) = 1] = L(a + u) - L(u)$$

and on the set where $z_t \to \infty$ it holds a.s. that

$$z_t^{\chi}/z_t \to \alpha \int_0^{\infty} e^{-\alpha u}\{L(a + u) - L(u)\}\,du/\{1 - \hat{L}(\alpha)\},$$

[under the assumptions of Theorem (6.10.1)]. By bounded convergence $E[z_t^{\chi}/z_t|z_t > 0]$ has the same limit as $t \to \infty$. $\quad \square$

Background

The important lemma (6.10.8) is from Harris's book. The ideas for rest are adapted from References [20] and [2], where the convergence of z_t^a was established. The relaxing of the condition $\sigma^2 < \infty$ to $\vartheta < \infty$ is from Reference [6]. It is worth pointing out that no martingale or other arguments are known, which would allow dispensing with the second moment of the reproduction distribution.* Let me also stress that it is the a.s. convergence that leads to asymptotic results of the same type for random fractions like z_t^a/z_t. It is hopeless to say anything about ratios of random variables, converging in distribution or to deduce the mean square behaviour of ratios from mean square convergence of variables. Compare Corollary (6.8.2), though.

6.11 THE STABLE AGE DISTRIBUTION AND PROCESSES WITH AN ANCESTOR DIFFERING FROM ITS PROGENY

For supercritical, non-lattice processes we proved that

$$m_t^a/m_t \to K(a) = \int_0^a e^{-\alpha t}\{1 - L(t)\}\,dt \Big/ \int_0^\infty e^{-\alpha t}\{1 - L(t)\}\,dt.$$

When the reproduction variance is finite, we saw further that the *actual age distribution* $z_t^a/z_t \to K(a)$ a.s. on the set where the population does not become extinct. This latter convergence can of course not be extended to the other cases. The ratio of means, however, still converges to $K(a)$ in the critical (with $\alpha = 0$) and subcritical (with $\alpha < 0$) Malthusian cases, provided the denominator of $K(a)$ converges. Since pointwise convergence of distribution functions towards a continuous proper distribution implies uniform convergence, we summarize:

Theorem (6.11.1)

In a branching process with non-lattice μ, *admitting a Malthusian parameter* α, *the* expected age distribution m_t^a/m_t *converges uniformly in* $0 \le a \le \infty$ *to* $K(a)$, *as* $t \to \infty$, *provided* $e^{-\alpha t}\{1 - L(t)\}$ *satisfies Condition (5.2.7) and*

$$\beta = \int_0^\infty ue^{-\alpha u}\mu(du) < \infty.$$

Changing the point of view somewhat, let us now assume that the population starts at $t = 0$ from an ancestor, whose age at that time is b. This leads to a process

$$z_t^a[b] = z_{t+b}^a(0) + \sum_{b \le \tau_0(n) \le t+b} z_{t+b-\tau_0(n)}^a \circ S_n, \qquad (6.11.1)$$

---

* Recall the footnote of p. 12, though.

where the probability law of $z_{t+b}^a(0)$ and the $\tau_0(n)$ is determined by the condition that $\lambda_0 > b$. Thus, define for $L(b) < 1$

$$m_t^a[b] = E[z_t^a[b] | \lambda_0 > b]$$

$$= 1_{[0,a)}(t+b)\{1 - L(t+b)\}/\{1 - L(b)\} + \int_b^{t+b} m_{t+b-u}^a \mu(du)/\{1 - L(b)\},$$

(6.11.2)

where we used $\mu_b(A) = E[\xi(A); \lambda > b] = E[\xi(A)] = \mu(A)$ for $A \subset (b, \infty)$.

We also consider processes starting from a b-aged ancestor whose reproduction process is given. Such processes (and the original one) we shall count by the random characteristic

$$\chi = 1_{[0,a \wedge \lambda)} 1_A(\xi),$$

$A \in \mathscr{B}(\mathcal{N})$ (see Section 5.4), so as to obtain information also about the reproduction processes in force at different times. We write $z_t^{a,A}$ for z_t^χ and $z_t^{a,A}[b,\rho]$ if the ancestor, of age b, has a given reproduction process ρ. Note that these processes are random measures in A and that Equations (6.11.1) and (6.11.2) take the forms, $\tau_\rho(n) = \inf\{t; \rho(t) \geq n\}$,

$$z_t^{a,A}[b,\rho] = z_{t+b}^{a,A}(0) + \sum_{b \leq \tau_\rho(n) \leq t+b} z_{t+b-\tau_\rho(n)}^{a,A} \circ S_n,$$

(6.11.3)

$$m_t^{a,A}[b,\rho] = 1_{[0,a)}(t+b) 1_A(\rho) P[\lambda > t + b | \lambda > b, \xi = \rho]$$

$$+ \int_0^{t+b} m_{t+b-u}^{a,A} \rho(du).$$

(6.11.4)

If in the original process we are given complete information about the ages of existing individuals at some time t and about their reproduction processes, then the future process is the sum of independent processes of type $\{z_u^{a,A}[b,\rho], u \geq 0\}$. Let us write Z_t for the process giving the information and note that Z_t is nothing but the random map $(b, B) \to z_t^{b,B}, 0 \leq b \leq \infty$, $B \in \mathscr{B}(\mathcal{N})$. Thus

$$E[z_{t+u}^{a,A} | Z_t] = \int_{R_+ \times \mathcal{N}(R_+)} m_u^{a,A}[b,\rho] z_t^{db, d\rho}.$$

Taking expectations on both sides results in

$$m_{t+u}^{a,A} = \int m_u^{a,A}[b,\rho] m_t^{db, d\rho}.$$

(6.11.5)

If the process is Malthusian and non-lattice, we multiply by $e^{-\alpha t}$. Let $t \to \infty$. Since the present characteristic satisfies

$$E[\chi(t)] = 1_{[0,a)}(t) P[\lambda > t, \xi \in A],$$

Theorem (6.9.2) implies that

$$e^{-\alpha t}m_t^{a,A} \to \int_0^a e^{-\alpha u}P[\lambda > u, \xi \in A]\,du/\beta = r(a, A).$$

Defining $R(a, A) = r(a, A)/r(\infty, \mathcal{N}(R_+))$ we thus have reason to suspect

Theorem (6.11.2)

For supercritical, non-lattice processes with Malthusian parameter α

$$e^{\alpha u}R(a, A) = \int m_u^{a,A}[b, \rho]\,R(db, d\rho).$$

If the process is not supercritical but admits a Malthusian parameter $\alpha \le 0$ the relation still holds provided the functions $e^{-\alpha t}\{1 - L(t)\}$ and $e^{-\alpha t}E[\xi(\infty); \lambda > t]$ satisfy Condition (5.2.7), and $\beta < \infty$.

Proof. According to Equation (6.9.1)

$$\int m_u^{a,A}[b, \rho]\,m_t^{db, d\rho} = E[m_u^{a,A}[t, \xi]; \lambda > t] + \int_0^t \mu(dv)\int m_u^{a,A}[b, \rho]\,m_{t-v}^{db, d\rho}.$$

As

$$E[m_u^{a,A}[t, \xi]; \lambda > t] \le P[\lambda > t] + \int_t^{u+t} m_{u+t-v}\mu_t(dv)$$

$$\le 1 - L(t) + E[y_u]\,E[\xi(\infty); \lambda > t],$$

by Equation (6.11.4), the renewal theorem applies to yield

$$\lim_{t \to \infty} e^{-\alpha t}\int_0^\infty m_u^{a,A}[b, \rho]\,m_t^{db, d\rho} = \int_0^\infty e^{-\alpha t} \cdot E[m_u^{a,A}[t, \xi]; \lambda > t]\,dt/\beta$$

$$= \int m_u^{a,A}[t, \rho]\,r(dt, d\rho),$$

where the last line just uses the definition of r. □

The function R emerges as a stable age-cum-reproduction distribution. Its marginal is the stable age distribution, $K(a) = R(a, \mathcal{N}(R_+))$. Taking $A = \mathcal{N}(R_+)$ in the theorem yields, in usual notation for conditional laws,

$$e^{\alpha u}K(a) = \int_0^\infty K(db)\int_n m_u^a[b, \rho]\,R(d\rho|b).$$

But from the definition (6.11.8) of R

$$R(B|b) = P[\lambda > b, \xi \in B]/P[\lambda > b] = P[\xi \in B|\lambda > b].$$

Since therefore

$$m_u^a[b] = E\left[m_u^a[b, \xi] \,|\, \lambda > b\right] = \int_n m_u^a[b, \rho] \, R(\mathrm{d}\rho|b)$$

we obtain

Corollary (6.11.3)
Under the conditions of Theorem (6.11.2)

$$e^{\alpha u} K(a) = \int_0^\infty m_u^a(b) \, K(\mathrm{d}b).$$

The corollary exhibits the sense in which the stable age distribution is precisely stable: Start the process from an ancestor whose age is random with distribution K. Then for any $0 \le a \le \infty$ the expected number of individuals aged a or less divided by the expected population size, the expected age distribution, remains $K(a)$ as time passes. (Note that the expected age distribution is not the same as the expectation of the actual age distribution). Furthermore, the expected size of the population at time t is in this case exactly $e^{\alpha t}$.

The converse statement is unfortunately not true; a population can grow exponentially (i.e. the mean population size is exactly a constant times an exponential) without having a stable age structure. This is the case e.g. with a Bellman–Harris process with exponential life lengths.

Let me also point at a simple fact which will be very useful in applications later. If an individual is sampled at random from a non-extinct population with known age structure at time t, then the probability distribution of its age is z_t^a/z_t and therefore the probability without conditioning upon the age structure that the age be $\le a$ is $E\left[z_t^a/z_t \,|\, z_t > 0\right]$. By Corollary (6.10.3) and dominated convergence this tends to $K(a)$ as $t \to \infty$. Thus, if individuals are sampled from old populations we may assume their ages to be stably distributed.

Example (6.11.4)
Consider binary splitting with $p > 1/2$. Then $L = (1 - p) F + pG$ and α is determined by $2p\hat{G}(\alpha) = 1$. Since

$$\int_0^\infty p e^{-\alpha t} G(t) \, \mathrm{d}t = p\hat{G}(\alpha)/\alpha = 1/2\alpha$$

$$K(a) = 2\alpha \int_0^a e^{-\alpha t}\{1 - (1 - p) F(t) - pG(t)\} \, \mathrm{d}t / \{1 - 2\hat{F}(\alpha)(1 - p)\}.$$

If there is no cell death this reduces to

$$K(a) = 2\alpha \int_0^a e^{-\alpha t} \{1 - G(t)\} \, dt.$$

Example (6.11.4)

In a process which has mean reproduction one and $L(0) = 0$ $K'(0) =$ $= 1/E[\lambda]$ by the definition of K [Equation (6.8.1)] whereas generally, as $\alpha \to 0$, $E[\lambda] = 1/K'(0) + O(\alpha)$, the last factor having the sign of α. This has some demographic appeal: In a stable population with mean reproduction essentially one suppose that you know the total population size n and the number of babies n_0 born during a short time interval, say a year. Then the mean life span of individuals is

$$E[\lambda] \approx 1/K'(0) \approx 1/K(1) \approx n/n_0.$$

Sweden has a fairly stable population though the age group around forty years is underrepresented. There is also a (very moderate) population increase. Thus we should expect n/n_0 to underestimate the life expectation. Indeed, using figures from 1968 n/n_0 is 67·84 for the female population and 63·89 for men. The life expectancies of newly borns that year were 76·13 and 71·72, respectively.   □

We have now considered processes where the (ξ_x, λ_x) are i.i.d. for $x \neq 0$ and (ξ_0, λ_0) has a specific distribution determined by the stable age distribution. More generally we might pose the question what happens if (ξ_0, λ_0) follows an arbitrary law, satisfying of course, $P[\xi_0(\infty) = \xi_0(\lambda_0)] = 1$. The process thus obtained, $\bar{z}_t^a$, can be written

$$\bar{z}_t^a = \bar{z}_t^a(0) + \sum_{n=1}^{\bar{\xi}_0(t)} z_{t-\bar{\tau}_0(n)}^a \circ S_n, \qquad (6.11.7)$$

where the bar indicates a changed distribution but otherwise things are as before. It follows that

$$\bar{\varphi}_t^a(s) = E[s^{\bar{z}_t^a}] = E[s^{\bar{z}_t^a(0)} \exp \int_0^t \log \varphi_{t-u}^a(s) \, \bar{\xi}_0(du)]. \qquad (6.11.8)$$

Similarly

$$\bar{m}_t^a = 1_{[0,a)}(t) \{1 - \bar{L}(t)\} + \int_0^t m_{t-u}^a \bar{\mu}(du) \qquad (6.11.9)$$

and $\bar{c}_u^{a,b}$ is obtained by barring the L, μ, μ_t and γ in Equation (6.4.3). Provided $\bar{\xi}_0(\bar{\lambda}_0)$ is a.s. finite, so is the new process. The asymptotics of the mean follows from Equation (6.11.9): if $m_t^a \sim ce^{\alpha t}$, $\alpha \in R$, then $\bar{m}_t^a \sim c\hat{\bar{\mu}}(\alpha)e^{\alpha t}$ if only $\hat{\bar{\mu}}(\alpha) < \infty$. From this and Equation (6.4.3) we may conclude that in the supercritical non-lattice case with finite reproduction variance

$$\bar{c}_u^{a,b}(t) \sim k(a)\,k(b)\,\{\bar{\vartheta} - \bar{\varphi} + (\vartheta - \varphi)\,\bar{\varphi}/(1 - \varphi)\}\,e^{2\alpha t + \alpha u},$$

where $\bar{\vartheta}$ and $\bar{\varphi}$ are defined in terms of $\bar{\gamma}$ and $\bar{\mu}$ as ϑ and φ from γ and μ. Hence also $e^{-\alpha t}\bar{z}_t^a$ converges in mean square to some limit variable. Its form can be seen from Equation (6.11.7), which indeed implies that a.s.

$$e^{-\alpha t}\bar{z}_t^a \rightarrow k(a) \sum_{n=1}^{\bar{\xi}_0(\bar{\lambda}_0)} e^{\alpha\bar{\tau}_0(n)} z^{(n)},$$

where the $z^{(n)}$ are independent with the distribution of z in Theorem (6.8.1). The asymptotic age distribution thus remains, the corresponding of course being true for processes z_t^χ counted by a random characteristic.

For critical processes the moment asymptotics are trivial. Extinction probabilities do not behave much worse; write $q_t = P[z_t = 0]$ and $\bar{m} = = E[\bar{\xi}_0(\infty)]$, assumed finite. Equation (6.11.8) with $s = 0$ implies that

$$tP[\bar{z}_t > 0] = t\left\{1 - E\left[\prod_{n=1}^{\bar{\xi}(\infty)} q_{t-\bar{\tau}(n)}; \bar{\lambda} \le t/2\right]\right\}$$

$$- tE\left[\prod_{n=1}^{\bar{\xi}(t)} q_{t-\bar{\tau}(n)}; \bar{\lambda} > t/2\right].$$

The latter term is $\le t\{1 - \bar{L}(t/2)\}$, which tends to zero provided $\bar{L}$ is taken to satisfy $t\{1 - \bar{L}(t)\} \to 0$. The former term equals

$$E\left[\sum_{n=1}^{\bar{\xi}(\infty)} t\{1 - q_{t-\bar{\tau}(n)}\} \prod_{k=n+1}^{\bar{\xi}(\infty)} q_{t-\bar{\tau}(k)}; \bar{\lambda} \le t/2\right]$$

$$= E\left[\sum_{n=1}^{\bar{\xi}(\infty)} \{t - \bar{\tau}(n)\}\{1 - q_{t-\bar{\tau}(n)}\} \prod_{k=n+1}^{\bar{\xi}(\infty)} q_{t-\bar{\tau}(k)} t/(t - \bar{\tau}(n)); \lambda \le t/2\right].$$

As $\bar{\tau}(n) \le \bar{\lambda}, t/\{t - \bar{\tau}(n)\} \le 2$ on the set where $\bar{\lambda} \le t/2$.

Dominated convergence yields

$$tP[\bar{z}_t > 0] \to 2\bar{m}\beta/\sigma^2$$

in the notation and under the conditions of Theorem (6.6.10) (a). Similarly, it can be shown that if the asymptotic exponentiality of Theorem (6.6.11) (c) or Theorem (6.6.17) is valid $\bar{z}_t/t$, given $\bar{z}_t > 0$, will have an exponential limit in distribution.

Finally a word about the non-lattice subcritical case with a Malthusian parameter and satisfying Equations (6.7.2) and (6.7.3): Theorem (6.7.2) leads to

$$\lim_{t \to \infty} e^{-\alpha t}P[\bar{z}_t > 0] = c\hat{\bar{\mu}}(\alpha)$$

if $c = \lim_{t \to \infty} e^{-\alpha t}P[z_t > 0]$. An analogue of Theorem (6.7.3) also holds.

Background
The stable age distribution has many applications e.g. to cell multiplication. It has been studied extensively by demographers. For references turn to Chapter 8.

6.12 THE TOTAL PROGENY

When $z_t \to 0$, i.e. always in not supercritical cases, the total number of individuals born by t, y_t, has a finite limit as $t \to \infty$. Indeed,

$$y_t \to \sum_{n=0}^{\infty} \zeta_n,$$

whose distribution was treated in Section 2.11. All results from there remain true; in particular the generating function of the total number of cells in a binary splitting process with cell disintegration probability $1 - p \geq 1/2$ remains that of Equation (2.11.6).

To study the time dependence and also the relations between z_t and y_t, we resort to the usual moment considerations, recalling that $y_t = z_t^\chi$ for $\chi_x(t)$ identically one. Equations (6.9.1), (6.9.2), and (6.9.3) turn into

$$E[y_t] = 1 + \int_0^t E[y_{t-u}] \mu(du), \qquad (6.12.1)$$

$$E[y_t y_{t+u}] = 1 + \int_0^{t+u} E[y_{t+u-v}] \mu(dv) + \int_0^t E[y_{t-v}] \mu(dv)$$

$$+ \int_{\substack{0 \leq v \leq t \\ 0 \leq w \leq t+u}} E[y_{t-v}] E[y_{t+u-w}] \gamma(dv, dw)$$

$$- \int_0^t E[y_{t-v}] E[y_{t+u-v}] \mu(dv) + \int_0^t E[y_{t-v} y_{t+u-v}] \mu(dv), \quad (6.12.2)$$

$$E[y_t z_t] = m_t + \int_0^t E[y_{t-u}] \mu_t(du) + \int_{0 \leq u, v \leq t} E[y_{t-u}] m_{t-v} \gamma(du, dv)$$

$$- \int_0^t E[y_{t-u}] m_{t-u} \mu(du) + \int_0^t E[y_{t-u} z_{t-u}] \mu(du). \qquad (6.12.3)$$

In the subcritical case this yields the same asymptotics as Equations (2.11.4), (2.11.5), and (2.12.3). In the critical (and throughout this section non-lattice) case, $\beta = \int_0^\infty u \mu(du)$,

$$E[y_t] \sim t/\beta \qquad (6.12.4)$$

as $t \to \infty$, by Corollary (5.2.14). By Equation (6.12.2) and the same corollary

$$E[y_t y_{t+u}] \sim \sigma^2 t^3/3\beta^3, \qquad (6.12.5)$$

where $\sigma^2 = \mathrm{Var}\,[\xi(\infty)]$, supposed finite, is the reproduction variance. And

$$E[y_t z_t] \sim \sigma^2 t^2 \int_0^\infty uL(du)/2\beta^3. \qquad (6.12.5)$$

Thus if $m = 1$, the correlation coefficient between y_t and z_t converges as time passes to

$$\left\{ 3 \int_0^\infty uL(du) \right\}^{1/2} \bigg/ 2\beta.$$

This is in full analogy with Equation (2.12.3).

The asymptotics of supercritical processes were displayed in Theorem (6.9.2.) and Corollary (6.10.2), $\beta = \int_0^\infty ue^{-\alpha u}\mu(du)$,

$$E[y_t] \sim e^{\alpha t}/\alpha\beta,$$

$$E[y_t y_{t+u}] \sim e^{2\alpha t + \alpha u}(\vartheta - \varphi)/(1 - \varphi)(\alpha\beta)^2,$$

$$E[y_t z_t] \sim e^{2\alpha t} \int_0^\infty e^{-\alpha t}\{1 - L(u)\}\,du(\vartheta - \varphi)/(1 - \varphi)\alpha\beta^2,$$

$$e^{-\alpha t}y_t \to z/\alpha\beta,$$

the last convergence holding a.s. and in mean square.

Background

For a discussion of similar problems in terms of convergence for supercritical Bellman–Harris processes (with a possibly infinite reproduction variance) see Reference [22].

6.13 INTEGRALS OF BRANCHING PROCESSES AND THEIR GENERALIZATIONS

Since a Galton–Watson process can be viewed as a general branching process with all life lengths equal to one, the sums

$$y_n = \sum_{k=0}^n z_k$$

of Sections 2.11 and 2.12 could be given the somewhat overambitions form

$$y_n = \int_0^n z_u\,du.$$

Thus, the total progeny by t, y_t, is not the sole imaginable counterpart of y_n in the general case. In this section we study

$$x_t = \int_0^t z_u\,du.$$

There is also a rationale for this from applications: if individuals (say virulent bacteria or viruses) produce toxins (or cells produce some secretion) at a constant rate, then the total amount of toxins produced up to time t should be proportional to x_t.

As was pointed out in Example (6.9.1), $x_t = z_t^\chi$ for $\chi(t) = t \wedge \lambda$. If different individuals produce their secretion independently and randomly we arrive at the more general model with χ a random measure with $P[\chi(\lambda, \infty) = 0] = 1$. We treat that situation, assuming that $E[\chi^2(\infty)]$ is finite. Note that the process y_t of the preceding section is a particular case of this, $\chi(t)$ placing always mass one at the origin.

Equations (6.9.1)–(6.9.3) yield, for subcritical processes, that

$$m_t^\chi \to E[\chi(\infty)]/(1 - m), \tag{6.13.1}$$

$$c_u^\chi(t) \to \{(1 - m)^2 E[\chi^2(\infty)] + 2(1 - m) E[\chi(\infty)]$$

$$E[\chi(\infty)\,\xi(\infty)] + (\sigma^2 + m^2 - m) E^2[\chi(\infty)]\}/(1 - m)^3, \tag{6.13.2}$$

$$c(t) \to 0. \tag{6.13.3}$$

To get more explicit expressions for the special case x_t, note that here

$$E[\chi(\infty)] = E[\lambda] = \int_0^\infty uL(\mathrm{d}u),$$

$$E[\chi^2(\infty)] = E[\lambda^2].$$

Of course, the processes are assumed non-lattice and with finite reproduction variance.

If $m = 1$, we must take recourse to Corollary (5.2.14): as $t \to \infty$

$$m_t^\chi \sim E[\chi(\infty)]\, t/\beta, \tag{6.13.4}$$

$$c_u^\chi(t) \sim \sigma^2 E^2[\chi(\infty)]\, t^3/3\beta^3, \tag{6.13.5}$$

$$c(t) \sim \sigma^2 E[\lambda]\, E[\chi(\infty)]\, t^2/2\beta^3, \tag{6.13.6}$$

from which an asymptotic correlation can be calculated, if desired.

The supercritical case has essentially been solved; we know the asymptotics of moments from Theorem (6.9.2), and the convergence $\mathrm{e}^{-\alpha t} z_t^\chi \to c_g z$ by Theorem (6.10.1). Let us only note that in the case of x_t,

$$g(t) = E[\lambda \wedge t] = \int_0^t \{1 - L(u)\}\, \mathrm{d}u,$$

and

$$c_g = \{1 - \hat{L}(\alpha)\}/\alpha^2\beta. \tag{6.13.7}$$

We return to the not supercritical case. Then only finitely many individuals are ever realized and since $\chi(t) \geq 0$ does not decrease in t, $z_t^\chi \uparrow$ some z^χ as $t \to \infty$. The Laplace transform

$$\varphi^\chi(s, t) = E[\mathrm{e}^{-sz_t^\chi}\},$$

$s \geq 0$, satisfies

$$\varphi^\chi(s, t) = E\left[e^{-s\chi_0(t)} \prod_{n=1}^{\xi_0(t)} \varphi^\chi(s, t - \tau_0(n))\right].$$

Let $t \to \infty$ in this. With

$$\varphi^\chi(s) = E[e^{-sz^\chi}]$$

monotone convergence implies that

$$\varphi^\chi(s) = E[e^{-s\chi(\infty)}\{\varphi^\chi(s)\}^{\xi(\infty)}]. \tag{6.13.8}$$

If now φ and ψ are two functions between zero and one both satisfying Equation (6.13.8), then

$$|\varphi(s) - \psi(s)| \leq E[e^{-s\chi(\infty)}|\varphi(s)^{\xi(\infty)} - \psi(s)^{\xi(\infty)}|]$$
$$\leq E[e^{-s\chi(\infty)}\xi(\infty)]|\varphi(s) - \psi(s)|.$$

Provided $\chi(\infty)$ is not identically zero this means that φ and ψ are the same. Hence the recursive schemes

$$\varphi_0 = 1,$$
$$\varphi_{n+1}(s) = E[e^{-s\chi(\infty)}\varphi_n^{\xi(\infty)}(s)],$$
$$\psi_0 = 0$$
$$\psi_{n+1}(s) = E[e^{-s\chi(\infty)}\psi_n^{\xi(\infty)}(s)]$$

define two sequences of functions, $\varphi_n \downarrow \varphi_\chi$ and $\psi_n \uparrow \varphi^\chi$.

Example (6.13.1)

Consider binary splitting with cell death frequency $1 - p \geq 1/2$ and $\chi(t) = t \wedge \lambda$, i.e.

$$z_t^\chi = x_t = \int_0^t z_u \, du.$$

Equation (6.13.8) reduces to

$$\varphi^\chi(s) = E[e^{-s\lambda}\{\varphi^\chi(s)\}^{\xi(\infty)}]$$
$$= (1 - p)\hat{F}(s) + p\{\varphi^\chi(s)\}^2 \hat{G}(s)$$

Hence

$$\varphi^\chi(s) = \{1 - \sqrt{4p(1 - p)\hat{F}(s)\hat{G}(s)}\}/2p\hat{G}(s).$$

It follows from this, or from monotonicity and Relation (6.13.1) that

$$E\left[\int_0^\infty z_u \, du\right] = \left\{(1 - p)\int_0^\infty uF(du) + p\int_0^\infty uG(du)\right\}\Big/(1 - 2p),$$

provided $p < 1/2$. Similarly Var $\int z_u \, du$ is expressible in terms of p and first and second moments of F and G. $\qquad\square$

Background

Integrals of birth and death processes were first treated by Puri [25]. His results were generalized to Bellman–Harris processes by myself [23] and by Waugh [27]. Asymptotics of the mean and second moments have been discussed by Weiner [28], who also studied the sum of ages of cells alive at t for $t \to \infty$. In the critical case Pakes's paper [24] gives a conditional result on the integral, given that $z_t > 0$, in the style of Theorem (2.12.3). All this is for the Bellman–Harris process.

6.14 MAXIMUM LIKELIHOOD ESTIMATION OF THE REPRODUCTION MEAN

Observe a Bellman–Harris process during a time interval $[0, t]$. A complete record of observation would give all life spans and all numbers of offspring, but also the age-distribution and number of individuals at time t. It could have the form of a vector (write ξ for $\xi(\infty)$) with elements $\sigma_x, \sigma_x \leq t$, and λ_x, ξ_x for $\sigma_x + \lambda_x \leq t$. The likelihood is

$$\prod_{\sigma_x + \lambda_x \leq t} L(d\lambda_x) \, p_{\xi_x} \prod_{\sigma_x \leq t < \sigma_x + \lambda_x} \{1 - L(t - \sigma_x)\}, \qquad (6.14.1)$$

and thus the log-likelihood term depending upon the reproduction law is

$$\sum_{\sigma_x + \lambda_x \leq t} \log p_{\xi_x} = \sum_{k=0}^{\infty} d_{tk} \log p_k.$$

Here d_{tk} is the number of individuals who have begotten exactly k children during $[0, t]$. Arguing as in Theorem (2.13.3), we obtain maximum likelihood estimators

$$\hat{p}_k = d_{tk}/d_t, \qquad k \in Z_+ \qquad (6.14.2)$$

(provided $d_t > 0$) and, by Lemma (2.13.2),

$$\hat{m}_t = (y_t - 1)/d_t = (y_t - 1)/(y_t - z_t).$$

This obviously generalizes to a start from z_0 ancestors of different ages,

$$\hat{m}_t = (y_t - z_0)/(y_t - z_t). \qquad (6.14.3)$$

The process $\{d_{tk}; t \geq 0\}$ is a branching process with the random characteristic $\chi_k(t) = 1_{\{\rho; \rho(\infty) = k\}}(\xi) \, 1_{[\lambda, \infty)}(t)$. Theorem (6.10.1) applies and since

$$E[\chi_k(t)] = P[\lambda \leq t, \xi(\infty) = k]$$

it follows that for any non-lattice supercritical process with finite reproduction variance a.s.

$$d_{tk}/d_t \to 1_{\{z_t \to 0\}} y(k)/y + 1_{\{z_t + 0\}}$$

$$\int_0^\infty e^{-\alpha u} P[\lambda \le u, \xi(\infty) = k]\, du \Big/ \int_0^\infty e^{-\alpha u} L(u)\, du, \quad (6.14.4)$$

as $t \to \infty$. Here y and $y(k)$ are the total progeny and the number of individuals in the whole population begetting exactly k children, respectively. In the Bellman–Harris case, and indeed as soon as $\xi(\infty)$ and λ are independent,

$$\int_0^\infty e^{-\alpha u} P[\lambda \le u, \xi(\infty) = k]\, du = p_k \int_0^\infty e^{-\alpha u} L(u)\, du.$$

Thus $d_{tk}/d_t \to p_k$ a.s. on the set where the process does not become extinct. On this same set a.s. $\hat{m}_t = (y_t - z_0)/d_t \to 1/\hat{L}(\alpha) = m$, where the last equality holds for Bellman–Harris processes. And further, since

$$y_t - md_t = \sum_{\sigma_x + \lambda_x \le t} \{\xi_x(\infty) - m\}.$$

Theorem A2 of the Appendix or, rather, its corollary applies exactly as to Galton–Watson processes, yielding that

$$(\hat{m}_t - m)\sqrt{d_t} \xrightarrow{d} 1_{\{z_t \to 0\}}(1 - z_0/\sqrt{y}) + 1_{\{z_t + 0\}} N(0, \sigma^2), \quad (6.14.5)$$

for Bellman–Harris processes, $\sigma^2 < \infty$ denoting the reproduction variance.

As the reader might have noted, the qualities of the estimators derived, d_{tk}/d_t and $\hat{m}_t$, hinge on the independence of life and reproduction. For binary splitting the likelihood can be written

$$\prod_{\sigma_x + \lambda_x \le t} \{pG(d\lambda_x)\}^{\xi_x/2} \{(1 - p) F(d\lambda_x)\}^{1 - \xi_x/2}$$

$$\prod_{\sigma_x \le t < \sigma_x + \lambda_x} \{1 - pG(t - \sigma_x) - (1 - p) F(t - \sigma_x)\}$$

and if the reader cares to differentiate the logarithm of this, he/she will see how the last factor complicates things. If $F = G$, so that the process is of the Bellman–Harris type, then $d_{t2}/d_t = (y_t - z)/2d_t = \hat{p}_t$ remains the maximum likelihood estimator of p. In any case its limit (on the set of non-extinction—and assuming the process non-lattice, of course) as $t \to \infty$ is a.s. by Equation (6.14.4)

$$\hat{p}_\infty = p \int_0^\infty e^{-\alpha u} G(u)\, du \Big/ \int_0^\infty e^{-\alpha u} \{pG(u) + (1 - p) F(u)\}\, du$$

$$= 1/\{1 + 2(1 - p)\hat{F}(\alpha)\}. \quad (6.14.7)$$

Here we used $P[\lambda \le u, \xi(\infty) = 2] = pG(u)$ and $2p\hat{G}(\alpha) = 1$.

Background

The topic is somewhat further elaborated in Section 9.2.

REFERENCES

1. Crump, K. S. and Mode, C. J., A general age-dependent branching process I and II. *J. Math. Anal. Appl.* **24**, 494–508, 1968 and **25**, 8–17, 1969. See also Mode's book, *Multitype branching processes*, Elsevier, New York, 1971.

2. Jagers, P., A general stochastic model for population development. *Skand. Aktuarietidskr.* **52**, 84–103, 1969.

3. Ryan, T. A. Jr., *On age-dependent branching processes*. Thesis. Cornell University, 1968.

4. Sevast'yanov, B. A., Vetvyaščiesya processy s prevraščeniyami, zavisyaščimi ot vozrasta častic. *Teor. Veroyatnost. i primenen.* **9**, 577–594, 1964. Translated as Age-dependent branching processes, *Theor. Prob. Appl.* **9**, 521–537, 1964. See also his book, *Vetvyaščiesya processy*, Mir, Moscow, 1971.

5. Sevast'yanov, B. A., Vetvyaščiesya processy or earlier papers in *Matematičeskie Zametki* **1**, 53–61, 1967 and **7**, 389–396, 1970.

6. Ganuza, E. and Durham, S. D., *Mean square and almost sure convergence of supercritical age-dependent branching processes*. University of South Carolina Mathematics Technical Report 60J80-2, 1973.

7. Jagers, P., Diffusion approximations of branching processes. *Ann. Math. Statist.* **42**, 2074–2078, 1971.

8. Baum, L. E., and Katz, M., Convergence rates in the law of large numbers. *Trans. Amer. Math. Soc.* **120**, 109–123, 1965.

9. Durham, S. D., Limit theorems for a general critical branching process. *J. Appl. Prob.* **8**, 1–16, 1971.

10. Feller, W., *An Introduction to Probability Theory and its Applications II*. John Wiley and Sons, New York, 1966.

11. Goldstein, M. I., Critical age-dependent branching processes: single and multitype. *Z. Wahrscheinlichkeitstheorie verw. Geb.* **17**, 74–88, 1971.

12. Holte, J. M., Extinction probability for a critical general branching process. *Stoch. Processes Appl.* **2**, 303–309, 1974.

13. Holte, J. M., *Limit Theorems for Critical General Branching Processes*. Thesis, Department of Mathematics, University of Wiscoisn, Madison, Wisc. 1974.

14. Sevast'yanov, B. A., *Vetvyaščiesya processy* (Branching Processes). Mir, Moscow, 1971.

15. Weiner, H. J., Critical age-dependent branching processes. *Proc. Sixth Berkeley Symp. Math. Statist. Prob.*, 1971.

16. Athreya, K. B. and Ney, P. E., *Branching Processes*. Springer, Berlin, 1972.

17. Ryan, T. A. Jr., *On age-dependent branching processes*. Thesis, Department of Mathematics, Cornell University, 1968.

18. Athreya, K. B., On the supercritical one-dimensional age-dependent branching process. *Ann. Math. Statist.* **40**, 743–763, 1969.

19. Doney, R. A., A limit theorem for a class of branching processes. *J. Appl. Prob.* **9**, 707–724, 1972.

20. Jagers, P., Convergence of general branching processes and functionals thereof. *J. Appl. Prob.* **11**, 471–478, 1974.

21. Weiner, H. J., Applications of the age distribution in age-dependent branching processes. *J. Appl. Prob.*, **3**, 179–201, 1966.

22. Doney, R. A., The progeny of a branching process. *J. Appl. Prob.* **8**, 407–412, 1971.
23. Jagers, P., Integrals of branching processes. *Biometrika* **54**, 263–272, 1967.
24. Pakes, A. G., A limit theorem for the integral of a critical age dependent branching process. *Math. Biosci.* **13**, 109–112, 1972.
25. Puri, P. S., On the homogeneous birth and death process and its integral. *Biometrika* **53**, 61–71, 1966.
26. Puri, P. S., Some limit theorems on branching processes and certain related processes. *Sankhyā* **31A**, 57–74, 1969.
27. Waugh, W. A. O'N., Age dependent branching processes under a condition of ultimate extinction. *Biometrika* **55**, 291–296, 1968.
28. Weiner, H. J., Sums of lifetimes in age dependent branching processes. *J. Appl. Prob.* **6**, 195–200, 1968.
29. Jagers, P., *Maximum likelihood estimation of the reproduction distribution in branching processes and the extent of disintegration in cell proliferation.* Report 1973-17, Department of Mathematics, Chalmers U. of Tech. and the U. of Göteborg, 1973.

Chapter 7

Neighbours of the General Process

7.1 IMMIGRATION

Many populations can be thought of as obtained from the appearance (immigration) of individuals at random times, each such individual initiating a branching process, maybe of the delayed type discussed towards the end of Section 6.11. This can be expressed in the following manner: there is given a point process η on R_+, the immigration process. At each $u \in R_+$ $\eta\{u\}$ i.i.d. branching processes are initiated. For ease of notation we take these to be not delayed, since by the arguments in Section 6.11 results can easily be generalized, for example to the particularly interesting case of a Poisson immigration process (immigrants coming from many independent in themselves unsignificant sources) with newcomers having stably distributed ages.

We write y_t^a for the process under consideration, the total number of individuals younger than a, and refrain from studying a y_t^χ process, counted by a random characteristic. Note that y_t has been the total progeny in Chapter 6. Set $\vartheta(n) = \inf\{t; \eta(t) \geq n\}$ and let $\{z_t^a(n), n \in N\}$ be i.i.d. branching processes. The basic—defining, if you please—relation is

$$y_t^a = \sum_{n=1}^{\eta(t)} z_{t-\vartheta(n)}^a(n) \tag{7.1.1}$$

The process will be referred to as supercritical etc. when so are the $z_t^a(n)$.

Some preliminary information may easily be obtained through the expectations $E[y_t^a]$. We write $v(t) = E[\eta(t)]$ and see that

$$E[y_t^a] = \int_0^t m_{t-u}^a v(\mathrm{d}u).$$

For a non-lattice, supercritical process with Malthusian parameter α. Lemma (5.2.10) implies that

$$\mathrm{e}^{-\alpha t} E[y_t^a] = \int_0^t \mathrm{e}^{-\alpha(t-u)} m_{t-u}^a \mathrm{e}^{-\alpha u} v(\mathrm{d}u) \to k(a)\,\hat{v}(\alpha),$$

190

provided $\hat{v}(\alpha) < \infty$. Even in non supercritical cases this convergence may obviously take place, but then the requirement that $\hat{v}(\alpha)$ must be finite implies that the number of immigrants must remain finite. Actually, if the process is critical and for some $c > \infty$

$$v(t + u) - v(t) \to u/c, \tag{7.1.2}$$

as $t \to \infty$, then Lemma (5.2.12) and Theorem (6.3.3) enhance that

$$E[y_t^a] \sim k(a)\, t/c.$$

Thus, if $v(t) \to \infty$ it is only in the subcritical case that we could hope for $\lim_{t \to \infty} E[y_t^a]$ to exist. And so it does when Relation (7.1.2) holds and the life expectation is finite. Indeed the reasoning leading to Theorem (5.2.6) results in

$$\lim_{t \to \infty} E[y_t^a] = \int_0^\infty m_u^a \, du/c, \tag{7.1.3}$$

since the Tauberian condition (5.2.7) is satisfied: Set $M_n = \sup_{0 \le t < 1} m_{n+t}$. Evidently, if we write $\mu(0) = 0$,

$$M_n \le 1 - L(n) + \sum_{k=0}^n M_{n-k}\{\mu(k + 1) - \mu(k)\}.$$

Therefore, if $m < 1$ and the life expectation is finite,

$$\sum_{n=0}^\infty M_n s^n \le \sum_{n=0}^\infty \{1 - L(n)\} s^n \Big/ \left\{1 - \sum_{n=0}^\infty s^n(\mu(n + 1) - \mu(n))\right\}$$

$$\to \sum_{n=0}^\infty \{1 - L(n)\}/(1 - m) < \infty$$

as $s \uparrow 1$, and

$$\sum_{n=0}^\infty \sup_{0 \le t \le 1} m_{n+t}^a \le \sum_{n=0}^\infty M_n < \infty. \tag{7.1.4}$$

The generating function $\psi_t^a(s) = E[s^{y_t^a}]$ is determined by

$$\psi_t^a = E\left[\prod_{n=1}^{\eta(t)} \psi_{t - \vartheta(n)}^a\right] = E\left[\exp \int_0^t \log \varphi_{t-u}^a \eta(du)\right]. \tag{7.1.5}$$

This means (5.4) that ψ_t^a is the Laplace transform of η evaluated at $-\log \varphi_{.-u}$ ($\varphi_t = 1$ for $t \le 0$). Hence if η is compound Poisson, say with the number of immigrants per bunch having generating function h and λ now the intensity of the non compound process, it follows that

$$\psi_t^a(s) = \exp \int_0^t \{h \circ \varphi_{t-u}^a(s) - 1\}\, v(du).$$

Since
$$0 \leq 1 - h \circ \varphi_t^a(s) \leq 1 - h \circ \varphi_t(s) \leq 1 - h \circ \varphi_t(0),$$

which does not increase, the arguments relied upon to prove the renewal theorem lead to:

Theorem (7.1.1)

Consider a general branching process with immigration, the immigration process being compound Poisson such that the underlying Poisson process has intensity λ satisfying Equation (7.1.2) and that the number of immigrants at a Poisson process epoch has generating function

$$h(s) = \sum_{n=0}^{\infty} a_n s^n. \tag{7.1.6}$$

Then, if
$$\int_0^{\infty} \{1 - h \circ \varphi_u(0)\} \, du < \infty, \tag{7.1.7}$$

$$\lim_{t \to \infty} E[s^{y_t^a}] = \exp \int_0^{\infty} \{h \circ \varphi_u^a(s) - 1\} \, du/c \tag{7.1.8}$$

exists and is a probability generating function.

Proof. We must only check the last assertion. But it follows from monotone convergence as $s \uparrow 1$ in the integral.   □

Corollary (7.1.2)

If the process is subcritical, $h'(1) < \infty$, and

$$\int_0^{\infty} t L(dt) < \infty,$$

then Equation (7.1.8) holds.

Proof. Here
$$1 - h \circ \varphi_u(0) \leq h'(1) \{1 - \varphi_u(0)\} \leq h'(1) m_u,$$

which was shown to be well behaved enough in Inequality (7.1.4).   □

Corollary (7.1.3)

If the processes $z_t^a(n)$ comply with the requirements of Theorem (6.7.2), including Condition (6.7.10), and

$$\sum_{n=2}^{\infty} a_n \log n < \infty,$$

then Equation (7.1.8) is true.

Proof. In this case

$$1 - h \circ \varphi_t(0) \le 1 - h(1 - ce^{-\alpha t})$$

for some $c > 0$ and t large enough. By Relation (2.5.2)

$$\int_0^\infty \{1 - h(1 - ce^{-\alpha t})\} \, dt < \infty \Leftrightarrow \sum_{n=1}^\infty a_n \log n < \infty. \qquad \square$$

The explicitness of the limit in Theorem (7.1.1) is very formal; little is known about φ_u^a. With immigrants arriving in bunches of i.i.d. size at the epochs of a renewal process still less is obtained, merely the existence of a limit.

Theorem (7.1.4)

Consider a branching process with immigration such that the immigration process is a compound renewal process with mass generating function h at each renewal epoch. Assume that the waiting time distribution between immigration epochs is non-lattice and has a finite expectation. Then $\lim_{t \to \infty} y_t^a$ exists in distribution provided Condition (7.1.7) holds [which is the case under the assumptions of either Corollary (7.1.2) or (7.1.3)].

Proof. Let J be the distribution function of waiting times in the immigration process. The generating function of y_t^a satisfies

$$\psi_t^a = 1 - J(t) + \int_0^t h \circ \varphi_{t-u}^a \psi_{t-u}^a J(du)$$

$$= 1 - J(t) + \int_0^t \{1 - h \circ \varphi_{t-u}^a\} \psi_{t-u}^a J(du) + \int_0^t \psi_{t-u}^a J(du).$$

This is a common renewal equation with maybe a somewhat strange constant function

$$f(t) = 1 - J(t) - \int_0^t \{1 - h \circ \varphi_{t-u}^a(s)\} \psi_{t-u}^a(s) J(du),$$

after insertion of the argument s. Not bothering about the fact that f involves ψ_t^a we conclude that

$$\lim_{t \to \infty} \psi_t^a(s) = 1 - \int_0^\infty \{1 - h \circ \varphi_u^a(s)\} \psi_u^a(s) \, du \bigg/ \int_0^\infty u J(du) \qquad (7.1.9)$$

must exist if f satisfies Condition (5.2.7). But so it does under Condition (7.1.7) if the expectation of J is finite. Finally we can use dominated convergence to conclude that

$$\lim_{s \uparrow 1} \int_0^\infty \{1 - h \circ \varphi_u^a(s)\} \psi_u^a(s) \, du = 0$$

and thus the limit in Equation (7.1.9) is a proper probability generating function. $\square$

We give a treatment of critical processes with immigration resting on Theorem (6.6.11) (b) or, rather, directly on the inequality of Theorem (6.6.7). Thus it is not powerful enough to deal with general processes.

Theorem (7.1.5)

Consider a critical Bellman–Harris process with finite reproduction variance σ^2 and an immigration of i.i.d. numbers of immigrants at the epochs of a point process η. Let $0 < a < \infty$ be the expected number of immigrants per immigration epoch and assume that

$$\lim_{t \to \infty} \eta(t)/t = y$$

exists a.s. and has generating function $g(s) = E[s^y]$, $0 \leq s \leq 1$. Further assume that

$$l = \int_0^\infty t L(\mathrm{d}t) < \infty.$$

Then, with $\alpha = 2al/\sigma^2$ and $\beta = \sigma^2/2l$, $v \geq 0$,

$$\lim_{t \to \infty} E[e^{-vy_t/t}] = g\{(1 + \beta v)^{-\alpha}\}.$$

In particular, if y constant, y_t/t tends in distribution, as $t \to \infty$, to a random variable with the gamma density

$$\left(1/\Gamma(\alpha y)\, \beta^{\alpha y}\right) u^{\alpha y - 1} e^{-u/\beta}, \qquad u \geq 0.$$

Note. *The condition $t^2\{1 - L(t)\} \to 0$, essential for the proof of Theorem (6.6.11), is not needed here. We use α and β like in Section 3.1 just here.*

Proof. In the present notation Equation (7.1.5) takes the form

$$\varphi_t = E\left[\prod_{n=1}^{\eta(t)} h \circ \varphi_{t - \vartheta(n)}\right],$$

if h as before is the generating function of the number of immigrants per immigration. By dominated convergence it suffices, thus, to prove that

$$\lim_{t \to \infty} \prod_{n=1}^{\eta(t)} h \circ \varphi_{t - \vartheta(n)}(e^{-v/t}) = (1 + \beta v)^{-\alpha y}$$

a.s. for any $v \geq 0$, or equivalently that

$$\lim_{t \to \infty} \int_0^t \log h \circ \varphi_{t-u}(e^{-v/t}) \, \eta(du) = -\alpha y \log(1 + \beta v).$$

By the Taylor expansion of $-\log h$ used in the proof of Theorem (3.1.2)

$$\int_0^t -\log h \circ \varphi_{t-u}(e^{-v/t}) \, \eta(du) = a \int_0^t \{1 - \varphi_{t-u}(e^{-v/t})\} \, \eta(du) +$$

$$+ \int_0^t \rho \circ \varphi_{t-u}(e^{-v/t}) \{1 - \varphi_{t-u}(e^{-v/t})\} \, \eta(du),$$

where $\lim_{s \uparrow 1} \rho(s) = 0$. If we can prove that the former of these integrals converges as $t \to \infty$, the latter must tend to zero, as $\varphi_{t-u}(e^{-v/t}) \geq \varphi_{t-u}(0) \to \to 1$.

Thus we shall show that

$$\lim_{t \to \infty} \int_0^t \{1 - \varphi_{t-u}(e^{-v/t})\} \, \eta(du) = (2yl/\sigma^2) \log(1 + \beta v)$$

a.s. According to Theorem $(6.6.7)$ $(L = \mu)$

$$1 - f_{n_t-u}(e^{-v/t}) - (1 - e^{-v/t}) \mu^{*n_t-u}(t-u) \leq 1 - \varphi_{t-u}(e^{-v/t})$$

$$\leq 1 - f_{k_t-u}(e^{-v/t}) + (1 - e^{-v/t}) \{1 - \mu^{*k_t-u}(t-u)\},$$

where $0 < \varepsilon < 1 \wedge \sigma^2/2$ is arbitrary and

$$n_t = [t(1 + \varepsilon)/l], \qquad k_t [t(1 - \varepsilon)/l]$$

as before. Insertion of this in the integral yields upper and lower bounds for

$$\int_0^t \{1 - \varphi_{t-u}(e^{-v/t})\} \, \eta(du).$$

We consider merely the former, the latter behaving very much the same. Fist, since $1 - \mu^{*k_t}(t) \to 0$ and $1 - e^{-v/t} \sim v/t$, as $t \to \infty$, it follows that a.s.

$$\lim_{t \to \infty} (1 - e^{-v/t}) \int_0^t \{1 - \mu^{*k_t-u}(t-u)\} \, \eta(du) = 0$$

from Lemma (5.2.12). Next we use Equation (3.1.6) to expand $f_{k_t-u}(e^{-v/t})$, choosing t_0 such that $|r_{k_t}| < \varepsilon$ for $t \geq t_0$ and $i = t_0(1 - \varepsilon)/l$ is an integer. Then

$$\int_0^{t-t_0} \{1 - f_{k_t-u}(e^{-v/t})\} \, \eta(du)$$

$$= \int_0^{t-t_0} \frac{(1 - e^{-v/t}) \, \eta(du)}{1 + k_{t-u}\{\sigma^2/2 + r_{k_t-u}(e^{-v/t})\}(1 - e^{-v/t})}$$

$$\leq \int_0^{t-t_0} \frac{(1 - e^{-v/t})\,\eta(du)}{1 + k_{t-u}(\sigma^2/2 - \varepsilon)(1 - e^{-v/t})}$$

$$= \int_{t_0}^{t} \frac{(1 - e^{-v/t})\,y\,du}{1 + k_u(\sigma^2/2 - \varepsilon)(1 - e^{-v/t})} + \int_0^{t-t_0} \frac{(1 - e^{-v/t})\,\{\eta(du) - y\,du\}}{1 + k_{t-u}(\sigma^2/2 - \varepsilon)(1 - e^{-v/t})}$$

The former integral is majorized by

$$\frac{yl}{1 - \varepsilon} \sum_{j=i}^{k_t} \frac{(1 - e^{-v/t})\,du}{1 + j(\sigma^2/2 - \varepsilon)(1 - e^{-v/t})}$$

which appeared in the proof of Theorem (3.1.2). By the arguments there it has, as $t \to \infty$, a superior limit not exceeding

$$yl(1 - \varepsilon)^{-1}(\sigma^2/2 - \varepsilon)^{-1} \log\{1 + v(\sigma^2/2 - \varepsilon)(1 + \varepsilon)/l\}.$$

The latter integrals equals

$$\frac{(1 - e^{-v/t})\,\{\eta(t - t_0) - y(t - t_0)\}}{1 + k_{t_0}(\sigma^2/2 - \varepsilon)(1 - e^{-v/t})}$$

$$+ \int_0^{t-t_0} \frac{(1 - e^{-v/t})\,\{\eta(u) - yu\}\,(\sigma^2/2 - \varepsilon)}{1 + k_{t-du}(\sigma^2/2 - \varepsilon)(1 - e^{-v/t})}$$

after an integration by parts. Here a.s.

$$\left| \frac{(1 - e^{-v/t})\,\{\eta(t - t_0) - y(t - t_0)\}}{1 + k_{t_0}(\sigma^2/2 - \varepsilon)(1 - e^{-v/t})} \right| \leq v\left|\eta(t - t_0) - y(t - t_0)\right|/t \to 0.$$

As to the integral, note that the measure k_t is concentrated on the lattice $jl/(1 - \varepsilon)$, $j \in N$, each such point getting mass one. Thus for any $c > 0$ and $t - u = jl/(1 - \varepsilon)$

$$1/(1 + ck_{t-du}) = 1/(1 + cj) - 1/\{1 + c(j + 1)\}$$

$$= c/(1 + cj)\{1 + c(j + 1)\} \leq c = ck_{t-du}.$$

Hence the integral is dominated by

$$(v\sigma^2/2t)^2 \int_0^{t} \left|\eta(u) - yu\right| k_{t-du} \to 0$$

as well. After noting that a.s.

$$0 \leq \int_{t-t_0}^{t} \{1 - f_{k_{t-u}}(e^{-v/t})\}\,\eta(du) \leq (1 - e^{-v/t})\{\eta(t) - \eta(t - t_0)\} \to 0$$

(recall that $f'_{k_t - u}(1) = 1$!), we conclude that

$$\limsup_{t \to \infty} \int_0^t \{1 - \varphi_{t-u}(e^{-v/t})\}\, \eta(\mathrm{d}u) \le yl(2/\sigma^2) \log \{1 + v\sigma^2/2l\},$$

since $\varepsilon > 0$ was arbitrarily small.

Similarly

$$\liminf_{t \to \infty} \int_0^t \{1 - \varphi_{t-u}(e^{-v/t})\}\, \eta(\mathrm{d}u) \ge yl(2/\sigma^2) \log \{1 + v\sigma^2/2l\}$$

and the proof is complete. $\qquad\qquad\square$

Turning, finally, to supercritical processes we see that Equation (7.1.1) implies

$$E[y_t^a y_{t+u}^b] = \int_{\substack{0 \le u \le t \\ 0 \le w \le t+u}} m_{t-u}^a m_{t+u-w}^b E[\eta(\mathrm{d}v)\,\eta(\mathrm{d}u)]$$

$$- \int_0^t m_{t-v}^a m_{t+u-v}^b E[\eta(\mathrm{d}v)] + \int_0^t c_u^{a,b}(t-v)\, E[\eta(\mathrm{d}v)]. \quad (7.1.10)$$

Hence, if $E[\hat{\eta}^2(\alpha)] < \infty$ (α now again the Malthusian parameter),

$$\lim_{t \to \infty} e^{-2\alpha t - \alpha u} E[y_t^a y_{t+u}^b]$$
$$= k(a)\,k(b) \{E[\hat{\eta}^2(\alpha)] + E[\hat{\eta}(2\alpha)](\vartheta - 1)/(1 - \varphi)\} \quad (7.1.11)$$

It follows that

$$\lim_{t \to \infty} E[\{e^{-\alpha t} y_t^a - e^{-\alpha(t+u)} y_{t+u}^a\}^2] = 0$$

and

$$\lim_{t \to \infty} e^{-\alpha t} y_t^a = y^a$$

exists in mean square. And since

$$\lim_{t \to \infty} E[\{e^{-\alpha t} y_t^a - K(a) e^{-\alpha t} y_t\}^2] = 0$$

the stable age distribution persists, $y^a = K(a)\,y$, y being the limit of $e^{-\alpha t} y_t$.

Without going into details, let us have a look at the a.s. convergence. For simplicity we consider just y_t and shall merely convince ourselves that

$$\int_0^\infty E[\{e^{-\alpha t} y_t - y\}^2]\, \mathrm{d}t < \infty \quad (7.1.12)$$

and then refer to Lemma (6.10.7). Writing $c(t) = e^{-\alpha t} E[z_t z]$ as in Section 6.10, we see that

$$E[\{e^{-\alpha t}y_t - y\}^2] = E[e^{-2\alpha t}y_t^2] - 2E[e^{-\alpha t}y_ty] + E[y^2]$$

$$= \int_{0 \le u,v \le t} e^{-\alpha(t-u)}m_{t-u}e^{-\alpha(t-v)}m_{t-v}e^{-\alpha(u+v)}E[\eta(du)\,\eta(dv)]$$

$$+ \int_0^t e^{-2\alpha(t-u)}\{c_0(t-u) - m_{t-u}^2\}\,e^{-2\alpha u}E[\eta(du)]$$

$$- 2\left\{ \int_0^t e^{-\alpha(t-u)}m_{t-u}k(\infty)\,e^{-\alpha u}E[\hat\eta(\alpha)\,\eta(du)] \right.$$

$$\left. + \int_0^t \{c(t-u) - e^{-\alpha(t-u)}m_{t-u}k(\infty)\}\,e^{-\alpha u}E[\eta(du)]\right\}$$

$$+ k^2(\infty)\{E[\hat\eta(\alpha)] + E[\hat\eta(2\alpha)](\vartheta - 1)/(1 - \varphi)\}.$$

From this Inequality (7.1.12) follows as in Section 6.10. We summarize:

Theorem (7.1.6)

Consider a general supercritical and non-lattice branching process with immigration process η. If the process has finite reproduction variance and, α the Malthusian parameter, $E[\hat\eta^2(\alpha)] < \infty$, then for some random variable y and $0 \le a \le \infty$

$$e^{-\alpha t}y_t^a \to K(a)\,y_\infty$$

almost surely and in mean square. Here

$$E[y] = k(\infty)\,E[\hat\eta(\alpha)]$$

and

$$\text{Var}\,[y] = k^2(\infty)\{\text{Var}\,[\hat\eta(\alpha)] + E[\hat\eta(\alpha)](\vartheta - 1)/(1 - \varphi)\}.$$

It goes without saying that there is a corresponding theorem for immigration processes counted with random characteristics.

Background

Except in Reference [1] beneath earlier only Bellman–Harris processes with renewal immigration have been considered. The difference is not very substantial, though. The methods used here follow References [3] and [7] closely. Pakes and Kaplan [5] give Inequality (7.1.7) as a necessary and sufficient conditon for convergence in the Bellman–Harris case of Theorem (7.1.4). During 1974 several papers on immigration have appeared in *J. Appl. Prob. and Stoch. Proc. Appl.*

7.2 INCREASING NUMBERS OF ANCESTORS

The results in Section 3.2 on Galton–Watson processes depended upon two types of arguments, the basic theorems for the super-, sub-, and just critical cases, and classical limit theory for sums of i.i.d. random variables, which entered through the independence of processes stemming from different ancestors. The latter property persists in the general case and thus we should expect results to follow from the basic theorems here much as in discrete time. Indeed, let $r_t \to \infty$ be integers and denote by $\{z_t^a(r_t)\}$ a process started from r_t ancestors, taken for simplicity all to be newly born at time zero.

Theorem (7.2.1)

Consider a non-lattice supercritical process with finite reproduction variance and $P[\hat{\xi}(\alpha) = 1] < 1$. Let r_t and $\{z_t^a(r_t)\}$ be as above. Then, for $0 \le a \le \infty$

$$\{z_t^a(r_t) - k(a) e^{\alpha t} r_t\} \sqrt{(1 - \varphi)/k(a) e^{\alpha t}} \sqrt{\vartheta - 1} \xrightarrow{d} N(0, 1) \qquad (7.2.1)$$

as $t \to \infty$.

The proof of this follows the proof of Theorem (3.2.1) literally.

Theorem (7.2.2)

If $m < 1$, $E[\hat{\xi}(\alpha) \log \xi(\infty)] < \infty$ in a Malthusian process and all other conditions of Theorem (6.7.2) are satisfied, then with $r_t \sim e^{-\alpha t}$,

$$\lim_{t \to \infty} E[s^{z_t^a(r_t)}] = \exp c\{g_a(s) - 1\},$$

where

$$c = \lim_{t \to \infty} e^{-\alpha t} P[z_t > 0]$$

and g_a is the generating function of the limit distribution in Theorem (6.7.3). If instead $e^{\alpha t} r_t \to \infty$ and $\sigma^2 < \infty$ a normal limit Equation (7.2.1) prevails.

Theorem (7.2.3)

Consider the case of a critical Bellman–Harris process with $\sigma^2 < \infty$ and $t\{1 - L(t)\} \to 0$. Then for all $u > 0$

$$\lim_{n \to \infty} E[e^{-uz_t(r_t)/t}] = \exp\{-2lu/(2l + u\sigma^2)\}$$

if only $r_t \sim t$. If $r_t/t \to \infty$, then again Equation (7.2.1) is true provided norming is by $b_t \sim \sigma t k(a)/l$, where $k(\infty)$ is assumed finite if we want the theorem to hold for $a = \infty$. As usual l is the expected span of life.

We omit the proofs.

7.3 CONVERGENCE TOWARDS CRITICALITY

If the reproduction is interwoven with the time structure in some complex manner, as might well be the case with general processes, the simultaneous passage to the limit of $m \to 1$ and $t \to \infty$ is difficult to get at. We contend ourselves with treating Bellman–Harris processes, where Goldstein's approach from Section 6.6 and Theorem (3.3.1) will cooperate fruitfully.

All the lemmata of Section 6.6 concerning generating functions at zero where true for arbitrary processes. As earlier, let f_k be the generating function of the kth generation. It follows that

$$-\mu^{*k}(t) \le f_k(0) - \varphi_t(0) \le (1 - L) * \sum_{j=0}^{k-1} \mu^{*j}(t),$$

and in the Bellman–Harris case, $\mu = mL$,

$$-m^k L^{*k}(t) \le f_k(0) - \varphi_t(0) \le (1 - L) * \sum_{j=0}^{k-1} m^j L^{*j}(t)$$

$$= 1 - m^{k-1} L^{*k}(t) + (m - 1) \sum_{j=1}^{k-1} m^{j-1} L^{*j}(t)$$

$$= m^{k-1} \{1 - L^{*k}(t)\} - (m - 1) \sum_{j=1}^{k-1} m^{j-1} \{1 - L^{*j}(t)\}$$

$$\le \{m^{k-1} + |m^{k-1} - 1|\} \{1 - L^{*k}(t)\} \le (1 + 2m^{k-1}) \{1 - L^{*k}(t)\}. \quad (7.3.1)$$

Let us now consider a set $\mathscr{K}_{\alpha,L}$ of Bellman–Harris processes, all with the same life span distribution L and a reproduction variance not less than α,* these variances also converging uniformly in the class. We also require that 1 be contained in the closure of all reproduction means of processes contained in $\mathscr{K}_{\alpha,L}$. Then Theorem (3.3.1) applies. So does Corollary (6.6.9) if only L satisfies the two requirements there. Thus, in the notation there

$$\liminf_{\substack{t \to \infty \\ m \to 1}} t\{f_t(0) - \varphi_t(0)\} \ge \liminf_{\substack{t \to \infty \\ m \to 1}} - tm^{n_t} L^{*n_t}(t) = 0,$$

provided the passage to the limit is such that m^t remains bounded. Under this same condition

$$\limsup_{\substack{t \to \infty \\ m \to 1}} t\{f_{k_t}(0) - \varphi_t(0)\} \le \limsup_{\substack{t \to \infty \\ m \to 1}} t(1 + 2m^{k_t - 1}) \{1 - L^{*k_t}(t)\} = 0.$$

By the concluding arguments of Theorem (6.6.11), the following has been shown at least partially:

---

* Any positive number, not a Malthusian parameter.

Theorem (7.3.1)

Let $\mathscr{K}_{a,L}$ be a class of Bellman–Harris processes as above. Assume that the expected life length is l and that $t^2\{1 - L(t)\} \to 0$. Write $a = f''(1)/2$ for any process is $\mathscr{K}_{a,L}$. In $\mathscr{K}_{a,L}$ let $m \to 1$ as time $t \to \infty$ so that $t(m - 1) \to c$. Then

$$atP[z_t > 0] \to c/(1 - e^{-c/l}) \tag{7.3.2}$$

to be interpreted as l if $c = 0$. In that case also

$$P[z_t/at \le u|z_t > 0] \to 1 - \exp -ul. \tag{7.3.3}$$

If L is not lattice,

$$E[z_t/at|z_t > 0] \to (1 - e^{-c/l})/c, \tag{7.3.4}$$

with the same interpretation as above for $l = 0$. In these expressions a should of course be taken as the entity of the $\mathscr{K}_{a,L}$-process in question.

Proof. Only details remain to justify the first and last claim. They are left out. The asymptotic exponentiality is more difficult to get at, since Lemma (6.6.5) cannot be applied. Reconsidering the arguments there, we find that

$$\varphi_t^{(1)}(s) = E[s^{z_t(0) + \xi_0(t)}] = f(s) + E[s^{z_t(0) + \xi_0(t)} - s^{\xi_0(\infty)}; \lambda_0 > t]$$

$$\le f(s) + E[s - s^{\xi(\infty)}; \lambda > t] \le f(s) + (1 - s)E[\xi(\infty) - 1; \lambda > t]$$

$$\le f(s) + |m - 1|(1 - s) = g(s).$$

This holds for all branching processes with $\xi(\infty)$ independent of the life span λ. It follows directly that

$$\varphi_t^{(2)}(s) \le g\{f(s) + |m - 1|(1 - s)\}.$$

To proceed we recall Lemma (3.3.3), implying that m can be chosen so close to one that $f'(0) > |m - 1|$. We assume that to be done. Since $g'(s) = = f'(s) - |m - 1|$, g is then increasing. Further $g'(1) = m - m + 1 = 1$ for $m \ge 1$ and $g'(1) = 2m - 1 < 1$ otherwise. Hence

$$\varphi_t^{(2)}(s) \le g \circ f(s) + g'(1)|m - 1|(1 - s) \le f_2(s) + |m - 1|\{1 - f(s)\} + $$

$$+ |m - 1|(1 - s) \le f_2(s) + (1 + m)|m - 1|(1 - s).$$

By induction

$$\varphi_t^{(k)}(s) \le f_k(s) + |m - 1|(1 - s)\sum_{j=0}^{k-1} m^j = f_k(s) + (1 - s)|m^k - 1|$$

for all $k \in N$. Joining this together with Lemmata (6.6.1) and (6.6.6) and the bound on $(1 - L) * \sum_{j=0}^{k-1} \mu^{*j}$ from Relation (7.3.1), we perceive that

$$-(1 - s)\{m^k L^k(t) + |m^k - 1|\} \le \varphi_t^{(k)}(s) - \varphi_t(s) - (1 - s)|m^k - 1| \le$$

$$\leq f_k(s) - \varphi_t(s) \leq \varphi_t^{(k)}(s) - \varphi_t(s) + (1 - s)(1 - L) * \sum_{j=0}^{k-1} m^j L^{*j}(t)$$

$$\leq (1 - s)(2m^{k-1} + 1)\{1 - L^{*k}(t)\}$$

for any s, k, t. If we let now $t \to \infty$ and $m - 1 = o(t)$, then for any $u > 0$

$$\liminf t\{f_{n_t}(e^{-u/t}) - \varphi_t(e^{-u/t})\} = \limsup t\{f_{k_t}(e^{-u/t}) - \varphi_t(e^{-u/t})\} = 0.$$

This in conjunction with Equation (7.3.2) proves Equation (7.3.3).   □

Background

Theorem (7.3.1) is new, though incomplete, Equation (7.3.3) being proved only for $c = 0$. For the lattice case some further results have been obtained by Reference [9].

7.4 DIFFUSION APPROXIMATIONS

This section serves to establish a counterpart to Theorem (3.4.1) for Bellman–Harris processes. Some background and heuristics were given in Section 3.4, so we manage without circumlocutions here.

Theorem (7.4.1)

Consider a sequence of Bellman–Harris processes $\{z_{n,t}; t \geq 0\}$, $n \in N$, each with reproduction generating function f_n,

$$f_n(s) = \sum_{k=0}^{\infty} p_{nk} s^k,$$

and life span distribution L. Write $m_n = f_n'(1)$, $2b_n = f_n''(1)$ and assume that as $n \to \infty$

$$m_n = 1 + a/n + o(1/n), \quad some \quad a \in R, \tag{7.4.1}$$

$$b_n \to \quad some \quad b > 0 \tag{7.4.2}$$

$$for\ all \quad \varepsilon > 0 \sum_{k > \varepsilon n} k^2 p_{nk} \to 0. \tag{7.4.3}$$

Further require that $L(0) = 0$ and

$$l = \int_0^\infty t L(dt) < \infty.$$

Then, for each $t \geq 0$ the process $z_{n,nt}(n)/n$, started from n ancestors, $\xrightarrow{d} x_t$, where $\{x_t; t \geq 0\}$ is the diffusion starting from $x_0 = 1$ and having infinitesimal drift ax/l and variance $2bx/l$, $x \geq 0$.

Note. *The reader may wish to replace n by $r_n \to \infty$ as in Theorem (3.4.1).*

The diffusion terminology is graphic, but has little bearing on the proof. Indeed, turning back to Equation (3.4.5) we realize, that the real matter is the convergence

$$E\left[e^{-uz_{n,nt(n)/n}}\right] = \left\{\varphi_{n,nt}(e^{-u/n})\right\}^n$$

$$\to \varphi(t, u) = \begin{cases} \exp\left\{-ue^{at/l}/(1 + ub(e^{at/l} - 1)/a)\right\}, & \text{if } a \neq 0, \\ \exp\left\{-u/(1 + ubt/l)\right\}, & \text{if } a = 0, \end{cases} \quad (7.4.4)$$

for $u, t \geq 0$. What we shall show is the equivalent assertion

$$\lim_{n \to \infty} n\left\{1 - \varphi_{n,nt}(e^{-u/n})\right\} = -\log \varphi(t, u), \quad (7.4.5)$$

for any $u, t \geq 0$.

Proof. The starting point is a Taylor expansion of the reproduction generating function in Equation (6.3.3),

$$\varphi_{n,nt}(e^{-u/n}) = e^{-u/n}\left\{1 - L(nt)\right\} + \int_0^{nt} f_n \circ \varphi_{n,nt-v}(e^{-u/n})\, L(dv)$$

$$= e^{-u/n}\left\{1 - L(nt)\right\} + L(nt) - m_n \int_0^t \left\{1 - \varphi_{n,n(t-v)}(e^{-u/n})\right\} L(n\, dv)$$

$$+ b_n \int_0^t \left\{1 - \varphi_{n,n(t-v)}(e^{-u/n})\right\}^2 L(n\, dv) - \int_0^t r_n \circ \varphi_{n,n(t-v)}(e^{-u/n})\, L(n\, dv).$$

If we introduce for fixed $u \geq 0$

$$g_n(t) = n\left\{1 - \varphi_{n,nt}(e^{-u/n})\right\},$$

this can be written

$$g_n(t) = n(1 - e^{-u/n})\left\{1 - L(nt)\right\}$$

$$+ m_n \int_0^t g_n(t - v)\, L(n\, dv) - (b_n/n) \int_0^t g_n^2(t - v)\, L(n\, dv)$$

$$+ n \int_0^t r_n \circ \varphi_{n,n(t-v)}(e^{-u/n})\left\{1 - \varphi_{n,n(t-v)}(e^{-u/n})\right\}^2 L(n\, dv). \quad (7.4.6)$$

Taking Laplace–Stieltjes transforms of this and multiplying by n, we obtain

$$b_n\hat{L}(s/n)\hat{g}_n^2(s) + n\left\{1 - m_n\hat{L}(s/n)\right\}\hat{g}_n(s)$$

$$-n(1 - e^{-u/n})n\left\{1 - \hat{L}(s/n)\right\} - \rho_n(s) = 0. \quad (7.4.7)$$

Here

$$\rho_n(s) = n^2\hat{L}(s/n)s \int_0^\infty e^{-st} r_n \circ \varphi_{n,nt}(e^{-u/n})\left\{1 - \varphi_{n,nt}(e^{-u/n})\right\}^2 dt.$$

We first show that this tends to zero as $n \to \infty$. By Theorem (6.5.5) either $\varphi_{n,nt}(e^{-u/n}) \leq q_n$ or $\varphi_{n,nt}(e^{-u/n}) \leq e^{-u/n}$, if q_n denotes the relevant extinction probability. It is however not difficult to check that $q_n = 1 - a/bn + o(1/n)$ in case $a > 0$, and we prove that in Lemma (7.4.2). Hence,

$$g_n(t) = n\{1 - \varphi_{n,nt}(e^{-u/n})\} \leq n(1 - q_n) + n(1 - e^{-u/n})$$

remains bounded as $n \to \infty$. Obviously $\hat{L}(s/n) \to 1$. If we can prove that $r_n \circ \varphi_{n,nt}(e^{-u/n}) \to 0$ boundedly, then $\rho_n \to 0$ would follow. However for $0 \leq v \leq 1$

$$0 \leq r_n(v) \leq f_n''(1) - f_n''(v) \leq \sum_{k=0}^{\varepsilon n} k^2 p_{nk}(1 - v^k) + \sum_{k > \varepsilon n} k^2 p_{nk}.$$

But for any $\varepsilon > 0$

$$\limsup_{n \to \infty} \sum_{k=0}^{\varepsilon n} k^2 p_{nk}(1 - e^{-ku/n}) \leq \limsup_{n \to \infty} u \sum_{k=0}^{\varepsilon n} k^3 p_{nk}/n$$

$$\leq \limsup_{n \to \infty} u \sum_{k=0}^{\varepsilon n} \varepsilon n k^2 p_{nk}/n = \varepsilon \sigma^2 u.$$

By Assumption (7.4.3) $r_n \circ \varphi_{n,nt}(e^{-u/n}) \to 0$, and the convergence is bounded since $r_n \leq \sup_n f_n''(1) < \infty$ by Equation (7.4.2).

As to the remaining terms in Equation (7.4.7),

$$b_n \hat{L}(s/n) \to b,$$

$$n\{1 - m_n \hat{L}(s/n)\} = n\{1 - \hat{L}(s/n)\} + \hat{L}(s/n) n(1 - m_n) \to ls - a,$$

and

$$n(1 - e^{-u/n}) n\{1 - \hat{L}(s/n)\} \to uls.$$

The functions g_n themselves do not increase in t if only $e^{-u/n} \leq q_n$, as Theorem (6.5.5) informs us. However by Lemma (7.4.2) beneath this is finally the case if only $u > a/b$. Further $g_n(0) = n(1 - e^{-u/n})$ and $g_n(\infty) = n(1 - q_n)$ [Theorem (6.5.5) again] and Helly's selection theorem (page 263 of Reference [10]) yields a subsequence n_k such that $g_{n_k} \to$ some limit g weakly. We let $n_k \to \infty$ in Equation (7.4.7), applying the continuity theorem for Laplace transforms to convince ourselves that g must satisfy

$$b\hat{g}^2(s) + (ls - a)\hat{g}(s) - uls = 0. \tag{7.4.8}$$

By the following lemmata this shows that g must be of the asserted form $-\log \varphi(t, u)$. $\qquad\qquad\qquad\qquad\qquad\qquad\qquad\qquad\square$

Lemma (7.4.2)

If f_n satisfies Relations (7.4.1) and (7.4.2) with $a \geq 0$, then $q_n = 1 - - a/bn + o(1/n)$, as $n \to \infty$.

Proof.

$$q_n = f_n(q_n) = 1 - m_n(1 - q_n) + \{b_n + r_n(q_n)\}(1 - q_n)^2$$

and

$$1 - q_n = \{a + o(1)\}/\{nb_n + nr(q_n)\}. \qquad \square$$

Lemma (7.4.3)

Equation (7.4.8) admits the solution $-\log \varphi(t, u)$, defined by Equation (7.4.4).

Proof. We consider only $a \neq 0$, the other case can be treated in the same manner. Note that Equation (7.4.6) has a sense also if f_n is not probability generating and consider it for

$$f_n(s) = 1 - (1 + a/n)(1 - s) + b_n(1 - s)^2,$$

$$L(t) = 1 - e^{-t/l}.$$

It reduces to the Riccati differential equation

$$g_n' = ag_n/l - bg_n^2/l$$

$$g_n(0) = n(1 - e^{-u/n}).$$

The solution (page 21 of Reference [13]) is

$$g_n(t) = an(1 - e^{-u/n})e^{at/l}/\{a + bn(1 - e^{-u/n})(e^{at/l} - 1)\},$$

which converges to $-\log \varphi(t, u)$, as $n \to \infty$. Since g_n satisfies Equation (7.4.7) we let $n \to \infty$ to obtain that g solves Equation (7.4.8). $\qquad \square$

Lemma (7.4.4)

For a given positive initial value Equation (7.4.8) cannot have more than one solution, which is bounded on finite intervals.

Proof. Assume $a \neq 0$ and consider two solutions, $g = -\log \varphi(\cdot, u)$ and $h, h(0) = u$. Then, for $s > a/l$

$$\hat{g}(s) - \hat{h}(s) = b\{\hat{h}^2(s) - \hat{g}^2(s)\}/(ls - a).$$

Since $b/(ls - a)$ is the transform of $be^{at/l}/a$, it follows that

$$g(t) - h(t) = \int_0^t \{h^2(u) - g^2(u)\} e^{-au/l} \, du \, b \, e^{at/l}/l.$$

The function h must be differentiable,

$$g'(t) - h'(t) = a\{g(t) - h(t)\}/l + b\{h^2(t) - g^2(t)\}/l$$

and for some constant c

$$g(t) - h(t) = c \exp\left\{\left\{ at - b\int_0^t (g(u) + h(u))\,du\right\} \Big/ l \right\}.$$

Since $g(0) = h(0)$ also $c = 0$ and $h = g$. $\qquad\square$

Background

This is taken from Reference [12], the condition (7.4.3) being formulated in Reference [11], where function space convergence is also discussed.

REFERENCES

1. Durham, S. D., A problem concerning generalized age-dependent branching processes with immigration. *Ann. Math. Statist.* **42**, 1121–1123, 1971.
2. Foster, J. H., *The asymptotic behaviour of critical age-dependent branching processes with immigration.* Mimeographed, 1972.
3. Jagers, P., Age-dependent branching processes allowing immigration. *Teor. Veroyatnost. Primenen.* **13**, 230–242, 1968. Correction, ibid. p. 755.
4. Kaplan, N., *Limit theorems for an age-dependent branching process with immigration.* Mimeographed, 1972.
5. Pakes, A. G. and Kaplan, N. L., On the subcritical Bellman–Harris process with immigration. *J. Appl. Prob.* **11**, 652–668, 1974.
6. Kawazu, K., The limit distribution of the age-dependent branching process with immigration in the supercritical case. *Tamkang J. Math.* **3**, 9–15, 1972.
7. Pakes, A. G., Limit theorems for an age-dependent branching process with immigration. *Math. Biosci.* **14**, 221–234, 1972.
8. Radcliffe, J., The convergence of a supercritical age-dependent branching process allowing immigration at the epochs of a renewal process. *Math. Biosci.* **14**, 37–44, 1972.
9. Nagaev, S. V., Perehodnye yavleniya dlya zavisyaščih ot vozrasta vetvyaščihsya processov s diskretnym vremenem I. (Transition phenomena for age dependent branching processes with discrete time). *Sibirsk. Mat. Ž.* **15**, 367–394, 1974.
10. Feller, W., *An Introduction to Probability Theory and its Applications II.* John Wiley and Sons, New York, 1966.
11. Grimvall, A., *Approximation of Bellman–Harris branching processes.* Report LiH-MAR-R-73-6, Department of Mathematics, Linköping University, Linköping, 1973.
12. Jagers, P., Diffusion approximations of branching processes. *Ann. Math. Statist.* **42**, 2074–2078, 1971.
13. Kamke, E., *Differentialgleichungen I.* 2. Auflage. Akademische Verlagsgesellschaft, Leipzig, 1943.

Chapter 8

Branching Processes and Demography

8.1 CLASSICAL CONTINUOUS TIME DEMOGRAPHY AND AGE DEPENDENT BIRTH AND DEATH PROCESSES

The growth of human and all bisexual populations is heavily dependent upon sex ratios and matching patterns. Classical demographic approaches disregard this and focus upon the female population. They are deterministic though assumptions are expressed in terms of individual probabilities. We shall see that they can be viewed as the study of expectations in certain branching processes.

Conventional assumptions for continuous time demographic theory can be summarized in four points (Reference [7]):

1. There is no migration.

2. There are, to all $a \geq 0$, probabilities $p(a)$ of an individual surviving to age a. Sometimes p is given as $l/l(0)$, where l is the number of individuals attaining age a in a cohort of size $l(0)$—the *radix*, say $= 100 \cdot 000$. The function p is usually taken to be continuous.

3. The probability that an individual, who is alive at age a, should beget one child during the age interval $(a, a + \Delta a]$ is $m(a)\Delta a + o(\Delta a)$, the probability of begetting none being $1 - m(a)\Delta a + o(\Delta a)$, independently of what occurred in lower ages.

4. The probability of survival p is zero for arguments $>$ some ω and m vanishes outside some interval contained in $[0, \omega]$.

This last requirement, though natural in a demographic context, is essentially an awkward way of guaranteeing the applicability of the renewal theorem. We shall not make use of it. The reader should interpret the words 'individual', 'child' etc. as referring always to females.

A further, though usually tacit, assumption is that individuals are independent of one another (though this is not needed all through as long as the aim is only conclusions on expectations). This leads to a branching process, started from several individuals with known ages. The reproduction process has the special property

$$P[\xi(a, a + \Delta a) = 0] = P[\lambda \leq a] + P[\lambda > a] P[\xi(a, a + \Delta a) = 0 | \lambda > a]$$

$$= 1 - p(a) m(a) \Delta a + o(\Delta a)$$

207

by Assumption 3. Since there it is also assumed (the independence) that

$$P[\xi(a + \Delta a) = 0] = P[\xi(a) = 0] P[\xi(a, a + \Delta a] = 0]$$

the usual differential argument implies that ξ is a Poisson process (5.4) with parameter measure (see Section 5.4) equal to the reproduction function μ,

$$\mu(a) = \int_0^a p(u) m(u) \, du. \tag{8.1.1}$$

The life distribution is $L = 1 - p$. The pair (ξ, λ) can be thought of as given by $\xi(t) = \eta(t \wedge \lambda)$ where η and λ are independent and η is Poisson with parameter

$$\int_0^a m(u) \, du.$$

The two functions m and the derivative $\mu' = pm$ are called the *gross* and *net maternity functions*, respectively, by demographers. Similarly, in the present context $\mu(\infty)$ is called the *net reproduction rate* and

$$E[\eta(\infty)] = \int_0^\infty m(u) \, du$$

the *gross reproduction rate*. Often also the existence of a *hazard function*,

$$d(t) = -\frac{d}{dt} \log p(t) = -p'(t)/p(t)$$

is required. The resulting branching process is then called an *age dependent birth and death process*. The functions m and d are referred to as *age specific birth* and *death intensities*.

Among branching processes the age dependent birth and death process is singled out by one important characteristic, which it shares with splitting processes, namely that given the number of individuals existing and their ages at any time point the future development of the process is independent of the past. Indeed, if z_t is made up of one individual aged a_1, one aged a_2 etc, then in the future the process evolves as the sum of z_t independent processes, started from one ancestor of age a_1, one of age a_2 and so forth. We shall call such processes *Markovian in the age structure*. To see that all branching processes need not have this property consider one where individuals beget just one child whose birth is uniformly distributed over the mother's life span. Assume that the latter is always of unit length and say that two individuals were observed at $t = 1/2$, the process being started as usual from a newly born ancestor at time zero.

The importance of Markovianness in the age structure is, of course, that it gives reason for basing predictions only on the age distribution at hand.

Background

In mathematical demography there are two trends, a discrete time theory, touched upon in Example (4.1.2), and the basic continuous time theory described here. The latter stems back to Lotka [8] and is often referred to as stable population theory, the reason being that a particularly important initial age distribution Q is precisely the stable one. The established demographic text is from Reference [6], more mathematical presentations being given in References [1] and [9]. A survey is Reference [2]. As pointed out in Section 1.1 the concept of an age dependent birth and death process is due to Kendall [5]. A fine recent mathematical presentation of birth and death processes and associated demographic problems is Reference [4]. The paper [3] develops parts of the theory in Chapter 6 specifically for birth and death processes.

8.2 LOTKA'S EQUATION

The basic relation in classical continuous time demography is known as Lotka's equation. It is a convolution equation similar in appearance to Equation (6.3.4) but still different. The typical derivation of it goes as follows: Let $b(t)$ be the density of births at t—not further explained—and assume that there is a continuous density q describing the distribution of the population over age intervals at time zero, the integral of q being the total number of individuals then. Let $g(t)$ be the density of births at t due to individuals existing at time zero. Then $b(t)$ consists of $g(t)$ and a part due to individuals coming from mothers who were themselves born during $[0, t]$. The latter will be

$$\int_0^t b(t - u)\, p(u)\, m(u)\, du$$

whereas g must satisfy

$$g(t) = \int_0^\infty q(a)\, m(a + t)\, p(a + t)/p(a)\, da,$$

since, in order to contribute, an individual of age a at time zero must be alive at time t, which she is with probability $p(a + t)/p(a)$, and give birth at time t with density $m(a + t)$. The result is *Lotka's integral equation*:

$$b(t) = \int_0^\infty q(a)\, m(a + t)\, p(a + t)/p(a)\, da + \int_0^t b(t - u)\, p(u)\, m(u)\, du. \quad (8.2.1)$$

Though this derivation is appealing it is far from complete. Also the Assumptions 1–4 from Section 8.1 were not all needed. And, actually, we can prove Lotka's equation much more generally for all branching processes satisfying a minor regularity condition. However then, if we do not require

Markovianness in the age structure, we must change the interpretation so that the process starts from just one individual with an arbitrary age distribution.

Theorem (8.2.1)

Consider a general branching process started from an ancestor whose age was random with a distribution Q such that

$$\int_0^\infty \frac{Q(da)}{1 - L(a)} < \infty.$$

Assume that the derivative μ' exists and is bounded. Then the derivative $b(t)$ of the total number of individuals born by t exists and satisfies

$$b(t) = \int_0^\infty \frac{\mu'(t + a)\, Q(da)}{1 - L(a)} + \int_0^t b(t - u)\, \mu(du). \tag{8.2.2}$$

For an age dependent birth and death process Equation (8.2.2) takes the form of Equation (8.2.1).

Proof. Consider a process started from an a-aged ancestor at $t = 0$. Let $y_t[a]$ be the total progeny stemming from it and born during the time interval $[0, t]$. As in Equation (6.11.3)

$$y_t[a] = 1 + \sum_{a \le \tau_0(n) \le t + a} y_{t + a - \tau_0(n)} \circ S_n, \tag{8.2.3}$$

where the decomposition is made in lines from the first generation individuals. However $y_t[a]$ can also be written (recall the definition of $y_t(x)$ from Section 6.2)

$$y_t[a] = 1 + \sum_{a \le \tau_0(n)} y_{t+a}(n) + \sum_{\substack{a \le \tau_0(n) \\ x \ne 0}} y_{t+a}(n, x). \tag{8.2.4}$$

If we define

$$y_{t,n} = \sum_{x \in N^n} y_t(x),$$

in analogy with $z_{t,n}$ in Section 6.7 this turns into

$$y_t[a] = 1 + \sum_{a \le \tau_0(n)} y_{t+a}(n) + \sum_{\substack{a \le \tau_0(n) \\ x \in I \\ k \in N}} y_{t+a}(n, x, k)$$

$$= 1 + \sum_{a \le \tau_0(n)} y_{t+a}(n) + \sum_{\substack{a \le \tau_0(n) \\ x \in I}} y_{t+a-\sigma(n,x),1} \circ S_{(n,x)}$$

by the argument in Lemma (6.1.1). Thus we have obtained a decomposition according to later lines. Taking expectations results in

$$E[y_t[a]|\lambda_0 > a] = 1 + \{\mu(t + a) - \mu(a)\}/\{1 - L(a)\}$$

$$+ \int_0^t E[y_{t-u,1}] E[y_{du}[a]|\lambda_0 > a] - E[y_{t,1}],$$

the last term arising since in the sum over all individuals (n, x), $n \in N$, $x \in I$, the ancestor did never appear. But as $E[y_{t,1}] = \mu(t)$ we have deduced that

$$E[y_t[a]|\lambda_0 > a] = 1 + \{\mu(t + a) - \mu(a)\}/\{1 - L(a)\}$$

$$+ \int_0^t \{E[y_{t-u}[a]|\lambda_0 > a] - 1\} \mu(du). \quad (8.2.5)$$

Now assume that the ancestor has a random age with distribution Q when the process starts. Integration of Equation (8.2.5) yields

$$E[\bar{y}_t] = 1 + \int_0^\infty \{\mu(t + a) - \mu(a)\}/\{1 - L(a)\} Q(da)$$

$$+ \int_0^t \{E[\bar{y}_{t-u}] - 1\} \mu(du),$$

the bar indicating the modified reproduction and life distribution of the ancestor. By the convergence and boundedness conditions in the theorem this can be differentiated into Equation (8.2.2).   □

Note. *It is worth observing that Equation (8.2.6) was derived for general branching processes without any further qualifications. Since $E[\bar{y}_t]$ does not decrease in t, its derivative certainly exists a.e. with respect to Lebesgue measure. It is to interchange differentiation and integratiation that we need the regularity conditions.*

Background

Lotka's equation [8] is a standard topic in demographic texts, like those quoted in the proceeding section.

8.3 THE GROWTH OF POPULATIONS

Consider a non-lattice branching process with the Malthusian parameter α, starting from an ancestor with age distribution Q. Assume Equation (8.2.2) satisfied and such that there are no difficulties in applying the renewal theorem, as is the case for a supercritical process. Then, as $t \to \infty$

$$b(t) \sim e^{\alpha t} \int_0^\infty v(a) Q(da) \Big/ \int_0^\infty t e^{-\alpha t} \mu(dt), \quad (8.3.1)$$

where

$$v(a) = e^{\alpha a} \int_a^\infty e^{-\alpha t} \mu(dt)/\{1 - L(a)\}. \qquad (8.3.2)$$

Thus $v(a)$ measures the reproductive power of an a-aged individual. It was introduced and termed the *reproductive value* by Fisher (Reference [11]).

The reproductive value has a further interesting property, also pointed out by Fisher:

Theorem (8.3.1)

Define

$$v_t = \int_0^\infty v(a) z_t^{da} \qquad (8.3.3)$$

for branching process with Malthusian parameter α. Then $E[v_t] = e^{\alpha t}$.

Proof. By Equation (6.3.4)

$$E[v_t] = \{1 - L(t)\} v(t) + \int_0^t E[v_{t-u}] \mu(du)$$

$$= e^{\alpha t} \int_t^\infty e^{-\alpha u} \mu(du) + \int_0^t E[v_{t-u}] \mu(du),$$

which is easily checked to have the unique solution $e^{\alpha t}$.   □

The counting of individuals by their v-values results in a martingale under suitable circumstances. To see this, define

$$v_t[b] = \int_0^\infty v(a) z_t^{da}[b]$$

and note that

$$E[v_t[b]|\lambda_0 > b] = v(t + b) \{1 - L(t + b)\}/\{1 - L(b)\}$$

$$+ \int_b^{t+b} E[v_{t+b-u}] \mu(du)/\{1 - L(b)\}$$

$$= e^{\alpha t} \left\{ e^{\alpha b} \int_{t+b}^\infty e^{-\alpha u} \mu(du)/\{1 - L(b)\} \right.$$

$$\left. + e^{\alpha b} \int_b^{t+b} e^{-\alpha u} \mu(du)/\{1 - L(b)\} \right\} = e^{\alpha t} v(b). \qquad (8.3.4)$$

Hence,

Theorem (8.3.2)

Let $\{z_t^a\}$ be a branching process which is Malthusian and Markovian in the age structure. Define $\mathscr{B}_t$ as the σ-algebra $\sigma\{z_u^a, 0 \le u \le t, a \in R_+\}$. Then $\{e^{-\alpha t}v_t; t \ge 0\}$ is a non-negative martingale with respect to $\{\mathscr{B}_t; t \ge 0\}$ and $v_\infty = \lim e^{-\alpha t}v_t$ exists a.s. (by continuous parameter martingale convergence which we have not discussed, see Reference [12]).

Rest of Proof.

$$E[v_{t+u}|\mathscr{B}_t] = \int_0^\infty E[v_u[a]|\lambda_0 > a]\, z_t^{da} = \int_0^\infty e^{\alpha u}v(a)\, z_t^{da} = e^{\alpha u}v_t. \qquad \square$$

The counting of individuals by their reproductive values thus means that inhomogeneities in population growth due to the age distribution disappear. The *crude birth* and *death rates*, on the other hand are the traditional means of expressing the expected rate of change in the conventionally counted population, i.e. to catch a little of these inhomogeneities. The former is defined by

$$\bar{b}(t) = \int_0^\infty m(a)\, m_t^{da}/m_t,$$

the latter by

$$\bar{d}(t) = \int_0^\infty d(a)\, m_t^{da}/m_t$$

in an age dependent birth and death process.

Provided the functions m and d remain bounded the well known renewal arguments apply to prove that

$$\lim_{t \to \infty} \bar{b}(t) = \int_0^\infty m(a)\, K(da) = 1/k(\infty) \qquad (8.3.5)$$

and

$$\lim_{t \to \infty} \bar{d}(t) = \int_0^\infty d(a)\, K(da) = 1/k(\infty) - \alpha \qquad (8.3.6)$$

for supercritical and non-lattice processes. From Section 6.10 it is also clear that for such processes the *actual birth* and *death rates*,

$$\bar{b}_t = \int_0^\infty m(a)\, z_t^{da}/z_t$$

and

$$\bar{d}_t = \int_0^\infty d(a)\, z_t^{da}/z_t,$$

converge to the same limit a.s. on the set where the population does not become extinct (if the second reproduction moment exists). Let us also note that the *crude* and *actual rates of increase*

$$\bar{\alpha}(t) = \bar{b}(t) - \bar{d}(t)$$

and

$$\bar{\alpha}_t = \bar{b}_t - \bar{d}_t$$

tend under the above circumstances to the Malthusian parameter.

The limits in Equations (8.3.5) and (8.3.6) have been given a name of their own, *intrinsic birth* and *death rates*. From Corollary (6.11.3) we see that if the process starts from an ancestor with a stably distributed age, then the crude rates are constant and equal to the intrinsic ones. By the Markovianness in the age structure it holds also that once the expected age distribution in a population of several individuals equals the stable one, then the crude rates are and remain the intrinsic ones.

Background

The reproductive value was introduced by R. A. Fisher in Reference [11]. It has been suggested repeatedly [10] that it be used to establish the a.s. convergence of branching processes. Apparently no progress has been made.* The crude and intrinsic rates defined are well established in demographic literature [6]. This last reference also gives several numerical applications.

8.4 THE AGE AT CHILDBEARING

In all applications of the renewal theorem we have made there appears the denominator

$$\beta = \int_0^\infty t e^{-\alpha t} \mu(dt).$$

In demographic literature it goes under the name of *the average age at childbearing*. The idea is that if newly born individuals are sampled at random (whatever this means) from the population, then the average age of their mothers should be precisely β. We shall give some basis for this.

Consider a non-lattice and supercritical branching process with finite reproduction variance. Define, for any $0 \le a \le \infty$, $\chi_a(t) = \xi(a \wedge t)$. Then

$$z_t^{\chi_a} = \sum_x \xi_x(a \wedge (t - \sigma_x)) \tag{8.4.1}$$

is the number of individuals born by t and by mothers aged a or less. Theorem (6.10.1) evidently applies and so a.s.

---

* See, however, the footnote of p. 12.

$$e^{-\alpha t}z_t^{\chi a} \to z\int_0^\infty e^{-\alpha t}E\left[\xi(a \wedge t)\right]dt/\beta = z\left\{\int_0^a e^{-\alpha t}\mu(t)\,dt + e^{-\alpha a}\mu(a)/\alpha\right\}\bigg/\beta$$

$$= z\int_0^a e^{-\alpha t}\mu(dt)/\alpha\beta.$$

Since

$$e^{-\alpha t}z_t^\chi\infty = e^{-\alpha t}(y_t - 1) \to z/\alpha\beta$$

—the ancestor is not counted by Equation (8.4.1)—we conclude that the fraction of individuals born by mothers not older than a

$$z_t^{\chi a}/y_t \to \int_0^a e^{-\alpha t}\mu(dt) \qquad (8.4.2)$$

a.s. on the set where $z_t \not\to 0$. Similarly the reader will concede that the expected age at childbearing satisfies

$$\int_0^\infty az_t^{\chi da}/y_t = z_t^\chi/y_t$$

for

$$\chi(t) = \int_0^t a\xi(da)$$

and hence coverges in the same sense to the average age at childbearing.

It is also clear that the expected fractions $m_t^{\chi a}/E[y_t]$ and $m_t^\chi/E[y_t]$ will have these limits under slightly more general circumstances. The function

$$\bar{\mu}(t) = \int_0^t e^{-\alpha u}\mu(du) \qquad (8.4.3)$$

may be referred to as the *stable distribution of the age at childbearing*.

Background

The basis for thinking of β as an average age of childbearing has been purely intuitive. Some aspects of it are discussed in a rather different vein by Keiding [4]. The given formulation is from Reference [13].

8.5 THE LENGTH OF GENERATIONS

The common sense concept of generation length is vague. In supercritical branching processes—we consider only non-lattice such processes in this section—it invites at least three different precise formulations. First, the length of a generation might be an individual's expected time from birth to childbearing,

$$b = \int_0^\infty t\mu(dt)/m. \qquad (8.5.1)$$

—Since we shall not use the gross maternity function of Section 8.1 any more we return to the notation $m = \mu(\infty)$.—This is the way the term 'generation time' is used by biologists, cf. Example (6.2.4) and Chapter 9.

Second the generation length could be the maternal age at which 'most' children are born, β, 'most' in the sense of the preceeding section. This yields the definition

$$\beta = \int_0^\infty te^{-\alpha t}\mu(dt).$$

Third, a sensible interpretation of the concept would be the time $\bar{t}$ it takes for a population with a stable age distribution to change its expected size by the factor m, being the expected contribution per individual. Then

$$e^{\alpha\bar{t}} = m$$

or

$$\bar{t} = \{\log m\}/\alpha. \qquad (8.5.2)$$

This definition is used in References [1] and [6]. It will turn out that b and β both have some good properties and that $\bar{t}$ is a sound compromise between them at least for α close to zero. At this juncture we only note that by Jensen's inequality

$$\alpha\beta = \int_0^\infty (\log e^{\alpha t})e^{-\alpha t}\mu(dt) \le \log \int_0^\infty e^{\alpha t}e^{-\alpha t}\mu(dt)$$

$$= \log m = -\log\left\{\int_0^\infty e^{-\alpha t}\mu(dt)/m\right\} \le \alpha \int_0^\infty t\mu(dt)/m = \alpha b. \qquad (8.5.3)$$

Let us define, as in Section 6.7,

$$z_{t,n} = \sum_{x\in N^n} z_t(x)$$

and

$$m_{t,n} = E[z_{t,n}]. \qquad (8.5.4)$$

Obviously

$$m_{t,0} = 1 - L$$

and

$$m_{t,n+1} = E\left[\sum_{k=1}^{\xi_0(t)} z_{t-\tau_0(k),n} \circ S_k\right] = \int_0^t m_{t-u,n}\mu(du).$$

Hence, for $n \in Z_+$

$$m_{t,n} = (1 - L)*\mu^{*n}(t) \tag{8.5.5}$$

and if g is the density of the stable age distribution,

$$g(a) = e^{-\alpha a}\{1 - L(a)\}/k(\infty),$$

then

$$e^{-\alpha t}m_{t,n} = g*\bar{\mu}^{*n}(t)\,k(\infty), \tag{8.5.6}$$

where $\bar{\mu}$ as defined in Equation (8.4.3) is the stable distribution of the age at childbearing.

Theorem (8.5.1)

For $-\infty < u < \infty$ *let* $t_n = \beta n + u\beta_2\sqrt{n} + o(\sqrt{n})$, β_2^2 *being the variance of* $\bar{\mu}$. *Let* φ *denote here the standardized normal density. Then, if the process is supercritical and not lattice,*

$$\lim_{n \to \infty} \sqrt{n}\,e^{-\alpha t_n}m_{t_n,n} = k(\infty)\,\varphi(u)/\beta_2.$$

Proof. First, the third moment of $\bar{\mu}$,

$$\beta_3 = \int_0^\infty t^3 e^{-\alpha t}\mu(dt),$$

is finite. By Esséen's expansion (page 512 of Reference [15])

$$\bar{\mu}^{*n}(\beta n + u\beta_2\sqrt{n} + o(\sqrt{n})) = \Phi(u) + \beta_3(1 - u^2)\,\varphi(u)/6\beta_2^3\sqrt{n} + o(1/\sqrt{n}),$$

Φ the standard normal distribution function. This holds as μ was assumed non-lattice. If we write $f(u) = \beta_3(1 - u^2)\,\varphi(u)/6\beta_2^3$, then

$$\begin{aligned}
g*\bar{\mu}^{*n}(t_n) &= \int_0^\infty \bar{\mu}^{*n}(\beta n + u\beta_2\sqrt{n} + o(\sqrt{n}) - t)\,g(dt) \\
&= \int_{-\infty}^{+\infty} \bar{\mu}^{*n}(\beta n + o(\sqrt{n}) + t\beta_2\sqrt{n})g\left(\beta_2\sqrt{n}(u - dt)\right) \\
&= \int_{-\infty}^{+\infty} g\left(\beta_2\sqrt{n}(u - t)\right)\varphi(t)\,dt + \\
&\quad + \int_{-\infty}^{+\infty} g\left(\beta_2\sqrt{n}(u - t)\right)f'(t)\,dt/\sqrt{n} + o(1/\sqrt{n}),
\end{aligned}$$

since g has bounded variation. It follows that

$$\beta_2\sqrt{n}\,g*\bar{\mu}^{*n}(t_n) \to \varphi(u)$$

as $n \to \infty$. By Equation (8.5.6) this ends the argument. $\qquad\square$

Corollary (8.5.2)

With t_n as above

$$\lim_{n \to \infty} \sqrt{n}\, m_{t_n,n}/m_{t_n} = \beta \varphi(u)/\beta_2.$$

In other words, at a time t the contribution to the population comes only from generations $[t/\beta] \pm O(\sqrt{t/\beta})$, the expected size of the contribution being approximately Gaussian bell shaped around $[t/\beta]$ and of the order of magnitude $1/\sqrt{t}$.

There is a stochastic analogue of this result. We shall deduce it by the well known L^2-considerations, assuming the reproduction variance to be finite. First,

$$E[z_{t,1}z_t] = E[z_t(0)\, z_t] = 1 - L(t) + \int_0^t m_{t-v}\mu_t(\mathrm{d}v)$$

as in Theorem (6.4.1). Recursively

$$E[z_{t,n+1}z_t] = E\left[\sum_{k=1}^{\xi_0(t)} z_{t-\tau_0(k),n} \circ S_k \sum_{k=1}^{\xi_0(t)} z_{t-\tau_0(k)} \circ S_k \right]$$

$$= \int_{0 \le v, w \le t} m_{t-v,n} m_{t-w}\gamma(\mathrm{d}v, \mathrm{d}w)$$

$$- \int_0^t m_{t-v,n} m_{t-v}\mu(\mathrm{d}v) + \int_0^t E[z_{t-v,n}z_{t-v}]\,\mu(\mathrm{d}v).$$

If we write

$$a_0(t) = 1 - L(t) + \int_0^t m_{t-v}\mu_t(\mathrm{d}v)$$

$$a_0(t) = \int_{0 \le v, w \le t} m_{t-v,n} m_{t-w}\gamma(\mathrm{d}v, \mathrm{d}w) - \int_0^t m_{t-v,n} m_{t-v}\mu(\mathrm{d}v), \qquad (8.5.7)$$

it follows that

$$E[z_{t,n}z_t] = \sum_{k=0}^{n-1} a_{n-k}*\mu^{*k}(t). \qquad (8.5.8)$$

Similarly, with

$$b_0(t) = 1 - L(t)$$

$$b_n(t) = \int_{0 \le v, w \le t} m_{t-v,n} m_{t-v,n}\gamma(\mathrm{d}v, \mathrm{d}w) - \int_0^t m_{t-v,n}^2 \mu(\mathrm{d}v), \qquad (8.5.9)$$

$$E[z_{t,n}^2] = \sum_{k=0}^{n-1} b_{n-k}*\mu^{*k}(t) \qquad (8.5.10)$$

for $n \in N$. We let t_n be as above, multiply by $e^{-2\alpha t_n}$, and note that

$$0 \le \sqrt{n}\, e^{-2\alpha(t_n - v)} a_n(t_n - v) \to k^2(\infty)\, \varphi(u)(\vartheta - \varphi)/\beta\beta_2$$

for any v. (Observe that φ is used to denote two different things). Since the integrand remains bounded it follows that

$$\sqrt{n}\, e^{-2\alpha t_n} E\left[z_{t_n,n} z_{t_n}\right] = \sum_{k=0}^{n-1} \int_0^{t_n} \sqrt{n}^{-2\alpha(t_n - v)} a_n(t_n - v)\, e^{-2\alpha v} \mu^{*k}(dv)$$

$$\to k^2(\infty)\, \varphi(u)(\vartheta - \varphi)/\beta\beta_2(1 - \varphi).$$

In the same manner

$$n\, e^{-2\alpha t_n} E\left[z_{t_n,n}^2\right] \to k^2(\infty)\, \varphi^2(u)(\vartheta - \varphi)/\beta_2^2(1 - \varphi)$$

and hence

$$E\left[\left\{\sqrt{n}\, e^{-\alpha t_n} z_{t_n,n} - \beta\varphi(u)\, e^{-\alpha t_n} z_{t_n}/\beta_2\right\}^2\right] = n\, e^{-2\alpha t_n} E\left[z_{t_n,n}^2\right]$$

$$- 2\beta\varphi(u)\sqrt{n}\, e^{-2\alpha t_n} E\left[z_{t_n,n} z_{t_n}\right]/\beta_2 + \beta^2\varphi^2(u)\, e^{-2\alpha t_n} E\left[z_{t_n}^2\right]\beta_2 \to 0$$

and we have proved:

Theorem (8.5.3)

Consider a supercritical and non-lattice branching process with finite reproduction variance. Let t_n and β_2 be as in Theorem (8.5.1). Then there is a random variable z such that, as $n \to \infty$

$$\sqrt{n}\, e^{-\alpha t_n} z_{t_n,n} \to k(\infty)\, \varphi(u)\, z/\beta_2$$

in mean square.

As before there is an immediate

Corollary (8.5.4)

Under the above conditions

$$\sqrt{n}\, z_{t_n,n}/z_{t_n} \to \beta\varphi(u)/\beta_2$$

in probability on the set where the process does not become extinct.

These results have all used a local central limit theorem requiring that third moments exist. Since the $\bar{\mu}$ has moments of any order this means no restriction but there are of course also global versions of the form

$$e^{-\alpha t} \sum_{n=0}^{n_t(u)} z_{t,n} \to k(\infty)\, \Phi(u)\, z/\beta \qquad (8.5.11)$$

in mean square for suitable $n_t(u)$, actually $= t/\beta + u\beta_2\sqrt{t}/\beta^{3/2} + o(\sqrt{t})$, cf. Reference [16], Φ the standardized normal distribution function. But it seems that the local statements give a more graphical picture of how an existing population is composed of generations.

When it comes to the dual problem of how individuals of one generation are distributed over time, we shall give only the global treatment. This is since here the rôle of $\bar{\mu}$ is taken by μ and so local statements do require a further moment.

Theorem (8.5.5)

Let b and $\sigma_o^2 < \infty$ be the expectation and variance of the normed reproduction function μ/m. Define

$$t_n = nb + u\sigma_o\sqrt{n} + o(\sqrt{n})$$

and the number of nth generation members born by t,

$$y_{t,n} = \sum_{x \in N^n} y_t(x).$$

If the process is supercritical, then

$$E[y_{t_n,n}]/m^n \to \Phi(u),$$

as $n \to \infty$. (Recall that $m^n = E[\zeta_n]$ is the expected size of the nth generation).

Proof.

$$E[y_{t,0}] = 1$$

and

$$E[y_{t,n+1}] = E\left[\sum_{k=1}^{\xi_0(t)} y_{t,n} \circ S_k\right] = \int_0^t E[y_{t-v,n}]\,\mu(dv).$$

Therefore

$$E[y_{t,n}] = \mu^{*n}(t)$$

for all $n \in Z_+$ and the assertion reduces to the central limit theorem for i.i.d. summands. $\square$

The corresponding local theorem starts from

$$E[z_{t,n}]/m^n = (1 - L)*\mu^{*n}/m^n,$$

divides this by the expected life length (assumed to be finite) and proceeds as in Theorem (8.5.1).

We turn to mean square convergence. As before

$$E[y_{t_0}\zeta_0] = 1,$$

$$E[y_{t,n+1}\zeta_{n+1}] = E\left[\sum_{k=1}^{\xi_0(t)} y_{t,n} \circ S_k \sum_{k=1}^{\xi_0(t)} \zeta_n \circ S_k\right]$$

$$= m^{n-1}\int_0^t E[y_{t-v,n}]\{\gamma(dv, \infty) - \mu(du)\} + \int_0^t E[y_{t-v,n}\zeta_n]\,\mu(dv).$$

Hence with

$$a_n(t) = \int_0^t E[y_{t-u,n}]\{\gamma(du, \infty) - \mu(du)\} \tag{8.5.12}$$

it holds for all $n \in N$ that

$$E[y_{t,n}\zeta_n] = \sum_{k=0}^{n-1} m^{n-k-1} a_{n-k-1} * \mu^{*k}(t).$$

Under the presumptions of Theorem (8.5.5)

$$a_n(t_n)/m^n \to \sigma^2 \Phi(u)$$

as $n \to \infty$, since

$$E[y_{t-u,n}] \le E[\zeta_n] = m^n,$$

and it can be directly checked that

$$E[y_{t_n,n}\zeta_n]/m^n \to \sigma^2 \Phi(u)/(m^2 - m). \tag{8.5.14}$$

In the same manner

$$E[y_{t_0,0}^2] = 1$$

$$E[y_{t,n+1}^2] = \int_{0 \le v,w \le t} E[y_{t-v,n}] E[y_{t-w,n}] \gamma(dv, dw)$$

$$- \int_0^t E^2[y_{t-v,n}] \mu(dv) + \int_0^t E[y_{t-v,n}^2] \mu(dv)$$

and

$$E[y_{t_n,n}^2]/m^n \to \sigma^2 \Phi^2(u)/(m^2 - m). \tag{8.5.15}$$

A glance at

$$E[\{y_{t_n,n} - \Phi(u)\zeta_n\}^2]/m^n$$

and a reference to Lemma (2.9.1) completes the proof of

Theorem (8.5.6)

Add to the assumptions of Theorem (8.5.5) that of requiring the reproduction variance to be finite. Then, in mean square,

$$y_{t_n,n} \to \Phi(u)\zeta$$

where

$$\zeta = \lim_{n \to \infty} \zeta_n/m^n$$

a.s. and in mean square.

There is, of course a sequel to this corresponding to Corollary (8.5.4) and also a local theorem. Instead of stating this let us summarize the results obtained by saying that though the $[t/\beta]^{th}$ generation dominates the population at time t, most of its members have not yet been realized. Indeed, the greatest part of that generation will be born later than $\bar{t}$, around time tb/β.

Finally, a word about the third suggested generation length, t. We consider supercritical processes such that

$$a = \int_0^\infty u^2 \mu(\mathrm{d}u) < \infty,$$

and α is close to zero, i.e. $m \approx 1$. Then, since

$$\bar{t} = (\log m)/\alpha = (m - 1)/\alpha - (m - 1)^2/2\alpha + O(m - 1)^3/\alpha,$$

and

$$m - 1 = \hat{\mu}(0) - \hat{\mu}(\alpha) = \alpha bm - \alpha^2 a/2 + O(\alpha^3),$$

$$\bar{t} = bm - \alpha a/2 - \alpha b^2 m^2/2 + O(\alpha^2).$$

However

$$\beta = \int_0^\infty e^{-\alpha u} u \mu(\mathrm{d}u) = bm - \alpha a + O(\alpha^2),$$

implying that

$$\bar{t} = bm + (\beta - bm)/2 - \alpha b^2 m^2 + O(\alpha^2)$$

$$= (b + \beta)/2 + b(m - 1)/2 - \alpha b^2 m^2/2 + O(\alpha^2)$$

$$= (b + \beta)/2 + \alpha b^2 m(m - 1)/2 + O(\alpha)$$

or

$$\bar{t} = (b + \beta)/2 + O(\alpha^2), \tag{8.5.16}$$

as $\alpha \downarrow 0$, whereas (recall Inequality (8.5.3))

$$0 \leq b - \beta = O(\alpha). \tag{8.5.17}$$

Background

The first result in this direction was Theorem (8.5.6) proved in Reference [16] for Bellman–Harris processes. The preceding results in the text are all from Reference [17], also restricting itself to the Bellman–Harris case. Relation (8.5.16) is given in various demographic texts. The reader might note the absense of results on a.s. convergence.

REFERENCES

1. Coale, A. J., *The Growth and Structure of Human Populations*. Princeton University Press, Princeton, N. J., 1972.
2. Costello, W. G. and Taylor, H. M., Deterministic population growth models. *Am. Math. Monthly* **78**, 841–855, 1971.
3. Doney, R. A., Age dependent birth and death processes. *Z. Wahrscheinlichkeitstheorie verw. Geb.* **22**, 69–90, 1972.
4. Keiding, N., *Lecture Notes on the Stochastic Population Model*. Institute of Mathematical Statistics, University of Copenhagen, Copenhagen, 1973.
5. Kendall, D. G., Stochastic processes and population growth. *J. Royal. Statist. Soc.* **B11**, 230–264, 1949.
6. Keyfitz, N., *Introduction to the Mathematics of Population*. Addison-Wesley, Reading, Mass., 1968.
7. Lopez, A., *Problems in Stable Population Theory*, Office of Population Research, Princeton University Press, Princeton, N.J., 1961.
8. Lotka, A., *Théorie analytique des associations biologiques II*. Hermann, 1939.
9. Pollard, J. H., *Mathematical Models for the Growth of Human Populations*. Cambridge University Press, Cambridge, 1973.
10. Athreya, K. and Ney, P., *Branching Processes*. Springer, Berlin, 1972.
11. Fisher, R. A., *The Genetical Theory of Natural Selection*. Dover, New York, 1958 (first published 1929).
12. Meyer, P. A., *Probabilités et potentiels*. Hermann, Paris, 1966. Translated as: Probability and Potentials, Blaisdell., Waltham, Mass., 1966.
13. Jagers, P., Convergence of general branching processes and functionals thereof. *J. Appl. Prob.* **11**, 471–478, 1974.
14. Bühler, W. J., The distribution of generations and other aspects of the family structure of branching processes. *Proc. 6th Berkeley Symp. Prob. Math. Statist. III*, University of California Press, Berkeley, 1972.
15. Feller, W., *An Introduction to Probability Theory and its Applications II*. John Wiley and Sons, New York, 1966.
16. Martin-Löf, A., A limit theorem for the size of the nth generation of an age dependent branching process. *J. Math. Anal. Appl.* **15**, 273–279, 1966.
17. Samuels, M. L., Distribution of the branching process population among generations. *J. Appl. Prob.* **8**, 655–667, 1971.

Chapter 9

Branching Models in Cell Kinetics

9.1 CELL PROLIFERATION AND BINARY SPLITTING

The problems of growth and division of cells lie in the very heart of of biology. However they have many aspects, ranging from the chemistry of DNA, proteins, or other substances to the complex growth of huge cellular systems. Though basic patterns remain the same, the proliferation can also take various forms. Mammalian cells differ from other types and within the same animals there are widely different systems. Some cell populations undergo a pure decay, like the adult ovary whose cells are continuously released and never divide. Neither do the neurons of the central nervous sytem in adult mammals divide, and since they die very seldomly they do in practice, for not too long time horizons, constitute an unchanging population. The cells of the bone marrow, the intestinal crypts and the skin epithelium provide more complex examples. Here a type of stem cell population is maintaining its size by division. The excess of cells produced become part of another, decaying, population, may be after migration. A well known example is given by the wandering of cells from inner to outer skin layers.

Tissue cultures and tumours, but also regenerating liver, constitute populations where cells continuously are born, grow, and divide or possibly die (disintegrate) or emigrate. This might be true for all cells but also here there may be non-proliferating cells. With each population is attached a *growth fraction*, the fraction of cells involved in proliferation. Further, so called differentiation may occur, daughters not being of the same cell type as their mothers. Or daughters might not proliferate, entering instead a rest stage (often called G_0 or prolonged G_1 cf. beneath). Still, often these complications are unsignificant, and tissue cultures or tumours being populations with an asexual reproduction and an exponential increase in size invite an analysis in terms of branching processes. [Example (6.2.4).]

A corner-stone of modern cell biology is the concept of the cell cycle. For some decades it has been known that there is in the life of a proliferating cell a well defined period during which DNA is being synthesized. A newly born cell enters first a presynthetic phase, the so called G_1 *phase*. During

224

this one proteins and RNA are being built up in preparation for the DNA synthesis. The G_1 period is usually long, longer than a third of the life and often substantially more. There appears to be great variability. G_1 is followed by the synthesis or S *phase*, also of considerable duration. In the subsequent phase, called G_2, RNA and proteins are preparing for the last and most dramatic stage of a cell's life, *mitosis*, (the M *phase*), during which cells clearly stand out against their neighbours and can be observed. In this short period ($<$ one hour in mammalian cells having a total life span of, say, between 10 and 20 hours) first chromosomes contract and unwind (prophase), the chromosomes line up at the equatorial plane (metaphase), they separate and move towards opposite poles (anaphase), reach these and the cell divides into two daughters (telophase), each entering a G_1 phase [See Example (4.1.1)]. Sometimes the phases G_1, S, and M are referred to jointly as interphase. The *cell cycle* (or generation time or mitotic cycle) is the totality of all the four phases. It is a random entity with an appreciable coefficient of variation. It is not necessarily the same as the life span of the cell, since cells might die before completing the cycle.

The model we shall use is that of binary splitting with cell death. Thus, consider a branching process where the life length of an individual x is with probability p a random variable T_x (the cell cycle) and with probability $1 - p$ a variable D_x (the time to the death of the cell; we use the words 'individual' and 'cell' interchangeably). If the cell dies, then $\lambda_x = D_x$ and x begets no children, whereas otherwise $\lambda_x = T_x$ and exactly two new individuals are born at the death of x. The reader may wish to reformulate this into exactly the terminology of Section 6.1 and also to check that it defines a splitting process.

The variable T (we delete the x's from now on) is the sum of four random variables, the duration of the respective phases,

$$T = G_1 + S + G_2 + M.$$

We denote phase duration by the same letter as the phase itself. There is no reason to assume phases independent of one another, except that sometimes this will ease calculations. Indeed there is some evidence of a negative correlation between them [12, 20].

Tissue cultures and tumours are often old and studied in circumstances of what biologists call a *balanced exponential growth*. This means that the population increases exponentially but also that its composition in phases and over ages, does not change. In our jargon, the population is stable. Thus we assume the process to be supercritical (i.e. $p > 1/2$), and also non-lattice, and starting at $t = 0$ from z_0 individuals with stably distributed independent ages. We shall use the attributes 'balanced', 'in balanced exponential growth', 'stable', 'with a stable age distribution' etc. synonymously.

Most easy to observe for the experimenter is usually the growth of the population. Since observed numbers are very great and we expect the growth to have the shape

$$m_t = z_0 e^{\alpha t}$$

an immediately appealing estimator of the Malthusian parameter is

$$\hat{\alpha}_t = (\log z_t - \log z_0)/t. \tag{9.1.1}$$

The parameter is closely related to the *doubling time* of a population t_d, defined by

$$e^{\alpha t_d} = 2, \tag{9.1.2}$$

i.e. $t_d = (\log 2)/\alpha$. If there were no cell death and T had zero variance then t_d would equal the *expected cycle time* $t_c = E[T]$. In literature there are many attempts at estimating "the cycle time" (without further specification) by equating it with t_d. By the definition of α and Jensen's inequality

$$1 = 2pE[e^{-\alpha T}] \geq 2pe^{-\alpha t_c}.$$

Thus,

$$e^{\alpha t_c} \geq 2p = pe^{\alpha t_d}$$

and

$$t_c \geq t_d(\log 2p)/\log 2. \tag{9.1.3}$$

In practice Var $T > 0$, the inequality is strict, and the difference $t_c - t_d$ might be considerable, in particular if p is close to one.

It is also possible to observe directly which cells are in mitosis, and thus to calculate the *mitotic index* defined as the ratio between the number of mitotic cells (i.e. cells in mitosis) and the total number of cells. Invoking Corollary (6.10.5) we see that the mitotic index should be (we take this as the definition)

$$mi = p \int_0^\infty e^{-\alpha t} P[T - M \leq t < T] \, dt \bigg/ \int_0^\infty e^{-\alpha t} \{1 - L(t)\} \, dt$$

$$= pE[e^{-\alpha T}(e^{\alpha M} - 1)]/\{1 - pE[e^{-\alpha T}] - (1 - p) E[e^{-\alpha D}]\}$$

$$= 2pE[e^{-\alpha T}(e^{\alpha M} - 1)]/\{1 - 2(1 - p) E[e^{-\alpha D}]\}. \tag{9.1.4}$$

Since mitosis is such a short part of the cycle time it holds that

$$mi \approx 2\alpha pE[Me^{-\alpha T}]/\{1 - 2(1 - p) E[e^{-\alpha D}]\}. \tag{9.1.5}$$

Similar expressions will meet repeatedly in the sequel; they arise in a manner like the average age at childbearing discussed in Section 8.4: Define a random characteristic $\chi(t)$, $t \geq 0$, to equal M if the cell has divided by t and M was the duration of its mitosis. Otherwise set $\chi(t) = 0$. If z_t^M is the total number of mitoses that have ended by t then z_t^χ/z_t^M is the average length of mitoses among those that have occurred during $[0, t]$. Letting $t \to \infty$ we obtain by Theorem (6.10.1) a.s. on the set of non-extinction that

$$z_t^\chi/z_t^M \to \int_0^\infty e^{-\alpha t} E[M; t > T]\, dt \bigg/ \int_0^\infty e^{-\alpha t} G(t)\, dt = 2pE[Me^{-\alpha T}].$$

Thus, $2pE[Me^{-\alpha T}]$ should be interpreted as the average mitotic time of individuals sampled at division. In concurrance with the terminology introduced by Quastler [11] we use this as the basis for a general definition: Let X be any random entity defined for each cell. Then

$$\tilde{E}[X] = 2pE[Xe^{-\alpha T}] = E[Xe^{-\alpha T}]/E[e^{-\alpha T}] \tag{9.1.6}$$

is the *flux expectation* of X. Clearly a flux distribution can be defined via the flux expectation of $1_{(-\infty, t]} \circ X$, the flux variance of X is $\tilde{E}[X^2] - \tilde{E}^2[X]$, etc. The concept is closely related to that of "carrier densities" [3, 17, 19] and is to be understood as the expected X-length of cells sampled at mitotic division.

Returning to Equation (9.1.5) we see that the mitotic index satisfies

$$mi \approx \alpha\tilde{E}[M]/\{1 - 2(1 - p)E[e^{-\alpha D}]\}.$$

Since many types of disintegration, if occurring, interrupt the cell cycle, it is, in certain cases, reasonable to assume that D is stochastically smaller than T. Then

$$1/2p = E[e^{-\alpha T}] \le E[e^{-\alpha D}] \le 1$$

and at least approximately

$$\alpha p\tilde{E}[M] \le (2p - 1)mi \le \alpha\tilde{E}[M]. \tag{9.1.7}$$

and when p is close to one

$$mi \approx \alpha\tilde{E}[M] = \tilde{E}[M] \log 2/t_d. \tag{9.1.8}$$

In other cases cell death occurs after a prolonged life span, implying that $E[e^{-\alpha D}]$ is little. Thus, anyhow Equation (9.1.8) holds.

If M were independent of the rest of the cycle, $E[M]$ would equal $E[Me^{-\alpha M}]/E[e^{-\alpha M}]$. In view of the littleness of αM, may be,

$$mi \approx \alpha E[M] = E[M] \log 2/t_d. \tag{9.1.9}$$

These two formulae are often met with in the form (page 123 of Reference [5] e.g.)

$$mi = (\log 2) M/T = 0{\cdot}693 t_m/t_c \tag{9.1.10}$$

valid approximately if $M = t_m$, $T = t_c$ are constant, and $p \approx 1$ or D is very large. (In the sequel we shall sometimes write $t_m = E[M]$ and similarly $t_s = E[S]$, $t_1 = E[G_1]$, $t_2 = E[G_2]$. Hooks above the entities indicate flux.)

Background

The concept of the cell cycle was first formulated in 1953 [7]. Before that knowledge about cells in interphase was rudimentary. Application of stochastic processes to cell proliferation goes back to, at least, Reference [9]. A more recent well written paper, studying the problems from the viewpoint of multi-type process is Reference [13], treating individuals in a phase as being of a particular type. Another general approach is Reference [4]. Many models have been formulated in connection with specific problems, in particular the study of so called fraction labelled mitoses, treated in Sections 9.4 and 9.6.

Experimentally oriented papers often analyze data as though cycle length and similar entities were not random. This may of course be helpful as a first step. Much of the contents in this section, and generally this chapter, can be viewed as discussing the limitations of such assumptions, and more sophisticated ones like taking the phases to be independent with some popular distribution (gamma, log-normal, normal) (Reference [14] gives a survey).

There are several good biological papers in the area. A classical book in the field, often referred to, is Reference [11]. A representative survey is Reference [14]. The book [2] contains several relevant articles and Reference [21] gives a lucid account of the chemical background, also surveyed in Reference [1]. Chapter 4 in Reference [5] is also commendable.

The binary splitting with cell death constitutes a model that is considerably more general than those of common occurrence in literature. Still it is a branching process having the limitations of such models. In reality some experimental findings indicate a positive correlation between the life spans of sisters, the continued expansion of populations evidently requires a very special nutritional and spatial environment, and in several experiments a decrease in proliferative activity with time has been noted.

9.2 ESTIMATION OF CELL DEATH

Tumorous cell populations increase differently from their normal counterparts, usually—but not always—quicker. Such a change can possibly depend on either of two factors, shorter cycle times or less risk for cell death, or may be both. Long it was taken for granted that changes in the cycle time distribution played the main rôle, indeed most models in the literature are valid only for $p = 1$, i.e. provided there is no cell disintegration. Recent results modify these conceptions [8, 16, 23, 24].

In Section 6.14 the maximum likelihood estimator

$$\hat{p}_t = (y_t - z_0)/2\, d_t$$

of p was derived for the case $F = G$, i.e. disintegration being possible only at the end of the cycle time. Recall that y_t and d_t are the total number of individuals, born and dead respectively, by t. By Equation (6.14.7) the estimator has an asymptotic bias

$$p - \hat{p}_\infty = p - 1/\{1 + 2(1 - p)\,E[e^{-\alpha D}]\}.$$

Invoking the argument preceding Inequality (9.1.7) we insert into this $1/2p \le E[e^{-\alpha D}] \le 1$, valid in case cell death tends to occur before division. We obtain.

$$1/(3 - 2p) \le \hat{p}_\infty \le p \tag{9.2.1}$$

or

$$\hat{p}_\infty \le p \le (3\hat{p}_\infty - 1)/2\hat{p}_\infty. \tag{9.2.2}$$

Hence $\hat{p}_t$ tends to underestimate p but the asymptotic error does not exceed $p - 1/(3 - 2p)$. As a function of p this is concave. It is zero at the points $1/2$ and 1, in between them first increasing and taking a maximum value of $(3 - 2\sqrt{2})/2 \approx 0.086$ at the argument $(3 - \sqrt{2})/2 \approx 0.793$ and then decreasing again.

In applications the initial cell number z_0 as well as the final cell number z_t can be observed. So can the number of mitoses z_t^M during the interval $[0, t]$ of observation. Clearly $y_t - z_0 = 2z_t^M$ and $d_t = 2z_t^M + z_0 - z_t$. For practical use we can thus estimate p by

$$\hat{p}_t = z_t^M/(2z_t^M + z_0 - z_t). \tag{9.2.3}$$

This has been done for a variety of *in vitro* tumorous and normal mammalian cells (human and Syrian hamster) [23]. Comparison shows that the probability of cell disintegration is considerably smaller in the tumorous than in the normal cell lines. For the normal cells the p-values lie, essentially, in the range 0.8–0.9 whereas for the tumorous counterparts they mostly exceed 0.9.

We can thus estimate cell death occurring during the life cycle or at its end by Equation (9.2.3). Situations where the time D to a possible death is long are more difficult to get at. Indeed, if $E[e^{-\alpha D}] \approx 0$, then $\hat{p}_\infty \approx 1$, whatever p is.

To see what can be said, let us assume, that newly born cells either— with probability p—complete the cell cycle, or in the contrary case they remain indefinitely, i.e. we take $D = \infty$. (We shall return to such models in Section 9.9). The likelihood formula (6.14.6) then reduces to

$$p^{d_t} \prod_{\sigma_x + \lambda_x \leq t} G(d\lambda_x) \prod_{\sigma_x \leq t < \sigma_x + \lambda_x} \{1 - pG(t - \sigma_x)\}. \tag{9.2.4}$$

If the last product is empty, clearly this is maximized by $p = 1$. Otherwise differentiation yields that the maximum likelihood estimator $\tilde{p}_t$ of p is the unique solution of $d_t = \sum_{\sigma_x \leq t < \sigma_x + \lambda_x} pG(t - \sigma_x)/\{1 - pG(t - \sigma_x)\}$. Expansion of the denominator provides the approximation

$$\tilde{p}_t \approx \frac{\sqrt{d_t \sum G^2(t - \sigma_x) + \sum \{G(t - \sigma_x)\}^2} - \sum G(t - \sigma_x)}{2 \sum G^2(t - \sigma_x)}.$$

This presumes knowledge of the cycle time distribution G, which is usually not available. See the following section.

Background

The experimental so called micro ciné technique used is exposed in References [24, 25]. Another model for cell proliferation with death was suggested by Horn [22].

9.3 THE CYCLE TIME DISTRIBUTION

In the preceding section we saw that decrease in the extent of cell death can have a substantial part in the changes of growth patterns when populations turn tumorous. Still it is possible that, say, a more rapid proliferation is not exhaustively explained by this but that there is also a shortening of cycle times. However these are difficult to measure. *In vivo* only indirect techniques, largely ad hoc (see later sections), are available. *In vitro* the period of possible observation. i.e. the time during which a population under observation, can maintain an exponential growth in the limited space available, is short as compared to plausible cycle times in human cells. When cells are followed from mitosis to mitosis, censored samples are obtained from which few conclusions can be drawn [16].

One resort is provided by the stable age distribution at hand during balanced exponential growth. If cells are sampled randomly and the time until division is recorded (in case of no disintegration), then in conjunction with the age distribution of sampled cells this should provide information about cycle length. Certainly these recorded times will be shorter than the time needed to observe a picked cell first to mitosis, then recording the time to mitosis of one of the daughters.

We consider at time $t = 0$ a cell aged u where $L(u) < 1$. The probability that it will not divide before t is

$$\{(1 - p)(1 - F(u)) + p(1 - G(t + u))\}/\{1 - L(u)\}.$$

Integrating this with respect to the stable age distribution we obtain the probability $R(t)$ that a cell, randomly sampled from a population in balanced exponential growth, will not divide during time t,

$$R(t) = 2\alpha \int_0^\infty \{(1 - p)(1 - F(u)) + p(1 - G(t + u))\} \, e^{-\alpha u} \, du/$$

$$\{1 - 2(1 - p) E[e^{-\alpha D}]\}$$

$$= 2\left\{1 - 2(1 - p) E[e^{-\alpha D}] - \alpha p e^{\alpha t} \int_t^\infty G(u) e^{-\alpha u} \, du\right\} \bigg/$$

$$\{1 - 2(1 - p) E[e^{-\alpha D}]\}. \tag{9.3.1}$$

This yields the differential equation

$$R' = \alpha R - 2\alpha \{1 - (1 - p) E[e^{-\alpha D}]\}/\{1 - 2(1 - p) E[e^{-\alpha D}]\}$$

$$+ 2\alpha p G/\{1 - 2(1 - p) E[e^{-\alpha D}]\}. \tag{9.3.2}$$

If $\bar{t}$ is the expected time to division, given that there is no disintegration, of a cell with a stably distributed age, then

$$\bar{t} = \int_0^\infty \{R(t) - R(\infty)\} \, dt/\{1 - R(\infty)\},$$

$$R(\infty) = 2(1 - p)(1 - E[e^{-\alpha D}])/\{1 - 2(1 - p) E[e^{-\alpha D}]\}$$

$$= 1 - (2p - 1)/\{1 - 2(1 - p) E[e^{-\alpha D}]\},$$

$R(\infty)$ being the probability of the cell going to disintegrate. Equation (9.3.2) turns into

$$R' = \alpha \{R - R(\infty)\} - 2\alpha p(1 - G)/\{1 - 2(1 - p) E[e^{-\alpha D}]\}$$

after little manipulation. Hence

$$R'/\{1 - R(\infty)\} = \alpha \{R - R(\infty)\}/\{1 - R(\infty)\} - 2\alpha p(1 - G)/(2p - 1)$$

$$\tag{9.3.3}$$

and integration from 0 to infinity yields the remarkably simple formula

$$t_c = \{(2p - 1)/2p\}(\bar{t} + 1/\alpha) \tag{9.3.4}$$

or, in terms of the doubling time,

$$t_c = \{(2p - 1)/2p\}\{\bar{t} + t_d/\log 2\}.$$

If there is no cell death this reduces to

$$t_c = (\bar{t} + 1/\alpha)/2 = (\bar{t} + t_d/\log 2)/2. \tag{9.3.5}$$

To estimate the mean cycle time *in vitro*, thus estimate α according to Equation (9.1.1), p according to Equation (9.2.3), select cells randomly and record the times until they divide (provided they do so), estimate $\bar{t}$ by the arithmetic mean of these, and finally insert into Equation (9.3.4).

Multiplication of Equation (9.3.3) by t before the integration provides a formula for $v_c^2 = \mathrm{Var}\,[\,T\,]$ in the same manner:

$$v_c^2 = \{(2p - 1)/2p\}\,(\bar{v}^2 - \bar{t}^2 + 2\alpha\bar{t}) - t_c^2. \qquad (9.3.6)$$

Here $\bar{v}^2$ is the variance of times to division among cells that divide in a stable population. Other entities, like the median of G, are also obtainable from Equation (9.3.3).

The formulae were applied to estimate expectation and variance of the cycle time of some human cells in Reference [26]. The result was that the tumorous cells, quite contrary to what had been suggested, seemed to have longer cycle times than their nomal counterparts. However the application was complicated by a rather extensive migration of cells (out of the field of observation) that had to be taken care of in formulae, and more definite conclusions might need further experimental basis. But in view of this and Section 9.2 it is tempting to end by a conjecture: Tumorous changes imply a strengthening of cells, involving that they become less prone to disintegration and have longer cycle times.

9.4 THE FRACTION LABELLED MITOSES

During the last two decades the method of fraction labelled mitoses has provided by far the most popular means of studying the cell cycle and its components. Experimentally it starts from introducing into a cell population tritiated thymidine. (Tritium is a radioactive hydrogene isotope, thymidine thymine with sugar). Cells which are synthesizing DNA at the moment, i.e. are in the S phase, and only these, incorporate the tritiated thymidine, getting a radioactive label that remains and can be observed. Not incorporated tritiated thymidine is rapidly (within one hour [35]) broken down to negligible concentrations. By washing or dilution with untritiated thymidine the uptake of radioactive labels can be made practically instantaneous. At mitotic division the labels are passed on to both daughter cells and so forth at later divisions, though in practice the amount of inherited tritiated thymidine becomes to small for detection after a few generations. The labels are detected by so called autoradiography [5, Chapter 1.6–1.10]. In this there are also statistical problems involved [29, 5, 12]. We do not enter upon them.

The fraction

$\mathrm{FLM}(t) =$ no. of labelled cells in mitosis at time t after the
tritiated thymidine pulse/total no. of mitotic cells
at the same time

is termed the *fraction labelled mitoses* (FLM) curve. (We shall use capital letters in this way when referring to empiric entities and Italic small letters for corresponding functions in models). An earlier name is the percentage labelled mitoses (PLM) curve.

Some of its properties to be expected are: The curve ought to start from zero, grow at a rate dependent upon the distribution of S and G_2 (the time from thymidine uptake to mitosis), approach one, decrease again after a period mainly influenced by S, increase again when daughters of initially labelled cells enter mitosis. This wave form should continue, though levelling off due to the variance of cycle times. The curve only considers cells in mitosis and therefore it should also be independent of the growth fraction and cell disintegration—except indirectly, through the age distribution. Some actual FLM curves are shown in Figure 9.4.1.

The purpose of the method is to infer properties of the cycle time and its phases from the shape of observed FLM curves. That presupposes some model and a highly simplified one is often relied upon, at least to convey basic ideas: It is assumed that all phases have constant durations, that there is no cell disintegration, that mitotic division results in exactly one proliferating daughter, and that the population has a stable age distribution at time zero, when the thymidine pulse is administered. As is easily checked, stable age distribution in this case means uniform distribution over the interval $[0, T]$. The probability of a randomly chosen cell in mitosis being

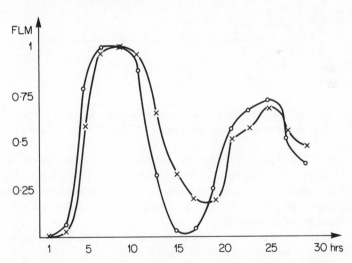

Figure 9.4.1 FLM-measurements on two cultivated normal human fibro-blast like cell populations. The curves connecting the two series of measurements have been drawn freehand just for the display. The data are due to K. Norrby (unpublished).

labelled t time units after the pulse, or equivalently the expected number of labelled mitotic cells at t divided by the total expected number of mitotic cells (which remains constant) then has the form

$$flm(t) = \begin{cases} 0 & 0 \leq t \leq G_2 \\ (t - G_2)/M & G_2 \leq t \leq G_2 + M \\ 1 & G_2 + M \leq t \leq G_2 + S \\ (G_2 + S + M - t)/M & G_2 + S \leq t \leq G_2 + S + M \\ 0 & G_2 + S + M \leq t \leq T \end{cases}$$

$$flm(nT + t) = flm(t), \qquad n \in N,$$

see Figure 9.4.2. Here M is taken as less than S.

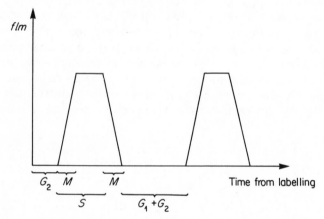

Figure 9.4.2   The fraction labelled mitoses in the imagined case of constant phases, one daughter per cell, and a uniform age of distribution over the interval $[0, T]$.

The curve thus obtained readily provides all information about the cycle; G_2 is the time until the fraction gets positive; S is the distance from the first point where $flm(t)$ takes the value $1/2$ to the second such point. Or, it is the area under the first wave. Similarly $G_2 + M/2$ and $G_1 + G_2 + M$ can be represented as areas.

Removal of the restriction to one child per individual does not change the function too much. Only, since the stable age distribution has now (for $1/2 < p \leq 1$) the form $K(t) = 2p\{1 - e^{-\alpha(t \wedge T)}\}$, the rises and falls of the curve will have a more concave shape. However, as the reader can easily convince him/herself, already this generalization invalidates the area method of measuring the duration of synthesis.

A more interesting generalization is that to stationary populations (i.e. each cell still begets exactly one daughter) with random phase durations and cycle times. This model has also been amply studied in literature [27, 28, 33, 43], and it yields the damped function, so typical of empiric FLM curves.

Such considerations of simple models have led to a variety of *ad hoc* methods in the analysis of actual curves. The original *boundary* or 0·5 *level* method, introduced in Reference [39], suggests estimating expected lengths of $G_2 + M/2$, S, $G_1 + G_2 + M$ from the distances between successive intersections of the curve with the level $1/2$, as motivated above. This breaks down if the second wave never reaches that height, as might well be the case. To avoid this a lower level in the boundary might be chosen or else the level $1/2$ might be replaced by half the height of the corresponding peak [38]. An alternative is provided by the *area* method, also hinted at in the preceding: estimate the expected duration in synthesis by the area under the first wave [30, 37]. One difficulty here is that the extension of a wave must be rather arbitrarily determined. If the convergence of the FLM curve to a plateau value is rapid (so that there is little or no disappearance of tritiated thymidine in divisions), then it has been suggested [27, 37], on the basis of the stationary model, that the asymptotic value estimates the ratio between the expected duration in S and the expected cycle time, the *plateau* method.

The quality of these proposed estimators has been investigated by simulation in special cases of less restrictive models [14]. The errors were generally found to be great. In the next section we shall give some explanation for this for the area and plateau methods, which are the ones that avail themselves to analysis in the general model of binary splitting with cell death.

Lately some more systematic approaches to inference on FLM curves have been made. But also these start from some specific model; it is assumed that phases have a joint normal distribution [12], which should be corrected for negativity, that phases are independent and lognormal [28, 43], or that phases are independent gamma variables [32, 33]. Cell disintegration is also taken into account in a simplified manner, not at all [32], with constant rate during the cycle time [33], or at division [33, 12]. Or only stationary populations are considered [28, 32, 43]. Macdonald [12] maximizes the likelihood of the observed FLM function, whereas the others minimize various sums of squares, to obtain parameter estimates.

Background

As is evident from the preceding, the literature on FLM abounds. There are several good survey articles [35, 38, 14, 47] and a very informative book [5]. The theoretical literature presents a somewhat confusing sight.

Often suggestions on the analysis of FLM curves are given in several sources, apparently independently. Decent procedures are advocated on shallow grounds. Deep facts are taken for granted and simple relations studied meticulously. But there are also fine analyses, displaying a very good feeling for the essentials and intricacies of population kinetics. Some of these are mentioned in the next section.

9.5 FLM FUNCTIONS IN BINARY SPLITTING

When it comes to more realistic models, like binary splitting with cell death, the fraction labelled mitoses is dependent upon individual cell properties in a complex and indirect manner. Still we shall be able to reach some conclusions. However instead of studying the random fraction that constitutes the real FLM, we shall limit our discussion to ratios of expectations.

Thus, consider a binary splitting process started from one newly born ancestor. Let $m_t^{(k)}(u, n)$ denote the expected number of cells in generation $k + n$ who are in mitosis at time $t + u$ and have an ancestor in the kth generation who was in synthesis at t. If $m_t^{SM}(u)$ denotes the expectation of the number of cells who have some ancestor synthesizing DNA at t and are themselves in mitosis at $t + u$, then

$$m_t^{SM}(u) = \sum_{k,n=0}^{\infty} m_t^{(k)}(u, n). \tag{9.5.1}$$

Since for $k \in Z_+$

$$m_t^{(k+1)}(u, n) = 2p \int_0^t m_{t-v}^{(k)}(u, n) \, G(dv),$$

it holds that

$$m_t^{SM}(u) = \sum_{n=0}^{\infty} m_t^{(0)}(u, n) + 2p \int_0^t m_{t-v}^{SM}(u) \, G(du).$$

However,

$$m_t^{(0)}(u, 0) = pP[G_1 \le t \le G_1 + S, T - M \le t + u < T],$$

which we shall denote by $pp_{SM}(t, u)$. Further

$m_t^{(0)}(u, 1)$

$$= 2p \int_t^{t+u} P[T - M \le t + u - v < T] \, P[G_1 \le t < G_1 + S, T \in dv].$$

If we write, for the probability of being in synthesis,

$$p_S(t, v) = P[G_1 \le t < G_1 + S, T \le t + v],$$
$$p_S(t) = P[G_1 \le t < G_1 + S] = p_S(t, \infty),$$

and

$$p_M(t) = P[T - M \le t < T]$$

for the probability of being in mitosis at age t, then

$$m_t^{(0)}(u, 1) = 2p \int_0^u p_M(u - v) \, p_S(t, dv) = 2p p_M * p_S(t, \cdot)(u).$$

Further

$$m_t^{(0)}(u, 2) = (2p)^2 \int_0^u p_S(t, dv) \int_0^{u-v} G(dw) \, p_M(u - v - w)$$
$$= (2p)^2 \, p_M * p_S(t, \cdot) * G(u)$$

and generally

$$m_t^{(0)}(u, n) = (2p)^n \, p_M * p_S(t, \cdot) * G^{*(n-1)}(u)$$

for $n \in N$. It follows that

$$m_t^{SM}(u) = p p_{SM}(t, u) + 2p p_M * p_S(t, \cdot) * \sum_{n=0}^{\infty} (2pG)^{*n}(u)$$

$$+ 2p \int_0^t m_{t-v}^{SM}(u) \, G(du) \qquad (9.5.2)$$

is the renewal equation that determines $m_t^{SM}(u)$ in terms of individual probabilities. But the expected number of mitoses at time t, m_t^M, satisfies

$$m_t^M = p p_M(t) + 2p \int_0^t m_{t-v}^M G(dv). \qquad (9.5.3)$$

thus

$$\lim_{t \to \infty} m_t^{SM}(u)/m_{t+u}^M$$

$$= \frac{\int_0^{\infty} e^{-\alpha t} \left\{ p_{SM}(t, u) + 2p_M * p_S(t, \cdot) * \sum_{n=0}^{\infty} (2pG)^{*n}(u) \right\} dt}{e^{\alpha u} \int_0^{\infty} e^{-\alpha t} p_M(t) \, dt} \qquad (9.5.4)$$

exists (recall that G is assumed non-lattice and that $p > 1/2$) and provides a determination of the expected FLM curve in a stable population. We define

$$flm(u) = \sum_{n=1}^{\infty} W_n(u), \tag{9.5.5}$$

where W_n, the nth wave, is determined by

$$W_1(u) = \int_0^{\infty} e^{-\alpha t} p_{SM}(t, u) \, dt \bigg/ e^{\alpha u} \int_0^{\infty} e^{-\alpha t} p_M(t) \, dt \tag{9.5.6}$$

and for $n > 1$

$$W_n(u) = (2p)^{n-1} p_M * \int_0^{\infty} e^{-\alpha t} p_S(t, \cdot) \, dt * G^{*(n-1)}(u) / e^{\alpha u} \int_0^{\infty} e^{-\alpha t} p_M(t) \, dt. \tag{9.5.7}$$

As $u \to \infty$, obviously $W_1(u) \to 0$. The sum $\sum_{n=2}^{\infty} W_n(u)$ is the solution of the renewal equation

$$x(u) = 2pG * p_M * \int_0^{\infty} e^{-\alpha t} p_S(t, \cdot) \, dt(u) \bigg/ \int_0^{\infty} e^{-\alpha t} p_M(t) \, dt$$

$$+ 2p \int_0^u x(u - v) G(dv),$$

normed by $e^{\alpha u}$. Hence $(2p\hat{G}(\alpha) = 1)$

$$\lim_{u \to \infty} \sum_{n=2}^{\infty} W_n(u) =$$

$$= \frac{\displaystyle\int_0^{\infty} e^{-\alpha u} \, du \int_0^u p_M(u - v) \int_0^{\infty} e^{-\alpha t} p_S(t, dv) \, dt}{\displaystyle\int_0^{\infty} e^{-\alpha t} p_M(t) \, dt \int_0^{\infty} u e^{-\alpha u} G(du)}$$

$$= \int_0^{\infty} e^{-\alpha t} \, dt \int_0^{\infty} e^{-\alpha v} p_S(t, dv) \bigg/ \int_0^{\infty} u e^{-\alpha u} G(du)$$

$$= E\left[\int_0^{\infty} e^{-\alpha t} \, dt \int_0^{\infty} e^{-\alpha v} 1_{[G_1, G_1 + S)}(t) \, 1_{[T-t, \infty)}(dv) \right] \bigg/ E[Te^{-\alpha T}]$$

$$= E\left[\int_0^{\infty} e^{-\alpha t} 1_{[G_1, G_1 + S)}(t) \, e^{-\alpha(T - t)} \, dt \right] \bigg/ E[Te^{-\alpha T}]$$

$$= E[Se^{-\alpha T}]/E[Te^{-\alpha T}] = \tilde{E}[S]/\tilde{E}[T]$$

in terms of flux expectations. We state this conclusion as

Theorem (9.5.1)

Consider a stable binary population with $p > 1/2$ and non-lattice G. As $t \to \infty$, the expected number of individuals in mitosis at time t with ancestors in synthesis at time zero, divided by the expected total number of mitotic cells at t, tends to the flux expected duration in synthesis, divided by the flux expected cycle time.

This gives some support to the plateau method, at least for p close to $1/2$, since then $\alpha \approx 0$. But it is difficult to assert anything general about the relation between $\tilde{E}[S]/\tilde{E}[T]$ and $E[S]/E[T]$. By Relation (8.5.3) $E[Te^{-\alpha T}] \leq E[T]E[e^{-\alpha T}]$ with equality if and only if $\alpha = 0$ or Var $T = 0$. If S and the rest of the cycle were independent, as is often assumed [3, 31, 44, 45, 46], then S and $e^{-\alpha T}$ would be negatively correlated, i.e. also $E[Se^{-\alpha T}] \leq E[S]E[e^{-\alpha T}]$ and conclusions about the ratios are difficult. Moreover, as pointed out in Section 9.1, the conventional assumption of independent phase times is questionable.

Example (9.5.2)

Assume that $T - S$ and S are independent. Then $E[Se^{-\alpha T}]/E[e^{-\alpha T}] = E[Se^{-\alpha S}]/E[e^{-\alpha S}]$. If S further is gamma distributed with parameters a, b, then

$$E[Se^{-\alpha S}]/E[e^{-\alpha S}] = a/(\alpha + b).$$

If T is also gamma A, B, then

$$E[Te^{-\alpha T}]/E[e^{-\alpha T}] = A/(\alpha + B),$$

and

$$\tilde{E}[S]/\tilde{E}[T] = a(\alpha + B)/A(\alpha + b) = (a/b)(B/A)(1 + \alpha/B)/(1 + \alpha/b)$$

$$= (E[S]/E[T])(1 + \alpha/B)/(1 + \alpha/b) < E[S]/E[T]$$

if and only if $1/B$, which is the coefficient of variation of T, is less than $1/b = \text{Coeff Var}[S]$. $\qquad\square$

After these comments on the plateau method, we turn to the first wave. Let us first note that

$$\int_0^\infty e^{-\alpha t}p_{SM}(t, u)\,dt = E\left[\int_0^\infty e^{-\alpha t}1_{[G_1, G_1+S)\cap[T-M-u, T-u)}(t)\,dt\right]$$

$$= E\left[\{e^{-\alpha G_1 \vee (T-M-u)} - e^{-\alpha(G_1+S)\wedge(T-u)}\}^+\right]/\alpha$$

$$= e^{\alpha u}E\left[\{e^{-\alpha(G_1+u)\vee(T-M)} - e^{-\alpha(G_1+S+u)\wedge T}\}^+\right]/\alpha,$$

second that

$$\int_0^\infty e^{-\alpha t} p_M(t) \, dt = E\left[\int_0^\infty e^{-\alpha t} 1_{[T-M,T)}(t) \, dt \right]$$
$$= E[e^{-\alpha T}(e^{\alpha M} - 1)]/\alpha. \qquad (9.5.8)$$

In conclusion, the first wave can be written

$$W_1(u) = E\left[\{ e^{-\alpha(G_1 + u) \vee (T-M)} - e^{-\alpha(G_1 + S + u) \wedge T} \}^+ \right] / E[e^{-\alpha T}(e^{\alpha M} - 1)],$$

$$(9.5.9)$$

a form that might be more suitable than Equation (9.5.6) for calculating special cases.

The area under the first wave can however be found directly:

$$\int_0^\infty e^{-\alpha u} \, du \int_0^\infty e^{-\alpha t} p_{SM}(t, u) \, dt = E\left[\int_{\substack{G_1 \le t < G_1 + S \\ T - M \le t + u < T}} e^{-\alpha(t+u)} \, du \, dt \right]$$

$$= E\left[\int_{\substack{G_1 \le t < G_1 + S \\ T - M \le v < T}} e^{-\alpha v} \, dv \, dt \right] = E[Se^{-\alpha T}(e^{\alpha M} - 1)].$$

Theorem (9.5.3)

The area A_1 under the first wave satisfies

$$A_1 = \int_0^\infty W_1(u) \, du = E[Se^{-\alpha T}(e^{\alpha M} - 1)]/E[e^{-\alpha T}(e^{\alpha M} - 1)]$$
$$= \tilde{E}[S(e^{\alpha M} - 1)]/\tilde{E}[e^{\alpha M} - 1] \approx \tilde{E}[SM]/\tilde{E}[M].$$

If S, the duration of mitosis M and the rest of the cycle $(G_1 + G_2)$ are independent then

$$A_1 = E[Se^{-\alpha S}]/E[e^{-\alpha S}] \le E[S]$$

with equality if and only if α or Var $S = 0$.

Completion of the proof is left for the reader. The approximate equality is of course supposed to be valid since M is considerably smaller than t_d.

Background

The most thorough analysis of the FLM curve in branching models was made in the serie of papers by Brockwell and co-authors [3, 31, 46], treating the case of binary Bellman–Harris processes with independent phases. Macdonald [12] gave results for dependent phases. Other references were mentioned in the preceding section.

9.6 CONTINUOUS LABELLING

As mentioned in Section 9.4, the FLM curve is independent of the growth fraction. The latter can be established by the related method of *continuous labelling*. Here tritiated thymidine is added not in a pulse but continuously from some time onwards (or at intervals certainly shorter than the S-phase). The fraction of labelled cells (i.e. among all cells) then increases, to one if all cells proliferate, to the growth fraction otherwise. The curve (the *CL curve*) has been used also to estimate the length of $G_1 + G_2$ [47, 49] and in companionship with the *continuous labelling mitoses* (CLM) curve, i.e. the fraction of labelled mitotic cells among cells in mitosis when labelling is continuous, to deduce further properties of the cell cycle. A third method used in continuous labelling is that of grain counts [5].

Experimentally, continuous labelling has the advantage over the pulse labelling of FLM studies that no washing is needed. However, there is the risk that radioactivity becomes so high that the population is affected.

Mathematically the method is treacherous. The simple model of a stationary process with constant phases leads to a CL curve starting from S/T and increasing linearly to one at time $T - S$. This is in astonishing conformity with experiments. Thus one estimator of $E[T] - E[S]$ advanced in literature is the time until the CL curve reaches one or some point close to it [5, 47, 49].

However, the corresponding curve in binary splitting is very messy, though it should, of course, have a fairly constant rate of increase. To calculate it, we have to elaborate the model somewhat, since the relation between phase times and the time to disintegration enters: cells which disintegrate in synthesis or later might well be labelled. (In the FLM-study the analogous problem was evaded by disregarding of the possibility of cell death during mitosis.) This means that we must consider the distribution of phase durations among disintegrating cells. Though it is enticing to assume that these are the same as among cells completing their life cycle, this can not be taken for granted and we shall write $P_D[G_1 \leq t]$ etc. for probabilities given disintegration and E_D for expectation. (Somewhat inconsistently we continue using $E[e^{-\alpha D}]$ instead of $E_D[e^{-\alpha D}]$.) Nothing seems to be known empirically about these phase distributions.

Proceeding as in Section 9.5 we define $m_t(u)$—in this section only—as the number of individuals alive at $t + u$, who have a mother in synthesis after time t or have themselves been in the S phase between t and $t + u$. Contributions to $m_t(u)$ accrue in four ways:

1. The ancestor is still alive at $t + u$ and she synthesized DNA in $[t, t + u)$. This has probability

$$pP[G_1 \leq t + u < T, G_1 + S > t] +$$

$$+ (1 - p) P_D[G_1 \leq t + u < D, G_1 + S > t].$$

2. The ancestor died by $t + u$, having synthesized some DNA between t and $t + u$. The expected number of individuals by this path is

$$2p \int_t^{t+u} m_{t+u-v} P[G_1 + S > t, T \in dv].$$

3. The ancestor died by $t + u$ having completed synthesis at t. This yields

$$2p \int_t^{t+u} E[z_{t+u-v}; \quad \text{the ancestor enters } S \text{ before} \quad t + u - v]$$

$$P[G_1 + S \leq t, T \in dv]$$

$$= 2p(1 - p) \int_t^{t+u} P_D[G_1 \leq t + u - v < D] P[G_1 + S \leq t, T \in dv]$$

$$+ 2p^2 \int_t^{t+u} P[G_1 \leq t + u - v < T] P[G_1 + S \leq t, T \in dv]$$

$$+ 4p^2 \int_t^{t+u} \int_0^{t+u-v} m_{t+u-v-w} G(dw) P[G_1 + S \leq t, T \in dv],$$

G as before the cycle time distribution among non-disintegrating cells.

4. Finally, the ancestor might have died before t. In this case we expect

$$2p \int_0^t m_{t-v}(u) G(dv)$$

labelled individuals at $t + u$.

Addition of the four terms results, as usual, in a renewal equation,

$$m_t(u) = pP[G_1 \leq t + u < T, G_1 + S > t]$$

$$+ (1 - p) P_D[G_1 \leq t \leq u < D, G_1 + S > t]$$

$$+ 2p(1 - p) \int_t^{t+u} P_D[G_1 \leq t + u - v < D] P[G_1 + S \leq t, T \in dv]$$

$$+ 2p^2 \int_t^{t+u} P[G_1 \leq t + u - v < T] P[G_1 + S \leq t, T \in dv]$$

$$+ 4p^2 \int_t^{t+u} \left\{ \int_0^{t+u-v} m_{t+u-v-w} G(dw) \right\} P[G_1 + S \leq t, T \in dv]$$

$$+ 2p \int_0^t m_{t-v}(u)\, G(dv) = f(t, u) + 2p \int_0^t m_{t-v}(u)\, G(dv). \qquad (9.6.1)$$

Assuming, as ever, supercriticality and no lattice life cycle, we obtain

$$\lim_{t \to \infty} m_t(u)/m_{t+u} = \alpha \int_0^\infty e^{-\alpha t} f(t, u)\, du / e^{\alpha u} \{1 - \hat{L}(\alpha)\} = cl(u), \qquad (9.6.2)$$

the last equality taken as the definition of the continuous labelling curve in balanced exponential growth. It is difficult to say much about this troublesome expression. However,

$$cl(0) = \alpha \int_0^\infty e^{-\alpha t} \{p P[G_1 \le t < G_1 + S]$$

$$+ (1 - p) P_D[G_1 \le t < G_1 + S]\}\, dt / \{1 - \hat{L}(\alpha)\}$$

$$= 2 \left\{ p E\left[\int_{G_1}^{G_1 + S} \alpha e^{-\alpha t}\, dt \right] + (1 - p) E_D\left[\int_{G_1}^{(G_1 + S) \wedge D} \alpha e^{-\alpha t}\, dt \right] \right\} \Big/$$

$$\{1 - 2(1 - p) E[e^{-\alpha D}]\} = 2p \{E[e^{-\alpha G_1}(1 - e^{-\alpha S})]$$

$$+ (1 - p) E_D[e^{-\alpha G_1} - e^{-\alpha D}; G_1 < D \le G_1 + S]$$

$$+ (1 - p) E_D[e^{-\alpha G_1}(1 - e^{-\alpha S}); G_1 + S < D]\} \Big/$$

$$\{1 - 2(1 - p) E[e^{-\alpha D}]\}, \qquad (9.6.3)$$

which is just the probability of a random cell in a stable population being in synthesis. Also $\lim_{u \to \infty} cl(u) = 1$.

For little u $cl(u)$ should consist essentially of first generation labelled cells. Thus

$$cl(u) \approx 2\alpha \left\{ p \int_0^\infty e^{-\alpha(t+u)} P[G_1 \le t + u < T, G_1 + S + u > t + u]\, du \right.$$

$$\left. + (1 - p) \int_0^\infty e^{-\alpha(t+u)} P_D[G_1 \le t + u < D, G_1 + S + u > t + u]\, du \right\} \Big/$$

$$\{1 - 2(1 - p) E[e^{-\alpha D}]\} = 2 \{p E[\{e^{-\alpha(G_1 \vee u)} - e^{-\alpha(T \wedge (G_1 + S + u))}\}^+]$$

$$+ (1 - p) E_D[\{e^{-\alpha(G_1 \vee u)} - e^{-\alpha(D \wedge (G_1 + S + u))}\}^+]\} /$$

$$\{1 - 2(1 - p) E_D[e^{-\alpha D}]\} \approx 2 \{p E[e^{-\alpha G_1} - e^{-\alpha(G_1 + S + u)}]$$

$$+ (1 - p) E_D[e^{-\alpha G_1} - e^{-\alpha D}; G_1 < D \le G_1 + S + u]$$

$$+ (1 - p) E_D[e^{-\alpha G_1} - e^{-\alpha(G_1 + S + u)}; D > G_1 + S + u]\} /$$

$$\{1 - 2(1 - p) E[e^{-\alpha D}]\}, \qquad (9.6.4)$$

for u so little that $P[u \le M + S]$ and $P[u \le G_1]$ are essentially one (say for $u \le 2$ hours in most cases). The approximation renders it tempting to suggest that the slope of the empiric CL line is the derivative *

$$cl'(0) \approx 2\alpha \{ pE[e^{-\alpha(G_1+S)}]$$
$$+ (1-p) E_D[e^{-\alpha(G_1+S)}; D > G_1 + S] \} / \{1 - 2(1-p) E[e^{-\alpha D}]\}$$

at zero. Then the line would hit one at time

$$t_1 = \{1 - cl(0)\}/cl'(0).$$

As $\alpha \to 0$ (i.e. $p \to 1/2$)

$$\{1 - 2(1-p) E[e^{-\alpha D}]\} cl'(0) \sim \alpha(1 + P_D[D > G_1 + S]).$$

Similarly, by $2pE[e^{-\alpha T}] = 1$

$$\{1 - 2(1-p) E[e^{-\alpha D}]\} \{1 - cl(0)\} = 2(1-p) E[1 - e^{-\alpha D}]$$
$$+ 2pE[1 - e^{-\alpha T}] - \{1 - 2(1-p) E[e^{-\alpha D}]\} cl(0) \sim \alpha\{E[D] + E[T]$$
$$- E[S] - E_D[D - G_1; G_1 < D \le G_1 + S] - E_D[S; G_1 + S < D]\}.$$

If cell death might only occur in G_1, this would mean that

$$t_1 \approx E[D] + E[T] - E[S]$$

for $p \approx 1/2$ and the proposed estimator would exaggerate $E[T-S]$ slightly. If, on the other extreme, cells can only die upon completed life cycle giving no birth with probability $1 - p \approx 1/2$, phases distributed as in case of mitotic division, (i.e. the Bellman–Harris setup)

$$t_1 \approx \{E[T] + E[T] - E[S] - E[S]\}/2 = E[T] - E[S].$$

In situations where there is no cell loss $p = 1$ and

$$t_1 = (1/\alpha) - \{2E[e^{-\alpha G_1}] - 1\}/2\alpha E[e^{-\alpha(G_1+S)}],$$

which has little connection with any sensible indicator of $T - S$. Thus, on the whole the Yamada–Puck estimator t_1 [49] seems difficult to justify.

Continuously labelled mitoses curves differ significantly from the fractions of the total population labelled by the same method. To begin with the fraction labelled mitoses must remain zero while the first labelled cells pass through G_2. Indeed, at time u after labelling started the expected fraction of labelled mitotic cells who were themselves alive when the thymidine was first added is a constant multiplied by

$$\alpha \int_0^\infty e^{-\alpha(t+u)} P[T - M \le t + u < T, G_1 + S > t] \, dt$$
$$= E[\{e^{-\alpha(T-M)\vee u} - e^{-\alpha T \wedge (G_1+S+u)}\}^+]$$
$$= E[e^{-\alpha(T-M)} - e^{-\alpha(G_1+S+u)}; G_2 \le u < G_2 + M]$$
$$+ E[e^{-\alpha(T-M)} - e^{-\alpha T}; G_2 + M \le u < T - M]$$
$$+ E[e^{-\alpha u} - e^{-\alpha T}; u > T - M].$$

---

* The term arising from the integration domain has been disregarded.

Thus, the instant when the CLM curve turns positive estimates a low quantile of the G_2 distribution, lower the larger is the population.

Background

A short introduction, rich in contents, is given on pages 116–118 of Reference [5]. The established survey is Reference [47]. References [40, 42, 48, and 49] report early experimentally oriented work with continuous labelling.

9.7 ARREST METHODS

The addition of colchicine or colcemide (or several other substances) to cell populations blocks division. Dependent upon the concentration of colchicine/colcemide, and upon the cell type, different behaviours are obtained: division may be immediately blocked upon administration of the blocking agent, or cells already in mitosis (suitably demarcated) at this moment may move on in the cell cycle unaffected, those entering the stage later however being arrested.

In the first case, occurring for example with Chinese hamster lung cells [51], the mitotic index at time t after blocking starts, $MI(t)$, has been calculated to

$$MI(t) = e^{\alpha(M+t)} - 1, \qquad (9.7.1)$$

for non random M and T [5, 52]. The logarithm

$$^{10}\log\{1 + MI(t)\} = \alpha(M + t)/\log 10$$

is called the *collection function*. Derivations are rather heuristic and it is obvious that Equation (9.7.1) can not hold for t large—the left hand side does not exceed one whereas $e^{\alpha t}$ is unbounded.

To find a more accurate formula we consider as usual a supercritical and non-lattice stable binary splitting process. The probability of a randomly chosen cell entering mitosis within time t is

$$\alpha p \int_0^\infty e^{-\alpha u} P[T - M \le t + u, T > u] \, du/\{1 - \hat{L}(\alpha)\}$$

$$= pE[e^{-\alpha(T-M-t)^+} - e^{-\alpha T}]/\{1 - \hat{L}(\alpha)\}$$

$$= \{2pE[e^{-\alpha(T-M-t)^+}] - 1\}/\{1 - 2(1-p)E[e^{-\alpha D}]\},$$

to be called A.

The probability of its disintegrating during the same time lapse is

$$\alpha(1 - p) \int_0^\infty e^{-\alpha u} P[u < D \le t + u] \, du/\{1 - \hat{L}(\alpha)\}$$

$$= 2(1 - p) E[e^{-\alpha(D-t)^+} - e^{-\alpha D}]/\{1 - 2(1 - p) E[e^{-\alpha D}]\} = B.$$

It follows that the expected number of mitotic cells divided by the expected total number of cells t time units after immediate blocking at mitosis must be

$$mi_b(t) = A/(1 - B)$$
$$= \{2pE[e^{-\alpha(T-M-t)^+}] - 1\}/\{1 - 2(1 - p)E[e^{-\alpha(D-t)^+}]\}. \quad (9.7.2)$$

The index b refers to the presence of a blocking agent. Here the numerator is

$$2pe^{\alpha t}E[e^{-\alpha(T-M)}; T - M > t] + 2pP[T - M \leq t] - 1$$
$$= 2pE[e^{-\alpha T}e^{\alpha(M+t)}] - 1 = e^{\alpha(M+t)} - 1$$

exactly if M is constant and $P[T - M > t] = 1$. We see that Equation (9.7.1) disregards both variation in mitotic times, cell death, and the behaviour for t of sizes approaching a cycle length. (After such a time the experimental picture is however also disturbed by the phenomenon of mitotic degeneration [52].)

If $p = 1$ and $T - M$ has the frequency function h, then Equation (9.7.2) yields

$$mi_b'(t) = \frac{d}{dt}\left\{2e^{\alpha t}\int_t^\infty e^{-\alpha u}h(u)\,du + 2\int_0^t h(u)\,du - 1\right\}$$

$$= 2\alpha e^{\alpha t}\int_t^\infty e^{-\alpha u}h(u)\,du - 2h(t) + 2h(t)$$

$$= \alpha\{mi_b(t) - 2P[T - M \leq t] + 1\}.$$

Hence,

$$\frac{d}{dt}\log\{1 + mi_b(t)\} = \alpha - 2\alpha P[T - M \leq t]/\{1 + mi_b(t)\} \quad (9.7.3)$$

and for t clearly less than any possible cycle time (minus mitosis) the collection function $\log\{1 + mi_b(t)\}$ is linear with slope α. Thus it can be used to estimate α or, equivalently, the doubling time $t_d = (\log 2)/\alpha$, a somewhat farfetched procedure as it might seem. However, consideration of the model with constant cycle times has led to the suggestion [52, 53] that the cycle time be estimated this way. By Inequality (9.1.3) this can obviously be rather misleading.

Blocking agents can be used together with tritiated thymidine, continuously added. In this procedure a fraction of labelled mitotic cells can be observed, besides the mitotic index. One might expect these two curves to be parallel with a horizontal distance determined by G_2 and the duration of the G_2 phase has also been thus estimated [52, 53]. We shall have a closer look at this:

The probability of a randomly chosen cell being both labelled and mitotic before t is given by

$$\alpha p \int_0^\infty e^{-\alpha u} P[T - M \le t + u, G_1 + S > u]\, du/\{1 - \hat{L}(\alpha)\}$$

$$= 2pE[e^{-\alpha(T-M-t)^+} - e^{-\alpha(G_1+S)}; t \ge G_2]/\{1 - 2(1 - p)E[e^{-\alpha D}]\}.$$

As in the derivation of Equation (9.7.2) it follows that the fraction labelled mitotic cells after immediate blocking is given by

$$lmi_b(t)$$

$$= \{2pE[e^{-\alpha(T-M-t)^+} - e^{-\alpha(G_1+S)}; t \ge G_2]\}/\{1 - 2(1 - p)E[e^{-\alpha(D-t)^+}]\}$$

$$= 2p\{E[e^{-\alpha(T-M-t)^+} - e^{-\alpha T}] + E[e^{-\alpha(G_1+S)}(e^{-\alpha(G_2+M)} - 1)]$$

$$- E[e^{-\alpha(T-M-t)^+} - e^{-\alpha(G_1+S)}; t < G_2]\}/\{1 - 2(1 - p)E[e^{-\alpha(D-t)^+}]\}$$

$$= mi_b(t) - 2p\{E[e^{-\alpha(G_1+S)}(1 - e^{-\alpha(G_2+M)})]$$

$$+ E[e^{-\alpha(T-M-t)^+} - e^{-\alpha(G_1+S)}; t < G_2]\}/\{1 - 2(1 - p)E[e^{-\alpha(D-t)^+}]\}.$$

$$(9.7.4)$$

Thus, $lmi_b(t) = 0$ for $P[t < G_2] = 1$, the function increases with t and if there is no cell death and $P[t \ge G_2] = 1$ the difference

$$mi_b(t) - lmi_b(t) = 2pE[e^{-\alpha(G_1+S)}(1 - e^{-\alpha(G_2+M)})] \qquad (9.7.5)$$

is constant. Indeed, since $\alpha(G_2 + M)$ is usually little we have, at least if the last two phases are independent of the former ones, that

$$mi_b(t) - lmi_b(t) \approx \alpha E[G_2 + M] \qquad (9.7.6)$$

for t a.s. not less than G_2. Since α is easily estimated (9.1.1) this provides a method of estimating the total expected post-synthetic sojourn.

Even in cases with rather substantial cell loss Equation (9.7.4) yields

$$\lim_{t \to \infty} lmi_b(t) = 2p\{1 - E[e^{-\alpha(G_1+S)}]\}/(2p - 1) \approx 2\alpha p E[G_1 + S]/(2p - 1)$$

$$(9.7.7)$$

yielding an estimator of the duration of the two first phases.

This formula and the possible exploitation (9.7.6) of the vertical distance between the two curves seem both to have passed unnoticed. On the other hand Puck and co-authors [51, 52, 53] have used the horizontal distance between collection functions as an estimator of precisely the G_2-phase, the basis for this being that $lmi_b(t) - 0$ for $t < G_2$ a.s. Approximate calculations like those preceding show however that this distance is $u \approx$ $\approx e^{-\alpha t}E[G_2 + M]$ if $mi_b(t) = lmi_b(t + u)$.

Finally, let us have a glance at the second case mentioned, of mitotic cells at the blocking agent supply moment not being arrested, while all cells entering mitosis later are. Obviously the mi_b-function here will start in the origin and increase slowly for a time of the length of mitosis, since then the population still grows. Afterwards the behaviour should be close to that of the mitotic index in circumstances of immediate blocking. It is not difficult (but somewhat messy) to verify this by calculation of the explicit curve. We leave it to the reader.

Background

The experimental methods described have been developed largely by T. T. Puck and co-workers, who also worked with the theoretical analysis under the assumption of no cell death and non-random phases. Reference [55] is a mathematical elaboration of their ideas. Reference [50] studies inference for some special phase distributions.

9.8 SYNCHRONY

In view of the mathematical difficulties in deducing properties of individual cells from the growth of stable populations, it is not strange that experimentalists have developed several methods of synchronizing mitotic cycles. If cells can be induced to march together from division to division then the population represents the individual and the cell cycle will be open to direct studies on a macroscale [62]. Such a synchronization is of course a futile hope: it would mean very much of interference with the object of study. But what can be obtained is a temporary synchronization, through selection of cells of the same age, or through blocking methods, for example like those discussed in the preceding section.

In this manner a branching process is obtained, most often starting from several newly born ancestors at $t = 0$. (There are also other methods, like accumulating cells in synthesis; we shall not enter upon that subject.)

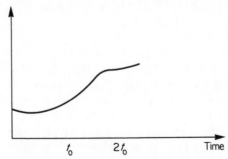

Figure 9.8.1 The development of a synchronized population.

Rather rapidly synchrony will disappear. However, if the increase in number of cells is little or none for a time t_0 and then more rapid, being again essentially zero by time $2t_0$ (Figure 9.8.1), then this means that only first generation individuals can have been born during $[0, 2t_0]$. Thus the expected number of cells, starting from m_0 newly born cells, will have the form

$$m_t = m_{t,0} + m_{t,1} = m_0(1 - p)\{1 - F(t)\} + m_0 p\{1 - G(t)\} + 2m_0 pG(t)$$

$$= m_0\{1 + pG(t) - (1 - p)F(t)\} \qquad (9.8.1)$$

at least for $0 \le t \le 2t_0$. If there is no cell death this offers an excellent way of estimating the cycle time distribution.

Methods of this type lead to the question of how to measure the degree of synchrony. Such a measure has been defined in Reference [57]: Consider a process with a differentiable expectation m_t, and define the normalized rate of cell division at r_t by

$$r_t = m_t'/m_t.$$

It can be viewed as the fraction of cells dividing per unit time. If the age distribution is stable, r_t is constant. On the other hand if r_t varies so does cell production, a peak in r_t representing a burst of mitotic activity. Further, if t is the first doubling time $m_{t_0} = 2m_0$, then the area under r_t $0 \le t \le t_0$,

$$\int_0^{t_0} r_t \, dt = \log m_{t_0} - \log m_0 = \log 2,$$

is independent of the integrand. Thus the portion of this area that lies above the level $\alpha_0 = (\log 2)/t_0$, which would be the normalized rate of production if the process were stable with the same doubling time, might be a reasonable measure of synchronity. Formally, we define, following Engelberg, the *degree of synchrony* during the first doubling period, ds, by

$$ds = \left\{ \int_0^{t_0} r_t \vee \alpha_0 \, dt - \log 2 \right\} \Big/ \log 2.$$

Clearly in a stable population $t_0 = t_d$ and $ds = 0$. Also $0 \le ds < 1$ and $ds = 0$ implies that $r_t = \alpha_0 = \alpha$. The degree of synchrony tends to one when r_t approaches one or several spikes.

The given definition is certainly not the only possible. For example it could be replaced by $\sup_{0 \le t \le t_0} r_t - \inf_{0 \le t \le t_0} r_t$. Also it has the disadvantage of relating to a fairly long time span. The degree of synchrony, or rather asynchrony, at a given time could be defined in various ways. However such definitions obviously require more knowledge than the number of individuals at the time. One possibility is to consider simply the variance of the empiric age distribution,

$$\int_0^\infty a^2 z_t^{da}/z_t - \left(\int_0^\infty a z_t^{da}/z_t \right)^2$$

or maybe the expectation analogue

$$\int_0^\infty a^2 m_t^{da}/m_t - \left(\int_0^\infty a m_t^{da}/m_t \right)^2.$$

Both of these converge—after the usual qualifications—to the variance of the stable age distributions. But the convergence need not at all be monotone. Another possibility of defining an instantaneous degree of synchrony is to measure the mitotic index, suitably normed (Zeuthen's index of synchrony [57].

A somewhat different aspect of synchrony is that of how individuals are distributed over generations, studied in Section 8.5. Restricting ourselves just for simplicity to the expectation formulation, we define

$$G_t(u) = \sum_{n \le u} m_{t,n}/m_t.$$

As usual we consider a supercritical and non-lattice process started from one zero aged ancestor. If η_t is a random variable with distribution G_t, then Theorem (8.5.1) or (8.5.11) imply that

$$(\eta_t - t/\beta)\, \beta^{3/2}/\beta_2 \sqrt{t} \to N(0, 1)$$

as $t \to \infty$ [β_2 defined in Theorem (8.5.1)]. If thus the variance of η_t is taken as a measure of asynchrony, it can be contended that the decay of synchrony over generations proceeds as $z_0 \beta_2^2 t/\beta^3$ for $t \to \infty$, if the process starts from z_0 ancestors.

Background

We have only scratched the surface of several intriguing problems: In what sense is a stable population the most "random"? How does this randomness evolve? How can the dependence of ages of different existing individuals and its decay as time passes be described? Burnett–Hall and O'N. Waugh [57] discuss several measures of synchrony. Sankoff [60] suggests making use of the age distribution.

9.9 THE COMPOSITION OF TWO-TYPE POPULATIONS: ENDOMITOSIS AND THE G_0 RESTING PHASE

In this section the topic and the model used, deviate somewhat from the main path of the chapter. The topic is not a method fairly generally used in the empiric study of cells but a specific problem, however basic and arising in at least three different contexts. The model resorted to is that of a two-type process.

The phenomenon of polyploidy was introduced in Example (4.1.2). It has been studied experimentally and mathematically in so called Ehrlich mouse ascites tumours [65, 66, 68]. There it has been observed that irrespective of the initial composition of the population, finally the tumour will always consist to 95% diploid cells. The main purpose here is to explain such facts.

In two-step models of carcinogenesis [67] it is imagined that first somehow there is a benign tumour initiated, which is interpreted as a subcritical process. The cells in this one may turn malign by chance, this being thought of as turning the ancestors of supercritical processes.

The third context in which the topic of this section is met is that of recent theories of tumour growth [63]. Here it is conceived that the growth fraction decreases by newly born cells having the possibility of entering a relatively long rest phase, called G_0 or sometimes prolonged G_1, from which they may or may not return to proliferation.

If this last possibility of return is disregarded, all the three examples display the same pattern: there is a main or first kind of individuals (type one), who can beget children of their own kind but also of an alternative, zeroth type. Type zero individuals can however only give birth to type zero cells and possibly (the last example) they may have eternal life. If a chance of return to the main cell type were allowed, the resulting model would be closely related to the so called stem cell hypothesis of growth [63].

Assume that cells of type one disintegrate with probability $1 - p$ after a time with distribution F, that their cycle time distribution is G, and that upon completed life cycle they give birth to i daughters of their own type and j cells of type zero with a probability p_{ij}. We write

$$a = p \sum_{i,j} i p_{ij},$$

$$b = p \sum_{i,j} j p_{ij}.$$

In the case of polyploidy and with q being the probability of endomitosis instead of mitotic division these reproduction means have the form

$$a = 2p(1 - q)$$
$$b = pq. \tag{9.9.1}$$

In the case where there are always two daughters, both of the first type with probability $1 - q$ both of the zeroth with probability q, the result is

$$a = 2p(1 - q)$$
$$b = 2pq, \tag{9.9.2}$$

these forms remaining if only division into two daughters is possible but these choose, say, independently, with probability q, to turn zero type, otherwise retaining their mother's type.

Cells of type zero are supposed to be lost (disintegrate) with probability $1 - p_0$, the time to this event then having distribution F_0. Otherwise, we presume they remain intact during a time with the possibly defective distribution G_0. After this, they can only beget children of type zero, their expected number being a_0/p_0.

As in earlier sections of this chapter we confine ourselves to the expected behaviour of the process. Results on almost sure and mean square convergence can be found in Reference [64]. Throughout G and G_0 are taken to be non-lattice and we consider first only proper G_0, i.e. $G_0(\infty) = 1$.

Let m_t, as usual, be the expected number of type one individuals at time t, provided the process was started at time zero from a newly born type one ancestor. Let n_t be the expected number of type zero cells in the same process and let m_t^0 denote the expectation of a process started from a type zero ancestor. The following equations follow as so many times before,

$$m_t = 1 - L(t) + a \int_0^t m_{t-u} G(du), \tag{9.9.3}$$

$$n_t = b \int_0^t m_{t-u}^0 G(du) + a \int_0^t n_{t-u} G(du), \tag{9.9.4}$$

$$m_t^0 = 1 - L_0(t) + a_0 \int_0^t m_{t-u}^0 G_0(du), \tag{9.9.5}$$

the first and the last being of the form of Equation (6.3.4), completely treated there. Here $L = pG + (1 - p)F$ and correspondingly for L_0.

We shall investigate the ratio $r_t = m_t/n_t$, assuming $b > 0$, of course. Define the Malthusian parameters α and α_0 by

$$a\hat{G}(\alpha) = 1,$$

$$a_0\hat{G}_0(\alpha_0) = 1$$

provided they exist. Then clearly, if $a_0 \geq 1$ and $\alpha < \alpha_0$ or $a < 1$, $r_t \to 0$ and the type zero cells will dominate completely (this is the carcinogenesis model). The rate of convergence is exponential, $r_t = O(e^{-(\alpha_0 - \alpha)t})$, assuming α exists. The same dominance occurs if $a = a_0 = 1$ and both L and L_0 have finite first moments. Because then

$$m_t \to p + (1 - p) E[D]/E[T]$$

$$m_t^0 \to p_0 + (1 - p_0) E[D_0]/E[T_0]$$

in an obvious notation, and by Lemma (5.2.12) and Corollary (5.2.14)

$$n_t \sim \{p_0 + (1 - p_0) E[D_0]/E[T_0]\} bt/E[T].$$

Also when $\alpha = \alpha_0 > 0$, this type of asymptotics obtains. The case when both processes are subcritical, $a < 1$, $a_0 < 1$, provides little interest (though some results can of course be established), whereas the case $a = 1$, $a_0 < 1$ reduces to a branching process with immigration, at least in so far expectations are concerned. And

$$r_t \to (1 - a_0) \{pE[T] + (1 - p) E[D]\}/\{p_0 E[T_0] + (1 - p_0) E[D_0]\}.$$

$$(9.9.6)$$

Also in the remaining case with both processes supercritical but $\alpha > \alpha_0 \geq 0$ or $\alpha > 0$ and $a_0 < 1$ a non-trivial limit can be obtained: Since

$$\int_0^\infty e^{-\alpha t} m_t^0 \, dt = \{1 - \hat{L}_0(\alpha)\}/\alpha\{1 - a_0 \hat{G}_0(\alpha)\}$$

from Equation (9.9.5), Equation (9.9.4) yields

$$n_t \sim e^{\alpha t} b\hat{G}(\alpha) \{1 - \hat{L}_0(\alpha)\}/\alpha\{1 - a_0\hat{G}_0(\alpha)\} a \int_0^\infty t e^{-\alpha t} G(dt)$$

$$= e^{\alpha t} b \{1 - \hat{L}_0(\alpha)\}/\alpha\{1 - a_0\hat{G}_0(\alpha)\} a^2 \int_0^\infty t e^{-\alpha t} G(dt)$$

and

$$r_t \to a\{1 - \hat{L}(\alpha)\} \{1 - a_0\hat{G}_0(\alpha)\}/b\{1 - \hat{L}_0(\alpha)\}. \qquad (9.9.7)$$

In the case of polyploidy (9.9.1) and no cell disintegration this is

$$r_t \to (1 - 2q)(1 - a_0 E[e^{-\alpha T_0}])/q(1 - E[e^{-\alpha T_0}]) \qquad (9.9.8)$$

where $1 \leq a_0 \leq 2$. A first conclusion from this is that the observed invariability of the ultimate composition of partially diploid cell populations can be viewed as a natural expression of the reproduction pattern. But then also an observed final fraction of type zero cells strictly less than one, as the 95 % in the Ehrlich mouse ascites tumour, must be interpreted as showing that these cells reproduce at a slower rate than the main cells do ($\alpha_0 < \alpha$ or may be $p_0 \leq 1/2$). If a_0 and the distribution of T_0 are known, and $p = p_0 = 1$ Equation (9.9.8) can be used for inference about q. If $E[e^{-\alpha T_0}]$ is very little, i.e. zero-type cells have considerably longer cycle times than have those of type one, then

$$r_t \approx (1 - 2q)/q \qquad (9.9.9)$$

an approximation that would yield a 4.77 % probability of endomitosis in the Ehrlich mouse ascites tumour.

We shall finally comment briefly upon the G_0 rest phase model. If it is assumed that type zero cells have an eternal life, provided they do not disintegrate, then Equations (9.9.4–5) reduce to

$$m_t = 1 - L(t) + a \int_0^t m_{t-u} G(du) \qquad (9.9.10)$$

$$n_t = b \left\{ G(t) - (1 - p_0) \int_0^t F(t - u) G(du) \right\} + a \int_0^t n_{t-u} G(du). \qquad (9.9.11)$$

If further $a < 1$, due to a comparatively great probability of new cells entering the G_0 stage, then a population starting from many proliferating cells first grows exponentially. The new cells are however resting to an increasing extent and after some time this displays itself in a slower population growth. Finally the total population levels off to

$$\lim_{t \to \infty} n_t = p_0 b / (1 - a),$$

which is $2q/(2q - 1)$ if Equation (9.9.2) is valid and $p = p_0 = 1$.

This behaviour conforms well to empiric data [63]. Also it implies that the growth fraction decreases from one to zero. Such a decrease is also supported experimentally [63].

In the critical case $a = 1$ m_t has a finite limit whereas n_t grows approximately linearly. If $a > 1$ however the amount of resting as well as of proliferating cells increases exponentially (α),

$$r_t \to a \{ 1 - \hat{L}(\alpha) \} / b \{ 1 - (1 - p_0) \hat{F}_0(\alpha) \}$$

which is just $r_t \to (a - 1)/b$ if cells never disintegrate. In case of a steadily expanding population such that Equation (9.9.2) holds the proportion of actively proliferating cells would be $1/2q - 1$.

Background

Though this has primarily stressed the problem of endomitosis the biologically basic topic touched upon here might well be that of a resting stage of cells. Reference [63] gives the biological theory. A lucid exposition in the pseudo-deterministic tradition is given in Reference [66].

REFERENCES

1. Baserga, R., Biochemical events in the cell cycle. In: Perry, S. (ed.) *Human Tumor Cell Kinetics.* National Cancer Institute Monograph 30. U.S. Department of Health, Education, and Welfare, Washington, 1969.
2. Baserga, R. (ed.), *The Cell Cycle and Cancer.* Marcel Dekker, New York, 1971.
3. Brockwell, P. J. and Trucco, E., On the decomposition by generations of the PLM-function. *J. Theoret. Biol.* **26**, 149–179, 1970.

4. Bronk, B. V., Dienes, G. J., and Paskin, A., The stochastic theory of cell proliferation. *Biophys. J.* **8**, 1353–1397, 1968.

5. Cleaver, J. E., *Thymidine Metabolism and Cell Kinetics*. North Holland, Amsterdam, 1967.

6. Harris, T. E., A mathematical model for multiplication by binary fission. In: *The Kinetics of Cellular Proliferation*, 368–381, Grume and Stratton, New York, 1959.

7. Howard, A. and Pelc, S. R., Synthesis of desoxiribonucleic acid in normal and irradiated cells and its relation to chromosome breakage. *Heredity Suppl.* **6**, 261–273, 1953.

8. Jagers, P., The composition of branching populations: A mathematical result and its application to determine the incidence of death in cell proliferation. *Math. Biosci*, **8**, 227–238, 1970.

9. Kendall, D. G., Stochastic processes and the growth of bacterial colonies. *Symposia of the Society for Experimental Biology* **7**, 55–65, 1953.

10. Kubitschek, H., Cell generation times; ancestral and internal controls. *Proc. 5th Berkeley Symp. Math. Statist. Prob.* **4**, 549–572, 1967.

11. Lamerton, L. F. and Fry, R. J. M. (eds.), *Cell Proliferation*. Blackwell, Oxford, 1963.

12. Macdonald, P. D. M., Statistical inference from the fraction labelled mitoses curve. *Biometrika* **57**, 489–503, 1970.

13. Mode, C. J., Multitype age-dependent branching processes and cell cycle analysis. *Math. Biosci.* **10**, 177–190, 1971.

14. Nachtwey, D. S. and Cameron, I. L., Cell cycle analysis. In: Prescott, D. M. (ed.), *Methods in Cell Physiology 3*. Academic Press, New York, 1968.

15. Nooney, G. C., Age distributions in stochastically dividing populations. *J. Theoret. Biol.* **20**, 314–320, 1968.

16. Norrby, K., Johannisson, G., and Mellgren, J., Proliferation in an established cell line. An analysis of birth, death and growth rates. *Exptl. Cell Res.* **48**, 582–594, 1967.

17. Painter, P. R. and Marr, A. G., Mathematics of microbial populations. *Ann. Rev. Microbiol.* **22**, 519–548, 1968.

18. Powell, E. D., Some features of the generation times of individual bacteria. *Biometrika* **42**, 16–44, 1955.

19. Powell, E. O., Growth rate and generation time of bacteria, with special reference to continuous culture, *J. Gen. Microbiol.* **15**, 492–511, 1956.

20. Steel, G. G., The cell cycle in tumours: An examination of data gained by the technique of labelled mitoses. *Cell Tissue Kinet.* **5**, 87–100, 1972.

21. Wilson, G. B., *Cell Division and the Mitotic Cycle*. Reinhold, New York, 1966.

22. Horn, M. Ein Beitrag zu stochastischen Modellierung der Kinetik von Zellverbänden. *Biom. Z.* **15**, 277–285, 1973.

23. Jagers, P., *Maximum likelihood estimation of the reproduction distribution and the extent of disintegration in cell proliferation*. Dept. of Mathematics, Chalmers U. of Tech., 1973.

24. Norrby, K., Population kinetics of normal, transforming and neoplastic cell lines. *Acta Path. Microbiol. Scand.* **78**, Suppl. No. 214, 1970.

25. Sisken, J. E., *Analysis of variations in intermitotic time. Cinemicrography in cell Biology*. Academic Press, New York, 1963.

26. Jagers, P. and Norrby, K., Estimation of the mean and variance of cycle times in cinemicrographically recorded cell populations during balanced exponential growth, *Cell Tissue Kinet.* **7**, 201–211, 1974.

27. Barrett, J. C., A mathematical model of the mitotic cycle and its application to the interpretation of percentage labelled mitoses data. *J. Nat. Cancer Inst.* **37**, 443–450, 1966.

28. Barrett, J. C., Optimized parameters for the mitotic cycle. *Cell Tissue Kinet.* **3**, 349–353, 1970.

29. Bartlett, M. S., Distributions associated with cell populations. *Biometrika* **56**, 391–400, 1969.

30. Bresciani, F., A comparison of the cell generative cycle in normal, hyperplastic and neoplastic mammary gland of the C3H mouse. In: *Cellular Radiation Biology,* 498–513. Williams and Wilkins, Baltimore, Maryland, 1965.

31. Brockwell, P. J., Trucco, E. and Fry, R. J. M., The determination of cell-cycle parameters from measurements of the fraction of labelled mitoses. *Bull. Math. Biophys.* **34**, 1–12, 1972.

32. Bronk, B. V., On radioactive labelling of proliferating cells; the graph of labelled mitoses. *J. Theoret. Biol.* **22**, 468–492, 1969.

33. Gilbert, C. W., The labelled mitoses curve and the estimation of the parameters of the cell cycle. *Cell Tissue Kinet.* **5**, 53–63, 1972.

34. Hartmann, N. R. and Pedersen, T., Analysis of the kinetics of granulosa cell populations in the mouse ovary. *Cell Tissue Kinet.* **3**, 1–11, 1970.

35. Lipkin, M., The proliferative cycle of mammalian cells. In Baserga, R. (ed.), *The Cell Cycle and Cancer,* 6–26, Marcel Dekker, New York, 1971.

36. Lipkin, M. and Descher, E., Comparative analysis of cell renewal in the gastrointestinal tract of newborn hamster. *Exptl. Cell Res.* **49**, 1–12, 1968.

37. Mendelsohn, M. L., The kinetics of tumor cell proliferation. In: *Cellular Radiation Biology,* 498–513. Williams and Wilkins, Baltimore, Maryland, 1965.

38. Mendelsohn, M. L. and Takahashi, M., A critical evaluation of the fraction of labelled mitoses method as applied to the analysis of tumor and other cell cycles. In: Baserga, R. (ed.), *The Cell Cycle and Cancer,* 58–95. Marcel Dekker, New York, 1971.

39. Quastler, H. and Sherman, F. G., Cell population kinetics in the intestinal epithelium of the mouse. *Exptl. Cell Res.* **17**, 420–438, 1959.

40. Stanners, C. P. and Till, J. E., DNA synthesis in individual L-strain mouse cells. *Biochim. Biophys. Acta* **37**, 406–419, 1960.

41. Steel, G. G., The cell cycle in tumours: an examination of data gained by the technique of labelled mitoses. *Cell Tissue Kinet.* **5**, 87–100, 1972.

42. Steel, G. G., Adams, K. and Barrett, J. C., Analysis of cell population kinetics of transplanted tumours of widely differing growth rate. *Brit. J. Cancer* **20**, 784–800, 1966.

43. Steel, G. G. and Hanes, S., The technique of labelled mitoses: analyses by automatic curve fitting. *Cell Tissue Kinet.* **4**, 93–105, 1971.

44. Takahashi, M., Theoretical basis for cell cycle analysis. I. Labelled mitosis wave method. *J. Theoret. Biol.* **13**, 202–211, 1966.

45. Takahashi, M., Theoretical basis for cell cycle analysis. II. Further studies on labelled mitosis wave method. *J. Theoret. Biol.* **18**, 195–209, 1968.

46. Trucco, E. and Brockwell, P. J., Percentage labelled mitoses curves in exponentially growing cell populations. *J. Theoret. Biol.* **20**, 321–337, 1968.
47. Wimber, D. E., Methods for studying cell proliferation with emphasis on DNA labels. In: Lamerton, L. F. and Fry, R. J. M. (eds.), *Cell Proliferation.* Blackwell, Oxford, 1963.
48. Robinson, S. H., Brecher, J., Lourie, I. S., and Haley, J. E., Leukocyte labeling in rats during and after continuous infusion of tritiated thymidine: implications for lymphocyte longevity and DNA reutilization. *Blood* **26**, 281–295, 1965.
49. Yamada, M. and Puck, T. T., Action of radiation on mammalian cells. IV. Reversible mitotic lag in the S3 HeLa cell produced by low doses of x-rays. *Proc. Nat. Acad. Sci.* (Wash.) **47**, 1181, 1961.
50. Macdonald, P. D. M., On the statistics of cell proliferation. In: Bartlett, M. S. and Hiorns, R. W. (eds.), *The Mathematical Theory of the Dynamics of Biological Populations,* 303–314. Academic Press, London and New York, 1973.
51. Puck, T. T., Studies of the life cycle of mammalian cells. *Cold Spring Harbour Symp. Quant. Biol.* **29**, 167–176, 1964.
52. Puck, T. T., Sanders, P., and Petersen, D., Life cycle analysis of mammalian cells II. Cells from the Chinese hamster ovary grown in suspension culture. *Biophys. J.* **4**, 441–450, 1964.
53. Puck, T. T., and Steffen, J., Life cycle of mammalian cells I. A method for localizing metabolic events within the life cycle, and its application to the action of colcemide and sublethal doses of X-irradiation. *Biophys. J.* **3**, 379–397, 1963.
54. Tannock, I. F., A comparison of the relative efficiencies of various metaphase arrest agents. *Exptl. Cell Res.* **47**, 345–356, 1965.
55. Trucco, E., A note on Puck and Steffen's equations for the life cycle analysis of mammalian cells. *Biophys. J.* **5**, 743–753, 1965.
56. Van't Hof, J., Experimental procedures for measuring cell population kinetic parameters in plant root meristems. In: D. M. Prescott (ed.), *Methods in Cell Physiology, III.* Academic Press, New York, 1968.
57. Burnett–Hall, D. G. and Waugh, W. A. O'N., Indices of synchrony in cellular cultures, *Biometrics* **23**, 693–716, 1967.
58. Engelberg, J., Measurement of degrees of synchrony in cell populations. In: Zeuthen, L., (ed.), *Synchrony in Cell Division and Growth,* 497–508. John Wiley and Sons, New York, 1964.
59. Stubblefields, E., Synchronization methods for mammalian cell cultures. In: Prescott, D. M. (ed.), *Methods in Cell Physiology III,* 25–44. Academic Press, New York, 1968.
60. Sankoff, D., Duration of detectible synchrony in a binary branching process. *Biometrika* **58**, 77–81, 1971.
61. Waugh, W. A. O'N., Models and approximations for synchronous cellular growth. *Proc. Sixth Berkeley Symp. Math. Statist. Prob., IV,* 137–145, 1972.
62. Zeuthen, E., Preface to Zeuthen, E. (ed.), *Synchrony in Cell Division and Growth.* John Wiley and Sons, New York, 1964.
63. Gavosto, F. and Pileri, A., Cell cycle of cancer cells in man. In: Baserga, R. (ed.), *The Cell Cycle and Cancer,* 99–128. Marcel Dekker, New York, 1971.

64. Jagers, P., The proportions of individuals of different kinds in two-type populations. A branching process problem arising in biology. *J. Appl. Prob.* **6**, 249–260, 1969.

65. Jansson, B., Absolute and relative growth of the total number of cells and of the number of cells of different ploidies in tumors. Mimeographed report, Stockholm 1967.

66. Jansson, B., Competition within and between cell populations. In: Sugaharo, T., Révész, L., and Scott, O. (ed.), *Fraction size in Radiobiology and Radiotherapy,* 51–72. Igaku Shoin, Tokyo, 1973.

67. Kendall, D. G., Birth-and-death processes and the theory of carcinogenesis. *Biometrika* **47**, 13–21, 1960.

68. Klein, G. and Révész, L., Quantitative studies on the multiplication of neoplastic cells in vivo I. *J. Nat. Cancer Inst.* **14**, 229–277, 1953.

69. Mode, C. J., *Multitype Branching Processes.* American Elsevier, New York, 1971.

Appendix

This appendix gives just two results. The first is wellknown and basic but not as easily available as you would wish, the second is a rather special theorem useful in inference for branching processes.

Theorem A1 (Dynkin)

Let Ω be a set. Any class $\mathscr{C}$ of subsets of Ω is called a d-system if

(a) $\Omega \in \mathscr{C}$
(b) $A, B \in \mathscr{C}, A \subset B \Rightarrow B - A \in \mathscr{C}$
(c) $\mathscr{C} \ni A_n \subset A_{n+1} \Rightarrow \bigcup_{n=1}^{\infty} A_n \in \mathscr{C}$.

If $\mathscr{C}$ is closed under intersection $(A, B \in \mathscr{C} \Rightarrow A \cap B \in \mathscr{C})$, then $d(\mathscr{C})$, which is the smallest d-system containing $\mathscr{C}$, coincides with $\sigma(\mathscr{C})$, the smallest σ-algebra containing $\mathscr{C}$.

Proof. Since any σ-algebra is a d-system, $d(\mathscr{C}) \subset \sigma(\mathscr{C})$. If $d(\mathscr{C})$ is closed under intersection, it is also a σ-algebra and the converse inclusion follows. To prove that $d(\mathscr{C})$ is thus closed define

$$\mathscr{D}_1 = \{B \in d(\mathscr{C}); B \cap A \in d(\mathscr{C}) \text{ for all } A \in \mathscr{C}\}.$$

Since $\mathscr{C}$ remains closed under intersection, $B \in \mathscr{C}$ implies that $B \cap A \in \mathscr{C} \subset \subset d(\mathscr{C})$ for all $A \in \mathscr{C}$. Hence $\mathscr{C} \subset \mathscr{D}_1$. But $\mathscr{D}_1$ is easily seen to be a d-system and so $d(\mathscr{C}) \subset \mathscr{D}_1$. As $\mathscr{D}_1$ is included in $d(\mathscr{C})$ by definition, actually $\mathscr{D}_1 = = d(\mathscr{C})$. Now allow A to belong to a larger class than $\mathscr{C}$, setting

$$\mathscr{D}_2 = \{B \in d(\mathscr{C}); B \cap A \in d(\mathscr{C}) \text{ for } A \in d(\mathscr{C})\}.$$

Again it is not difficult to show that $\mathscr{D}_2$ is a d-system. And if $B \in \mathscr{C}$, we have, for all $A \in \mathscr{D}_1 = d(\mathscr{C})$ that $B \cap A \in d(\mathscr{C})$ by the definition of $\mathscr{D}_1$. Hence $\mathscr{C} \subset \mathscr{D}_2$ and $d(\mathscr{C}) \subset \mathscr{D}_2$. But if $d(\mathscr{C}) \subset \mathscr{D}_2$, then for any A, B in $d(\mathscr{C})$ also $A \cap B \in d(\mathscr{C})$. Thus $d(\mathscr{C})$ contains with two sets always their intersection. $\qquad\square$

Here is a useful

259

Corollary

Let $\mathscr{C}$ be closed under intersection and $\mathscr{H}$ a linear function space satisfying

(a) $1 \in \mathscr{H}$ *and* $1_A \in \mathscr{H}$ *for all* $A \in \mathscr{C}$,
(b) *if* $0 \le f_j \le f_{j+1} \in \mathscr{H}$ *and* $f = \sup f_j$ *is finite, then* $f \in \mathscr{H}$.
Then $\mathscr{H}$ contains all $\sigma(\mathscr{C})$-measurable functions.
If (b) *holds only if f is bounded, then all bounded $\sigma(\mathscr{C})$-measurable functions are in $\mathscr{H}$.*

Proof. Let $\mathscr{D} = \{A ; 1_A \in \mathscr{H}\}$. Since $1_\Omega = 1 \in \mathscr{H}$, $A, B \in \mathscr{D}$, $A \subset B \Rightarrow 1_{B-A} = = 1_B - 1_A \in \mathscr{H}$, $\{A_j\}_1^\infty$ an increasing sequence in $\mathscr{D} \Rightarrow 1_{\cup A_j} = \sup_j 1_{A_j} \in \mathscr{H}$, we see that $\mathscr{D}$ is a d-system containing $\mathscr{C}$. By the preceding theorem $\sigma(\mathscr{C}) \subset \mathscr{D}$.

If $f \ge 0$ is $\sigma(\mathscr{C})$ measurable, it is the limit of an increasing sequence of linear combinations of indicator functions 1_A, $A \in \sigma(\mathscr{C}) \subset \mathscr{D}$. Each such linear combination is in $\mathscr{H}$ and by (b) f is in $\mathscr{H}$. If f may take negative values, we can write $f = f^+ - f^-$ and the proof is complete. □

The second result will be formulated and proved in terms of weak convergence on the function space $D[0, 1]$ with its customary Skorohood J_1-topology. So here $\overset{d}{\to}$ denotes weak convergence on that space. For our applications we need only finite-dimensional, or indeed one-dimensional, convergence.

The setup is the following:
On some basic probability space $(\Omega, \mathscr{S}, P\}$ are defined

(a) a sequence $\{\xi_k\}_1^\infty$ of independent, identically distributed random variables with mean zero and variance one;

(b) a sequence $\{v_n\}_1^\infty$ of not negative integer-valued random variables. There are positive numbers $a_n \to \infty$ such that $v_n/a_n \to$ some random variable $\eta \ge 0$ in probability. $P[\eta > 0]$ is not null. Define the sums

$$s_0 = 0, \qquad s_n = s_{n-1} + \xi_n,$$

$n = 1, 2, \ldots$ and the corresponding normed step functions,

$$X_n(t) = s_{[nt]}/\sqrt{n}$$

$0 \le t \le 1, n \ge 1$.

Theorem A2

Endow the basic probability space with the conditional measure, given that $\eta > 0$. Then, as $n \to \infty$,

$$X_{v_n} \overset{d}{\to} B,$$

where B is the Brownian motion on $[0, 1]$.

Corollary

If $v_n \to$ some random variable v in probability on the set where $\eta = 0$, then, as $n \to \infty$

$$X_{v_n} \overset{d}{\to} 1_{\{\eta > 0\}} B + X_v$$

where the one is the indicator function of its suffix, $X_\infty = 0$ and B is independent of the pair (η, X_v).

Proof of the theorem. By Theorem 16.3 of Reference [1], Donsker's theorem holds not only for i.i.d. summands but also if the joint distribution of the summands is absolutely continuous with respect to some product of identical measures on the line, having first and second moments zero and one respectively. Let Q be the conditional distribution,

$$Q(A) = P(A \cap \{\eta > 0\}) / P[\eta > 0],$$

$A \in \mathscr{S}$. Q is absolutely continuous with respect to P and the ξ_k are i.i.d. mean zero, variance one under P.

Hence also under Q

$$X_n \overset{d}{\to} B.$$

Let $r_n \to \infty$ but slowly enough, so that $r_n/\sqrt{n} \to 0$. Define

$$Y_n(t) = \begin{cases} \sum_{k=r_n}^{[nt]_*} \xi_k/\sqrt{n} & \text{if } nt \geq r_n \\ \\ 0 & \text{otherwise.} \end{cases}$$

Obviously

$$|X_n(t) - Y_n(t)| \leq \sum_{k=1}^{r_n} |\xi_k|/\sqrt{n} \to 0$$

a.s. with respect to P by the law of large numbers. By absolute continuity the convergence holds also a.s. with respect to Q. Hence

$$Y_n \overset{d}{\to} B.$$

Next, if $\varphi: R^k \to R$ is bounded and measurable, $\varphi(\xi_1, \ldots, \xi_k)$ and Y_n are independent under P for n large enough. Therefore if $\psi: D[0, 1] \to R$ is J_1-continuous and bounded

$$\int \psi \circ Y_n \varphi(\xi_1, \ldots, \xi_k) \, dP \to E[\psi \circ B] \cdot \int \varphi(\xi_1, \ldots, \xi_k) \, dP.$$

By the corollary to Dynkin's theorem this can be so extended that

$$\int \psi \circ Y_n \xi \, dP \to E[\psi \circ B] \int \xi \, dP$$

for any integrable random variable ξ measurable with respect to the σ-algebra $\mathscr{A}$ generated by the sequence $\{\xi_k\}$. Since each Y_n is measurable in $\mathscr{A}$, it holds generally for any integrable random variable ξ that

$$\int \psi \circ Y_n \xi \, dP = \int \psi \circ Y_n E[\xi | \mathscr{A}] \, dP \to E[\psi \circ B] \int \xi \, dP,$$

$E[\cdot | \mathscr{A}]$ denoting conditional expectation for P. In particular we can choose ξ of the form $\zeta 1_{\{\eta > 0\}}$ and obtain

$$\int \psi \circ Y_n \zeta \, dQ \to E[\psi \circ B] \int \zeta \, dQ$$

for all Q-integrable ζ.

By assumption $v_n/a_n \to \eta$ in P-probability and hence also in Q-probability. Define random elements Z_n by

$$Z_n(t) = \{(v_n/a_n) \wedge 1\} t$$

$0 \leq t \leq 1$. Clearly, with $Z(t) = (\eta \wedge 1) t$,

$$Z_n \to Z$$

in P- and Q-probability. Using the product topology we can invoke Reference [1], Theorem 4.5 to conclude that

$$(Y_{a_n}, Z_n, v_n/a_n) \overset{d}{\to} (B, Z, \eta)$$

where B and η are independent, and the converging triples are given their distribution under Q. (It is also not difficult to give a direct proof relying upon the preceding paragraph.) From this convergence, since $Y_{a_n} - X_{a_n} \to 0$ in Q-probability also, still using Q,

$$(X_{a_n}, Z_n, v_n/a_n) \overset{d}{\to} (B, Z, \eta).$$

The rest of the proof goes exactly as the upper half of page 148 in Reference [1]: Assume first that $\eta < 1$. The map T, $T(f, g, t) = f \circ g/\sqrt{t}$ is continuous at f if f is continuous, $t > 0$ and g is restricted to non-decreasing functions with range in the unit interval. Further

$$T(X_{a_n}, Z_n, v_n/a_n) = X_{v_n}.$$

Hence by the continuous mapping theorem, under Q,

$$X_{v_n} \overset{d}{\to} B \circ Z/\sqrt{\eta} = B$$

in distribution since $\eta > 0$. $Z(t) = \eta t$ and η and B are independent. If η is bounded by some constant, a change in the a_n brings it below 1. If it is unbounded approximation through $\eta \wedge k$ and

$$v_{n_k} = \begin{cases} v_n & \text{if } \eta \le k \\ a_n k & \text{if } \eta > k \end{cases}$$

and the strong Theorem 4.2 in Reference [1] completes the proof.

The Corollary is a direct sequel to the theorem: Let φ be a bounded continuous function $D[0, 1] \to R$. Then

$$\int \varphi \circ X_{v_n} dP = P[\eta > 0] \int \varphi \circ X_{v_n} dQ$$

$$+ \int_{\{\eta = 0\}} \varphi \circ X_{v_n} dP \to P[\eta > 0] E[\varphi \circ B] + \int_{\{\eta = 0\}} \varphi \circ X_v dP$$

$$= E[\varphi(1_{\{\eta > 0\}} B + X_v)]. \qquad \square$$

Background

Theorem A2 is due to Dion [2].

REFERENCES

1. Billingsley, P., *Convergence of Probability Measures*. Wiley, New York, 1968.
2. Dion, J.-P., *Estimation des probabilités initiales et de la moyenne d'un processus de Galton–Watson*. Mimeographed thesis. Département de Mathématiques, Université de Montréal à Quebec, 1972.

Index

Actual age distribution 176
Actual birth (death) rate 214
Adams, K. 257
Age dependent birth and death process 208
Age dependent branching process 8, 124
Age specific birth (death) intensity 208
Agresti, A. 85
Alive 8
Anaphase 225
Ancestor 6
Area method 235
Athreya, K. B. 12, 17, 85, 86, 167, 188, 223
Average age at child-bearing 10, 15, 214

Badalbayev, I. S. 53
Balanced exponential growth 225
Ballot theorem 103
Barrett, J. C. 256, 257
Bartlett, M. S. 256
Baserga, R. 254
Baum, L. E. 188
Bellman. R. 5, 17
Bellman–Harris process 8, 124
Benoiston de Châteauneuf, L. F. 1, 16
Bienaymé, I. J. 1, 2, 16
Billingsley, P. 122, 263
Binary Galton–Watson process 20
Binary splitting 129
Birth and death process 208
Boundary method 235
Bounded variation 115
Brecher, J. 257

Bresciani, F. 256
Brockwell, P. 240, 254, 256
Bronk, D. V. 254, 256
Brown, B. M. 53
Burnett-Hall, D. G. 250, 258
Bühler, W. J. 53, 223

Cameron, I. L. 255, 256
Candolle, A. de 2, 16
Carrier density 227
Cell cycle 129, 225
Cell death 129
Chiang, C. L. 96
Choquet–Deny theorem 106
Chow, Y. S. 122
Church, J. D. 85
Cleaver, J. E. 255, 256, 257
Coale, A. J. 223
Collection function 245
Continuous labelling (mitoses) curve 241
Costello, W. G. 223
Cox, D. 122
Critical 10, 23, 131
Critical (for multi-type processes) 93
Crude birth (death) rate 213
Crude rate of increase 214
Crump, K. S. 17, 52, 53, 135, 188

Daley, D. J. 122
Darling, D. A. 52
Defective renewal equation 104
Degree of synchrony 249
Descher, E. 256
Dienes, G. J. 255
Dion, J. P. 49, 53, 263

265

Directly Riemann integrable   108
Disintegration   129
Dmitriev, N. A.   5, 17
Doney, R. A.   167, 188, 189, 223
Doob, J. L.   35, 104
Doubling time   226
Dubuc, S.   52
Durham, S. D.   156, 188, 206
Dwass, M.   53
Dynkin's theorem   259

Endomitosis   90
Engelberg, J.   258
Environmental process   81
Erlang, A.   3, 17
Euler, L.   4, 16
Everett, C. J.   18
Excessive renewal equation   104
Expected age distribution   176
Expected cycle time   226
Extinction   22, 140
Extinction probability   9, 22, 140

Fahady, K. S.   85
Fearn, D. H.   81, 85
Fecund   140
Feichtinger, G.   96
Feller, W.   5, 17, 85, 104, 114, 122, 188, 206, 223
Fisher, R. A.   2, 17, 214, 223
FLM   233
Flux expectation   227
Foster, J. H.   84, 206
Fraction labelled mitoses   233
Fry, R. J. M.   255, 256
Functional of branching process   8
Furry, W. H.   5, 17, 52

Galton, F.   1, 16
Galton–Watson process   7
Ganuza, E.   188
Gavosto, F.   258
General branching process   8, 124
Generation   6
Generation time   129, 225
Gilbert, C. W.   256
Good, I. J.   76

Goldstein, M. I.   156, 188
G_0 phase   224
G_1 phase   224
G_2 phase   225
Grimwall, A.   85, 206
Gross maternity function   208
Gross reproduction rate   208
Growth fraction   224

Haldane, J. B. S.   2, 17
Haley, J. E.   257
Hanes, S.   257
Harris, T. E.   5, 17, 36, 52, 53, 166, 176, 255
Hartmann, N. R.   256
Hawkins, D.   35, 52
Hazard rate   208
Heathcote, C. R.   62, 84
Herbelot, L.   4, 16
Hewitt's and Savage's zero-one law   101
Heyde, C. C.   18, 37, 52, 53, 84
Hiorns, R. W.   86
Holte, J. M.   152, 156, 188
Horn, M.   230, 255
Howard, A.   255
Howe, R. B.   53

Imbedded Galton–Watson process   124
Individual   6, 7
Intensity   208
Interphase   225
Intrinsic birth (death) rate   214

Jagers, P,   17, 85, 122, 188, 189, 206, 255, 258
Janson, B.   258
Johannisson, G.   255

Kamke, E.   206
Kaplan, N.   12, 86, 206
Karlin, S.   85, 86
Katz, M.   188
Kawazu, K.   206
Keiding, N.   86, 223
Kendall, D. G.   2, 5, 17, 18, 209, 223, 255, 258

Kesten, H. 35, 52
Keyfitz, N. 86, 223
Kingman, J. F. K. 143
Klein, G. 258
Kolmogorov, A. N. 5, 17, 30, 52
Kolmogorov's zero-one law 101
Kubitschek, H. 255

Labkovskiy, V. A. 85
Lamerton, L. F. 255
Lamperti, J. 85
Lattice 10, 107
Leontovič, A. M. 85
Leslie, J. R. 53
Levina, L. V. 85
Lewis, P. A. W. 122
Life-length 123
Lindvall, T. 52, 85, 129
Lipkin, M. 256
Locally bounded variation 115
Loève, M. 86
Lopez, A. 223
Lotka, A. 4, 17, 209, 223
Lourie, I. S. 257

M phase 225
Macdonald, P. D. M. 235, 240, 255, 256
Malthus, T. 1, 16
Malthusian parameter 10, 132
Malthusian process 10
Markovian in the age structure 208
Marr, A. G. 255
Martin-Löf, A. 223
Martingale 97
Maternity function 208
McKendrick, A. G. 5, 16
Mellgren, J. 255
Mendelsohn, M. L. 256
Metaphase 225
Meyer, P.-A. 122, 223
Mitosis 175, 225
Mitotic index 175, 226
Mitotic time 175, 225
Moser, L. 4, 16
Mountford, M. D. 86
Muhamedhanova, R. 85

Multi-type (*r*-type) Galton–Watson process 7

Nachtwey, D. S. 255
Nagayev, A. V. 49, 53
Nagayev, S. V. 85, 206
Net maternity function 208
Net reproduction rate 208
Neveu, J. 122
Ney, P. 5, 17, 52, 188
Nielsen, J. E. 86
Non-lattice 10, 107
Nooney, G. C. 255
Norrby, K. 233, 255
Number of children 7

Otter 18

Painter, P. R. 255
Pakes, A. G. 45, 53, 84, 189, 206
Parameter measure 121
Paskin, A. 255
Pedersen, T. 256
Pelc, S. R. 255
Petersen, D. 257
Pileri, A. 258
Plateau method 235
Point process 121
Poisson process 121
Pollard, J. H. 96, 223
Polyploidy 90
Positively regular 89
Powell, E. O. 255
Predecessor 8
Prescott, D. M. 256, 258
Process generating function 129
Progeny 39
Puck, T. T. 247, 248, 257, 258
Puri, P. S. 186, 189
Pyateckiy-Šapiro, I. I. 85

Quastler, H. 227, 256
Quételet, A. 4
Quine, M. P. 84

Radcliffe, J. 206
Radix 207

Radon measure   104
Random characteristic   14
Random environment   81
Rao, K. M.   104, 122
Realized   7
Renewal equation   104
Renewal process   121
Reproduction   123
Reproduction function   9
Reproduction generating function   9, 21, 129
Reproduction law   7
Reproduction mean   10, 21
Reproduction process   8
Reproduction variance   11, 21
Reproductive value   212
Révész, L.   258
Robbins, H.   122
Robinson, S. H.   257
Royden, H. L.   122
Ryan, T. A.   5, 17, 188

S phase   225
Samuels, M. L.   223
Sanders, P.   257
Sankoff, D.   258
Savits, T.   12
Schröder, E.   2, 16
Seneta, E.   18, 52, 53, 84, 85
Sevast'yanov, B. A.   5, 17, 96, 129, 156, 188
Sevast'yanov process   8, 124
Sharpe, F. R.   16
Sherman, F. G.   256
Siegmund, D.   122
Simons, G.   104
Singular   89
Siraždinov, S. H.   85
Sisken, J. E.   255
Slack, R. S.   26, 52
Smith, W. L.   86, 122
Span   107
Spitzer, F.   52
Splitting process   8, 124
Stable age distribution   12, 164
Stanners, C. P.   257
Steel, G. G.   255, 257

Steffen, J.   258
Steffensen, J. F.   3, 17
Stigler, S. M.   52, 53
Stiglum, B.   35, 52, 53
Stopping time   97
Stubblefield, E.   258
Subcritical   10, 23, 131
Subcritical for multi-type processes   93
Submartingale   97
Supercritical   11, 23, 132
Supercritical for multi-type processes 93
Supermartingale   97
Süssmilch, J. P.   3, 16

Takács, L.   104
Takahashi, M.   256, 257
Tannock, I. F.   258
Telophase   225
Till, J. E.   257
Trucco, E.   254, 256, 257, 258
Turnball, B. W.   86
Type   87

Ulam, S   18, 35
Uniformly integrable   101
Urbanik, K.   18

Van't Hof, J.   258
Varying environment   70
Vere-Jones, D.   52, 122
Volterra, V.   5, 17

Watson, H. W.   2, 16
Waugh, W. A. O'. N.   86, 189, 250, 258
Weiner, H. J.   169, 186, 188, 189
Wilkinson, W. E.   86
Williamson, L. A.   84
Wilson, G. B.   255
Wimber, D. E.   257

Yamada, M.   257
Yaglom, A. M.   30, 35
Yaglom's theorem   29
Yule, G. U.   5, 17

Zero-one laws   101
Zeuthen   258

Applied Probability and Statistics (Continued)

HOEL · Elementary Statistics, *Third Edition*
HOLLANDER and WOLFE · Nonparametric Statistical Methods
HUANG · Regression and Econometric Methods
JAGERS · Branching Processes with Biological Applications
JOHNSON and KOTZ · Distributions in Statistics
         Discrete Distributions
         Continuous Univariate Distributions-1
         Continuous Univariate Distributions-2
         Continuous Multivariate Distributions
JOHNSON and LEONE · Statistics and Experimental Design: In Engineering and the Physical Sciences, Volumes I and II
LANCASTER · The Chi Squared Distribution
LANCASTER · An Introduction to Medical Statistics
LEWIS · Stochastic Point Processes
MANN, SCHAFER and SINGPURWALLA · Methods for Statistical Analysis of Reliability and Life Data
MILTON · Rank Order Probabilities: Two-Sample Normal Shift Alternatives
OTNES and ENOCHSON · Digital Time Series Analysis
PRENTER · Splines and Variational Methods
RAO and MITRA · Generalized Inverse of Matrices and Its Applications
SARD and WEINTRAUB · A Book of Splines
SEAL · Stochastic Theory of a Risk Business
SEARLE · Linear Models
THOMAS · An Introduction to Applied Probability and Random Processes
WHITTLE · Optimization under Constraints
WONNACOTT and WONNACOTT · Econometrics
YOUDEN · Statistical Methods for Chemists
ZELLNER · An Introduction to Bayesian Inference in Econometrics

Tracts on Probability and Statistics

BILLINGSLEY · Ergodic Theory and Information
BILLINGSLEY · Convergence of Probability Measures
CRAMÉR and LEADBETTER · Stationary and Related Stochastic Processes
JARDINE and SIBSON · Mathematical Taxonomy
KINGMAN · Regenerative Phenomena
RIORDAN · Combinatorial Identities
TAKACS · Combinatorial Methods in the Theory of Stochastic Processes